SCIENCE, POLITICS AND SOCIAL PRACTICE

BOSTON STUDIES IN THE PHILOSOPHY OF SCIENCE

*Editor*

ROBERT S. COHEN, *Boston University*

*Editorial Advisory Board*

THOMAS F. GLICK, *Boston University*
ADOLF GRÜNBAUM, *University of Pittsburgh*
SAHOTRA SARKAR, *McGill University*
SYLVAN S. SCHWEBER, *Brandeis University*
JOHN J. STACHEL, *Boston University*
MARX W. WARTOFSKY, *Baruch College of the City University of New York*

VOLUME 164

# BOSTON STUDIES IN THE PHILOSOPHY OF SCIENCE

Editorial Committee for the Robert S. Cohen *Festschrifts*:

KOSTAS GAVROGLU, *National Technical University, Athens, Greece*
ADOLF GRÜNBAUM, *University of Pittsburgh*
JÜRGEN RENN, *Max-Planck-Institut für Wissenschaftsgeschichte, Berlin*
SAHOTRA SARKAR, *McGill University*
SYLVAN S. SCHWEBER, *Brandeis University*
JOHN J. STACHEL, *Boston University*
MARX W. WARTOFSKY, *Baruch College, The City University of New York*

Volume I   *Physics, Philosophy, and the Scientific Community*
           Essays in the philosophy and history of the natural sciences and mathematics
Volume II  *Science, Politics and Social Practice*
           Essays on Marxism and science, philosophy of culture and the social sciences
Volume III *Science, Mind and Art*
           Essays on science and the humanistic understanding in art, epistemology, religion and ethics

ROBERT S. COHEN 1987
Courtesy of Boston University Photo Services

# SCIENCE, POLITICS AND SOCIAL PRACTICE

Essays on Marxism and science, philosophy of culture
and the social sciences
In honor of Robert S. Cohen

*Edited by*

KOSTAS GAVROGLU
*National Technical University, Athens*

JOHN STACHEL
*Boston University*

MARX W. WARTOFSKY
*Baruch College, The City University of New York*

KLUWER ACADEMIC PUBLISHERS
DORDRECHT / BOSTON / LONDON

**Library of Congress Cataloging-in-Publication Data**

```
Science, politics, and social practice : essays on Marxism and
  science, philosophy of culture and the social sciences : in honor of
  Robert S. Cohen / edited by Kostas Gavroglu, John Stachel, Marx W.
  Wartofsky.
      p.   cm. -- (Boston studies in the philospphy of science ; v.
  164)
    Includes index.
    ISBN 0-7923-2991-0 (set). -- ISBN 0-7923-2989-9 (alk. paper)
    1. Science--Philosophy.  2. Science--Social aspects.  3. Cohen, R.
  S. (Robert Sonné)   I. Cohen, R. S. (Robert Sonné)  II. Gavroglou,
  Kōstas.  III. Stachel, John J., 1928-    .  IV. Wartofsky, Marx W.
  V. Series.
  Q175.S4225  1994
  501--dc20                                                    94-3703
```

ISBN 0-7923-2989-9
ISBN Set: 0-7923-2991-0

Published by Kluwer Academic Publishers,
P.O. Box 17, 3300 AA Dordrecht, The Netherlands.

Kluwer Academic Publishers incorporates
the publishing programmes of
D. Reidel, Martinus Nijhoff, Dr W. Junk and MTP Press.

Sold and distributed in the U.S.A. and Canada
by Kluwer Academic Publishers,
101 Philip Drive, Norwell, MA 02061, U.S.A.

In all other countries, sold and distributed
by Kluwer Academic Publishers Group,
P.O. Box 322, 3300 AH Dordrecht, The Netherlands.

*Printed on acid-free paper*

All Rights Reserved
© 1995 Kluwer Academic Publishers
No part of the material protected by this copyright notice may be reproduced or
utilized in any form or by any means, electronic or mechanical,
including photocopying, recording or by any information storage and
retrieval system, without written permission from the copyright owner.

Printed in the Netherlands

TABLE OF CONTENTS

EDITORIAL PREFACE ix

### A. RELATIONS OF MARXISM TO SCIENCE AND PHILOSOPHY

| | |
|---|---|
| HERBERT APTHEKER / On Bob Cohen | 1 |
| LIN CHUN / Love and Hate: Learning "Human Nature" under Communism | 5 |
| DAINIAN FAN / Marxist Philosophy and Humanism in Mainland China – An Outline | 15 |
| BURTON DREBEN / Cohen's Carnap, or Subjectivity is in the Eye of the Beholder | 27 |
| WŁADYSŁAW KRAJEWSKI / Stalinism and Soviet Science | 43 |
| PETER MARCUSE / Herbert Marcuse on Real Existing Socialism: A Hindsight Look at *Soviet Marxism* | 53 |
| JOHN STACHEL / Marx on Science and Capitalism | 69 |
| REN-ZONG QUI / Karl Popper and Karl Marx | 87 |

### B. SCIENCE, TECHNOLOGY AND SOCIETY

| | |
|---|---|
| PAUL T. DURBIN / Technological Praxis: Reflections | 99 |
| SHELDON KRIMSKY / Science, Society and the Expanding Boundaries of Moral Discourse | 113 |
| MIHAILO MARKOVIĆ / The Concept of Scientific Law | 129 |
| GYORGY MARKUS / After the 'System': Philosophy in the Age of the Sciences | 139 |
| MARIA MARKUS / Civil Society and the Politisation of Needs | 161 |

## C. METHODOLOGICAL, ONTOLOGICAL AND PHENOMENOLOGICAL ISSUES IN THE PHILOSOPHY OF THE SOCIAL SCIENCES

ABNER SHIMONY / Cybernetics and Social Entitles 181
KRISTIN SHRADER-FRECHETTE / Risk Models and Geological Judgments: The Case of Yucca Mountain 197
DOROTHY WEDDERBURN / Some Reflections on Inequality and Class Structure 215
KURT H. WOLFF / From "Dualism of Human Nature" to "Human Being as a Mixed Phenomenon" 235
DANILO ZOLO / The Tragedy of Political Science 247

## D. CULTURE, REASON AND ETHICS

RICCA EDMONDSON / Reasoning in the Social World: Prolegomenon to a Sociology of Argument 267
CHARLES L. GRISWOLD, JR. / Rhetoric and Ethics: Adam Smith on Theorizing about the Moral Sentiments 295
PATSY HALLEN / Ecofeminism as Reconstruction: Making Peace with Nature 321
DAVID B. ZILBERMAN / On Cultural Relativism and "Radical Doubt" 359
PAOLA ZAMBELLI / From the *Quaestiones* to the *Essais*: On the Autonomy and Methods of the History of Philosophy 373

INDEX OF NAMES 391

# EDITORIAL PREFACE

The essays in this *Festschrift* are celebrations of the human mind in its manifold expressions – philosophical, scientific, historical, aesthetic, political – and in its various modes – analytical, systematic, critical, imaginative, constructive. They are offered to Robert S. Cohen on the occasion of his 70th birthday, in acknowledgment of his own extraordinary participation in the life of the mind, and of his unfailing encouragement and facilitation of the participation of others. It is fitting that these volumes should appear in the *Boston Studies in the Philosophy of Science*, the series which he co-founded so many years ago, and of which he has been the principal editor for more than three decades. (These are perhaps the only volumes of that series which he has not edited or co-edited!)

The three volumes that constitute this *Festschrift* cover the range of Cohen's interests as a philosopher/scientist/humanist, as they also represent the spectrum of his professional and personal friendships. (Regretfully, the editors could not include contributions from more of them here.) The first volume centers around the philosophy and history of the natural sciences and mathematics; Volume Two collects essays related to Marxism and science, philosophy of culture and the social sciences; and the third volume focuses on science and the humanistic understanding in art, epistemology, religion and ethics.

The editors and the editorial committee express their thanks to Annie Kuipers, our editor, conscience and guide at Kluwer Academic Publishers, who has been guardian angel of the *Boston Studies* these many years, and a good friend of Robert's; to her able assistant, Evelien Bakker; to Carolyn Fawcett for apt translation and editorial assistance; and of course, to Robin Cohen for all around enthusiasm, timely revelations and steady support.

KOSTAS GAVROGLU   *National Technical University, Athens*
JOHN STACHEL   *Boston University*
MARX WARTOFSKY   *Baruch College and the Graduate Center of The City University of New York*

HERBERT APTHEKER

## ON BOB COHEN

I want to recall two moments in my long association with Bob.

The mid-fifties marked the nadir of McCarthyism. One of its prime manifestations was the persecution of Communists. In 1955 the Government chose as one of its victims, Junius Scales, the Chairman of the Communist Party in North Carolina. Scales was indited under the membership clause of the Smith Act, facing, upon conviction, up to ten years in prison.

For the defense, two professors at the University of North Carolina testified – bravely – that they knew Scales as a student characterized by honesty and sincerity but had no awareness of his politics. A minister in the university town of Chapel Hill offered similar testimony.

The substantive defense witness was the 32-year old Bob, then an assistant professor of physics and philosophy at Wesleyan University in Connecticut. He appeared for the defense as a non-Communist but an authority on Marxism – a description with which the prosecution did not quarrel.

Speaking very much as in a classroom (one of the courses he taught at Wesleyan was on Marxism), Professor Cohen testified quite briefly but clearly as to the universal scope of Marxism, the various sources from which it derived, its scientific character, and the fact that it was partisan towards democracy. He touched on the growth of this body of thought, how it had changed in the course of time, and how its application depended decisively upon specific place and time. He stressed – in accordance with the occasion – that the advocacy of violence was hostile to Marxism. This, he added, was one of Marx's differences with the anarchists. He sought also to convey to the court and the all-male, all-white jury something of the profound respect with which Marxism was regarded among all serious scholars and thinkers.

The cross-examination of Bob was conducted in an exceedingly hostile manner – including hardly veiled anti-Semitism. Bob retained his calm and replied as substantively as he could to the raving of the government's lawyer.

Bob, being asked by the defense why he, a non-Communist, had testified on behalf of the defendant, said that he believed that any citizen who possessed knowledge that might be helpful to the furtherance of justice ought to be willing to offer it; that he was disturbed by the grossly incorrect characterization of Marxism that he had read about in other cases as coming from prosecution witnesses – and again in this case – and that he felt that it should be corrected, especially where the possible imprisonment of people was involved. Further, in his opinion, these prosecutions of Communists – as in the present instance – were a major part of the growing restrictions against civil liberties in our country. These restrictions troubled him, as a citizen and a scholar. He thought that aiding Communists, as among those most persecuted, was crucial to the defense of the Bill of Rights.

The prosecution's summation took two and a half hours; it was marked by an illiterate, demagogic harangue including an openly anti-Semitic attack upon this Northern "perfesser." He added that he expected the jury to return a guilty verdict quickly, thus helping preserve Carolina's way of life. The jury was out for ninety minutes, declared Scales guilty; he was sentenced to six years in prison.

\* \* \*

After McCarthy's alcoholism brought him to an early grave, the poison he personified was somewhat muted. With the war upon Vietnam, however, its worst aspects reappeared. In an effort to help counter this reaction, what became known as the American Institute for Marxist Studies (AIMS) was founded, early in the 1960's. This originated at the initiative of the Communist Party; however, it affirmed and maintained throughout its existence that it was not a Party organization but rather an instrument seeking to further serious, civilized discourse on Marxism, defined in the most catholic manner.

It was hoped that AIMS might help counter the prevailing anti-communist hysteria and to suggest that Marxism was a system of thought rather than a criminal conspiracy.

Harry F. Ward agreed to serve as its honorary chairman and the early launching of the effort did bring an encouraging response – especially from scholars – throughout the nation. Tragically, Dr. Ward died soon after AIMS' establishment; Professor Dirk Struik agreed to serve as Ward's successor.

Bob agreed with AIMS' purposes and of course had the highest respect for Ward and Struik. He accepted the position of Chairman and played, in its two decades of existence, a modest part in furthering its purposes.

It hosted several symposia in various parts of the country. It sponsored debates and discussion – some lasting several days – with scholars of many viewpoints (including that of a Catholic priest) which attracted audiences reaching into the hundreds. It published about twenty books – several of lasting consequence – as well as some thirty Occasional Papers.

AIMS also maintained a good library – specializing in subjects relevant to its purpose. This was open, without charge, to the public; it served as a major source for more than one published dissertation. Also AIMS issued a bi-monthly newsletter, finally reaching a subscription list of over four thousand; this contained news relevant to its objective, as well as lists of books and articles that would interest its readers. The latter reached an international audience; by far, however, most subscribers resided in the United States.

Bob was an active Chairman throughout the years. He attended all semi-annual meetings of its Board, chaired some of its public events and, especially, helped keep AIMS on its non-sectarian course.

AIMS played a modest effort for several years in furthering scholarship, counteracting vicious baiting of radicalism and offering assistance to young partisans of reasonable discourse – several of whom went on to distinguished careers.

Alas, in the 1950's and 1960's few were the people of Bob's position and distinction who had the courage to try to counteract the deliberately concocted hysteria of the Government and the dominant press. But the vindication of Bob's effort has come with the passage of time. His life-time commitment to the role of philosopher-activist is memorable and has played a part in the continuing struggle to bring decency to our nation.

LIN CHUN

# LOVE AND HATE: LEARNING "HUMAN NATURE" UNDER COMMUNISM*

> ... there are three essential moral qualities of human beings to which science has no direct relation. The first is tenderness. The second is kindliness. . . . The third is human intimacy and love.
> – Robert S. Cohen[1]

I was twelve when I first learnt that it was Marx's conviction that "Nothing human is alien to me." I never really understood it in the subsequent twenty years. Along with my peers living in Communist China, we grew up to believe that one of the basic Marxist tenets was that there was no such thing as universal or abstract human nature. Not that the nature of men, if it does exist at all, must be historically and sociologically specific, but simply that people of different social classes have different natures. (How about in a classless society? I don't remember if I had ever been puzzled by questions like this.) Indeed, for a long while, to us anything that claimed to be above classes, either psychological phenomena or aesthetic tastes or personal characters, were illusory or deceitful bourgeois ideologies. Within that framework of thinking, what Marx meant by the phrase was absolutely beyond comprehension.[2]

I'm talking about the experience of a generation that was born right after the victory of the Chinese revolution; but to a great extent of an earlier generation too, the one immediately before us which also received its primary political education in the post-1949 period. The formative influence on these two generations of the unending official campaign against the "bourgeois/revisionist theory of human nature" contributed a great deal to the basic characteristics of culture and politics in ordinary Chinese life prior to the post-1978 reforms. This is not to say that lately there has been any attempt on the side of the government propaganda machinery to clear away the damages of that long-standing anti-humanism. Nor is there any autonomous intellectual movement resembling the earlier east European reformism under the now seemingly outdated slogan, "socialism with a human face." But the economic reforms have brought about some ambiguous yet lively changes in the popular attitude toward "humanity", in terms of "individuality" along with an emerging civil society. I will return to this new development.

The only legitimate political science in the age I am recalling was the theory of class struggle. But even Mao's doctrine of "continuous revolution" did not offer us any scientific criterion to define or identify antagonistic classes in a society which was claimed to be the "historical period of transition to communism". We were taught, and more or less came naturally, to love our socialist motherland (or more precisely, depend on it), our people, our collectives (i.e., study and/or work communities), the Communist Party and, especially during the cultural revolution, our great leader Chairman Mao. Yet proletarian love, which meant utter devotion of oneself to the communist cause, would only be completed by proletarian hatred in order to make a revolutionary. And being a qualified successor to the revolution was the purpose of life. So we were also taught to hate the "class enemies," a category vaguely including members of the overthrown exploiting classes and those who were in one way or another connected to the (US) imperialist, (USSR) "modern revisionist", and other domestic and/or foreign reactionary projects.

It was not at all easy in practice, though, first to spot these enemies and then hate them, especially because hidden enemies, according to the logic of "class struggle under the conditions of the proletarian dictatorship," could be anywhere. But also because, as I gradually realized, neither hate nor love can be inculcated merely with reasons or ideological arguments. There was an ex-landlord in our production team where for five years I lived with the peasants. I still remember very well my confused feelings whenever I saw that old man who was always working hard and always alone. I knew what the correct class stand should be but couldn't help pitying him. In our supposed Marxist education, members of an enemy class and that class itself were the same thing, although the possibility of some individual members being reformed was not denied. In any case, the more we tried to understand our own political emotions, the muddier the situation became. Even love was not simple. Consider, for example, the people as a whole whom we loved (as an abstraction) were, as a matter of fact, divided into the "left," "center," and "right," whatever these categories might imply; the Party to which we gave our all had forever been engaged in two-line ("correct" v. "erroneous") struggles, and highly placed party officials were often found guilty of the most evil crimes.

The earlier middle school political classes proved not to be helpful. We had lots of heated debates in the classroom over such topics as what was the moral justification for saving a drowning person before knowing his or her class status (obviously there wouldn't be time to do an investigation), since the rescue action might cost a comrade's life. (I wonder if patients in our hospitals had received treatment regardless of their class origins. Maybe the doctors were not as brutally naive as we were.) We passionately criticized 20th-century anti-war literature, western as well as Chinese, for its pacifist messages. Any narrative in any literary or art form that had a ring of telling human rather than class characters was viewed as a lie of class compromise. The impression of these criticisms on us seems not to be erasable so that more than 20 years later, my first reaction to reading Erich Fromm was ironically a sense of regret for not having his work (especially *The Art of Loving*) then as an ideal object of criticism! I don't think we were entirely wrong though, for I still believe that reason is not enough and in our time there are still just wars or necessary violence. But surely those one-sided and extremely oversimplified discussions only strengthened our blindness to reality, and our tendency to arbitrary hatred. In a more rational manner, we wondered, if a class heritage was no longer economically definable but only handed on from generation to generation in political terms, then how and when and on what ground could one expect final abolition of enemy classes? Our teachers were quite paralysed during those discussion sessions. They did not foresee, of course, nor did anybody else, that they were themselves soon to fall victims of the storms of class struggle.

Teachers and intellectuals in general were among the first targets of the violent cultural revolution. That revolution was also a big test of our (we as the revolutionary youth) being clear about what to love and what to hate. I have long tried to answer the following questions which I asked myself with hindsight: Did we really hate our teachers who hardly hurt us in any way before the event? Why could we be so cruel to them that most of them were humiliated and many beaten? Why were the first activists of the movement usually top students with "correct" family background and therefore hitherto well treated rather than those who should have had more reasons to rebel?[3] Why was the violence in fact more fierce in the all-girls schools (in Beijing, at least) where several principals were beaten to death or committed suicide? These are hard and very painful questions. It finally becomes clear to me that we not so much hated our teachers as felt superior to them, because we were "born red" to carry out the historical mission of the revolution, fighting against the bourgeois educational system for which they were practically responsible. Still our "class consciousness" kept pushing us further to hate, to eliminate any compassion for these people to whom we owed much but who now by nature, so we were convinced, differed from us because of their different political identities. One of the most unforgettable moments in my life was when our history teacher's eyes and mine met while I was giving orders to the teachers (I was only in my early teens). I must have looked so fierce that he couldn't recognize me, once his best student who truly respected him.

In a frenzied atmosphere of struggle which was said to be a matter of life and death for revolutionary China, tenderness was such a sign of weakness or betrayal that one must avoid it. The more serious we were, the more we felt guilty and the harder we tried to overcome it. And the only way to prove one's steadfastness and militancy was to be ruthless. We needed this proof also very much for fear of isolation, of being expelled from where we ought to belong. A number of my classmates actively joined others in attacking their own grandparents or parents who were either ex-capitalists or newly determined "capitalist roaders". More than one of those who were attacked died from physical injury. I know that on the part of my classmates the pain only came later (and would stay forever): as they were overwhelmed by what was seen then as an imperative and by fear of discrimination and exclusion, they believed in what they did with moral justifications. Moreover, to explain the explosive social behaviors of the time, it was also true that Mao's call for "great democracy" and actual mass mobilization from below had an immediate effect of liberation. Our excitement (how wonderful to have no more classes, exams, disciplines . . . , but only take our own initiatives!) was real, energy released.

> Take up the pens as guns,
> Concentrate fire on the reactionary gang!
> Whoever dares to say a bad word about the Party,
> Send him to the Kingdom of Hell at once!

This was a popular song of the Red Guards, and it took the place in our young hearts of flowers and birds and pink fantasies of a loving world. We were ready to move to destroy anything in the way, by the "weapons of criticism", yes, but by "the criticism of weapons" too if necessary, be it the people we once loved, or the earlier make-up of our own characters. Such a release of complications can be seen also in the lasting hatred and in many cases violent clashes among competing rebel factions who in fact shared the same language and ideology of Mao's thought. In the name of the revolution, they fought one another for power and control, honor and survival. As soon as the old bureaucratic repression was largely removed, and the freedom for political participation was granted to the masses without an institutionally supported citizenry, human impulses, the reality as well as the confusions and distortions of them, were exposed. (In this connection, I tend to think that the wars and destructions being waged in the former Soviet empire are something similar to the explosion of barbarism during the Chinese cultural revolution in terms of such negative liberating effect.)

Do we see some resemblance here to the fascist youth movements? The late historian Li Shu and the late social theorist Lin Wei, both old Communists and honored friends of mine, first suggested this resemblance to me as a serious research topic in 1979. However, it was not until later when I began to read Theodor Adorno (*The Authoritarian Personality*), Fromm ("the fascist personality"), and Wilhelm Reich (*The Mass Psychology of Fascism*), that I realized how deep the psychological dimension of the matter was, where, rather than at the phenomenal surface, the comparisons should be made. It turned out to be plain that there was little difference between a loyal Red Guard, a committed Nazi, and, say, a religious fundamentalist, in their mental conflicts and psychic mechanisms. Each subordinates himself to a lofty ideal, a great cause, a god or sacred leader, a higher power; each has a constant sense of guilt for what remains of his self and therefore an urge to release his anxiety and guilt feeling by destructive and self-destructive actions. The larger social dimensions and conditions of this personal irrationality are conceptualized in such key notions as "organized mysticism" which is the socially penetrating patriarchal authoritarianism of the family and the church under National Socialism (Reich), the "social unconscious" in terms of rationalization of repression or ideology against true knowledge about a repressive society (Fromm, *Beyond the Chains of Illusion*), "surplus repression" which denotes specific socio-historical interests and institutions of domination in addition to the general quality of civilized (i.e. repressed) life (Herbert Marcuse, *Eros and Civilization*), and most pessimistically, "instrumental reason" which explains some basic qualities of modern and late modern human existence, including systematic automatization of men (the Frankfurt school). I find the sociopsychological approach useful for any inquiry into the movements of the Chinese masses under Communism during extraordinary periods such as the Great Leap Forward and the Cultural Revolution. But that approach is especially helpful to me in understanding the origin and social functions of communist puritanism in which a possible explanation of the cruelty of teenage school girls might lie.

The emancipatory spirit of the populist and anti-bureaucratic upsurge incited by Mao in the second half of the 1960s was profoundly manipulated by the Mao cult. Quite different from "1968" in the west, the upsurge in China was by no means a general liberation movement against the existing morality of the status quo, including sexual repression. On the contrary, since the proclaimed goal of the cultural revolution was to counter "capitalist restoration" and hence the "bourgeois" values rather than what was called the "feudalist olds", the traditional asceticism of the old Chinese ethics (especially the Song-Ming Confucian School)[4] was not only left intact but also combined with the military-Communist tradition to strengthen a mentality that was anti-individualist in general and anti-sexual in particular. For many years our celebrated slogan was "Fear neither hardship nor death" (which actually reflected the country's material backwardness and the need of hard work and sacrifice). Indeed the most serious of us, then working in the factories and the countryside or serving in the army, deliberately sought hardship and even death, for self-sacrifice was the highest value of life for a revolutionary.

In such a culture, one that essentially denied pleasure and enjoyment (thus love, play, art . . .), it is not surprising to observe an absence of privacy and (conscious) interpersonal love, especially among educated young people who sincerely believed in the propagandized morals. The love for Mao, to be sure, to a great extent filled our emotional life. There was hardly a sense of emptiness or loneliness so long as he was there. Beyond this religious-like love, there were certainly friendship relations and comradely solidarity, but we did not see affection for a friend as "love" anyway. My own experience, which was ordinary (in comparison to those who had their family members marked as class enemies), was that I felt too ashamed of being "petty-bourgeois sentimental" to acknowledge that I loved my parents and sister, for example. As to sexual love, we knew so little about it, not only because sexuality became a taboo in the official language and education, but also because we ourselves avoided any knowledge of it; it was thought incompatible with a noble-minded Communist. We were not, or too horrified to be, even curious. This attitude also very much shaped our aesthetic outlook. The varieties of our judgment over what was beautiful and what was ugly aside, rarely had any of us, to my knowledge, learnt to appreciate such things, for example, as the beauty of the human body (!).

Where Erich Fromm saw an "escape from freedom", we escaped from intimacy and love. For many of us, it was not too long. But for thousands of others like my elder sister whose "self-revolution" was achieved more successfully in their youthful days, that escape has become a life-long exile, a life tragedy.[5] The Freudian concept of sublimation in terms of desexualization or transformation of sexual energy is perfectly applicable here, although it seemed to need to be supplemented by a substantially greater "surplus" (to borrow from Marcuse) repressive operation than culture or civilization as such would require. Communist puritanism, specifically in China, suppressed love and sexuality for the sake (unconsciously?) of the very survival of the system and the regime, which wanted selfless workers to build socialism and obedient fighters to win the class struggle. Sex repression was only a way, if not the only way, to achieve this, namely, to make the citizens machine-like or to eliminate independent individuals. The sublimation process was therefore much more thorough and bitter there, and essentially political. It must be also said that that process was not a universally shared experience: the degenerate life style of so many of our leaders has gradually been disclosed since the late 1970s, which was a further revelation of the hypocritical nature of puritanical moralism and helped the "spiritual crisis" or mass disillusion about Chinese Communism to spread widely.

To me what is most revealing is sublimation as alteration of the instinctual structure. Losing the balance of that structure means, Freud wrote, destructive elements are unbound from the erotic component and "released in the form of inclinations to aggression and destruction".[6] This might be exactly what happened to the Red Guards. Speaking with Freud, Norman Brown points out that the fundamental polarity in human nature is "love and hate, love and aggression, love and the will to power".[7] The disease called man is his suppressed self, hence his hostility toward life in which the more his love is repressed, the more he tends to the other pole. Maybe this is why the girls were more violent than the boys in some prestigious Beijing schools? Consider first the general environment wherein both societal and self repression were heavier on the female sex (even in China where the Communist revolution achieved a victorious women's liberation movement), in the forms of formal ethical education as well as continuing traditional norms and attitudes. Secondly, for a dedicated female Red Guard, in order to meet the "proletarian standards", her self-criticism could be endless, her wish to atone for her weaknesses desperate. (One example that I know of is that seeing blood from violence was seen as a punishment for one's having lived like a "bourgeois young lady"). Her inner tensions and contradictions needed an expression. She was less confident of her own strength than her male comrades and motivated to gain and prove that strength in an action. Thirdly, the glorification of her mystical conviction about the indecency and wrong of sexual love was at the same time a dehumanization of her sensibilities and feelings. She found a sense of satisfaction and pride in attacking others, which demonstrated her revolutionary determination. Finally, at least Wilhelm Reich's approach would suggest actual bodily suffering from suppressed (female) puberty desires to be also a source of her irrational behavior. I'm not in a position to argue either for or against this on a scientific basis and I leave it to other researchers.

The age of blind class struggle, of vast-scale social irrationalism and human tragedies, was not merely a nightmare. It was part of our history and made us think. It nurtured and tempered a "thinking generation" and eventually catalyzed a new path toward enlightenment and demystification, toward humanism. What we should have learnt from those years, among many lessons, is that we, not the "class enemies", are also capable of persecution and destruction. It is indeed the shared biological needs and potentialities of the human individuals that in the end makes sense of our quest for human emancipation. In more concrete terms, it must be pointed out that under the Chinese condition of the "social unconscious", much of our love and hate was unnaturally politicized. The nature of the love for Mao, as can be judged now, was submission; and "class hatred", abstract as it often was, was a matter of identity: by, and only by, hating whoever was on the other side could we secure ourselves a right place to stand. Today, as arbitrary class politics, together with the special phraseology and language that powerfully sustained it, has been radically relaxed, in the vacancy created by dissolution of the old ideological emotions our once crippled and misled senses and passions might have a chance to develop healthily. On the other hand, we see such reemerging and rapidly growing practices as corruption and prostitution, among all kinds of new social problems. I have no excuse for any of my generation's misdeeds, nor nostalgia for our being naively idealist. Yet living now with the rising market and prevailing money fetishism it is a significant contrast to remember that once upon a time, without money, life was meaningful.

However, unlike some left-wing western friends who have been upset by newly developed Chinese capitalism, I believe that the economic reform is necessary, even if only for its tremendous emancipatory outcomes seen in new initiatives from below and the self-determining and self-realizing activities of the ordinary people. This can be for us an age of construction, the time to really desire a civilized but non-repressive future: man as the end, the naturalization of moral values, the free development of each and all, and progress under conditions most favorable to and worthy of human nature.[8] In a letter to her lover Leo Jogiches, Rosa Luxemburg wrote:

> My soul is bruised . . . Would I not be happier instead of looking for adventure to live with you somewhere in Switzerland quietly and closely, to take advantage of our youth and to enjoy ourselves . . . in fact I have a cursed longing for happiness.[9]

"Cursed", what a familiar word and heart. This passage is very personal but not untypical. She restrained her longing for her own happiness not because of the "reality principle", nor the "surplus repression" of an authoritarian regime, but purely and willingly to work for the socialist struggle. Even love is a basic necessity of the human essence, it can transcend individual fulfillment. My point is then simply that if there is anything precious left from our communist education, it is that idealist heroism and self-sacrifice are indispensible on the road to true happiness and a better world. This remains so, however, only where and when the movement that involves these is itself firmly and evidently toward the coincidence, the harmony of personal and social achievement, toward flourishing of the best human qualities, and toward the triumph of love over hate, or life over death.

*Chinese Academy of Sciences and Boston University*

## NOTES

\* Professor Bob Cohen has been the first reader and commentator of whatever I wrote in the last three years when I was in the US as a visiting scholar. Fortunately the present paper is not an exception although he did not know that I was to submit it to his *Festschrift*. As before, I am grateful to him for his help with the paper, but I also and especially thank him for encouraging me to write on this topic.

I use 'man' and 'him' as gender–free noun and pronoun in the paper.

[1] 'Reflections on the ambiguity of science', in L.S. Rouner (ed.), *Foundations of Ethics* (Notre Dame, 1983), p. 233.

[2] By contrast, another maxim that Marx favored, 'Be skeptical of everything', was fairly understood and popularly inspiring. It strangely survived in the peak of the personality cult of Mao who himself encouraged youthful challenges to any authority and orthodoxy.

[3] The fact that the latter people were further discriminated against at the beginning of the movement can only explain their momentary passivity but not the motives of those privileged activists.

[4] It is well-known that there was no lack of pornography in pre-19th century popular Chinese literature (and culture in general). But it had been purely entertaining and never played a role, comparable to the hedonistic literature of Renaissance humanism, in the modern social and cultural transformations.

[5] However, of course most people had not been really hostile to sex. Also notice the huge gap between urban and rural experiences. That gap and the different standards in accordance with it were taken for granted, and the peasants seemed to view their own lives as naturally outside the territory where the Party morality had the power of control.

[6] *The Ego and the Id* (Hogarth Press, London, 1950), p. 80

[7] *Life Against Death* (Wesleyan University Press, 1959), p. 53.

[8] Cf., Reich on 'natural work-democracy' in *The Mass Psychology of Fascism* (New York, 1946); R.S. Cohen, following Marcuse, on the moral ideal becoming a dimension of the natural in 'Individuality and common purpose: the philosophy of science', in *The Science Teacher, 31* (1964), reprinted in *Science and Education 4*, 1994; Marx, beside *The Communist Manifesto* in *Capital* (Moscow, 1962), pp. 799–800.

[9] Cited in J.P. Nettl, *Rosa Luxemburg* (Oxford University Press, 1966), p. 163.

DAINIAN FAN

# MARXIST PHILOSOPHY AND HUMANISM IN MAINLAND CHINA – AN OUTLINE

Professor Robert Cohen is a Marxist philosopher and a cordial friend of Chinese philosophers. On the occasion of the celebration of his seventieth birthday, I would like to make a brief, historical sketch of the situation of Marxist philosophy and humanism in mainland China.

## I

At the time of the May Fourth New Cultural Movement, in 1919, Marxist philosophy was introduced to China from Japan and Russia, after earlier anticipatory exposure from the 1890's on.

In 1923, Chinese academic circles witnessed a major debate on the issue of science and the philosophy of life, or science and metaphysics. During the debate, Marxist Chen Duxiu (1880–1942) advocated the Marxist materialist conception of history, claiming "only objective and material causes can change society, explain history, and govern the philosophy of life", and maintaining that the economic base can determine the superstructure. (Chen, 1923)

In 1934, the philosopher Zhang Dongsun (1887–1972) and his colleagues initiated a polemic against Marxist philosophy, comprehensively and systematically criticized the Marxist proposition "philosophy is partisan", and surmized it would be disadvantageous to the development of philosophy. But their criticism had no significant influence on Chinese society at that time.

Meanwhile, from 1934 to 1935, Marxist philosopher Ai Si-qi (1910–1966) successively published his "philosophical talks" in a journal, *Study and Life*, which expounded Marxist philosophy in popular language. He stated that "philosophical ideology has a class character", and maintained that the "revolutionary class has to adhere to materialism, that is the partisanship of philosophy" (Ai, 1979). These talks were published as a book entitled *Mass Philosophy* in 1936. After that, up to 1948, 32 editions were printed; and tens of thousands of copies were circulated. This philosophical book was very influential and was enthusiastically welcomed by the Chinese people, especially by the young intellectuals at that period.

In 1937, in Yenan, the revolutionary base of the Chinese Communist Party (CCP), Mao Zedong (1893–1976) published two famous philosophical articles, *On Practice*, and *On Contradiction*, in order to criticize the dogmatists inside the CCP. In the article *On Practice*, Mao stressed that:

> The Marxist philosophy of dialectical materialism has two outstanding characteristics. One is its class nature: it openly avows that dialectical materialism is in the service of the proletariat. The other is its practicality: it emphasizes the dependence of theory on practice, emphasizes that theory is based on practice and in turn serves practice. (Mao, 1965, v. 1, 297)

In the book *On New Democracy*, published in 1940, Mao suggested:

> On this point, the possibility exists of a united front against imperialism, feudalism and superstition between the scientific thought of the Chinese proletariat and those Chinese bourgeois materialists and natural scientists who are progressive, but in no case is there a possibility of a united front with any reactionary idealism. (Mao, 1965, v. 2, 381)

In 1942, in *Talks at the Yenan Forum on Literature and Art*, Mao criticized the theory of human nature, and stated that:

> There is only human nature in the concrete, no human nature in the abstract. In class society there is only human nature of a class character; there is no human nature above classes. (Mao, 1965, v. 3, 90)

As Mao's leadership in the CCP was consolidated, Mao Zedong thought became the guiding ideology inside the Party.

## II

After the Chinese Communist Party seized state power throughout mainland China in 1949, Marxism, Leninism and Mao Zedong thought became an ideology with a dominant position there. According to the Marxist viewpoint that philosophy has a class character and partisanship, all the contemporary non-Marxist philosophical doctrines were condemned as bourgeois (or feudalist) reactionary philosophy, and criticized. Traditional Chinese philosophy and Western philosophy could only be taught in a course on the history of philosophy. Non-Marxist philosophers became objects to be educated and remolded, and they mostly stopped their teaching, research and writing.

From 1951 to 1955, Mao Zedong personally initiated criticism of the film "The Life of Wu Hsun" (Mao, 1977, v. 5, 57–58), criticism of Professor Liang Shuming's 'reactionary' ideas (*Ibid.*, 121–130), criticism of Yu Ping-po's study of *The Dream of the Red Chamber* (*Ibid.*, 150–151), criticism of Hu Shi's (1891–1962) philosophy, and criticism of Hu Feng's thought about literature (*Ibid.*, 176–183). Idealism, pragmatism, humanism and theories of human nature repeatedly became targets of criticism.

In 1956, by Mao Zedong's direction, the director of the Propaganda Department under the Central Committee of the CCP, Lu Dingyi (1906–) gave a lecture entitled "Let a Hundred Flowers Blossom, Let a Hundred Schools of Thought Contend". Although Lu still affirmed the criticism of Hu Shi's philosophical ideas and other idealistic doctrines, he proposed:

Among the people, there is not only the freedom to propagandize materialism, but also to propagandize idealism. The debate between the two sides can also proceed freely. (Lu, 1956)

Inspired by the Double Hundred Policy, philosophers gradually became active in mainland China. Some philosophers suggested that the non-Marxist doctrines of the contemporary Western philosophical schools should be reevaluated and allowed to be taught, discussed and studied. But soon after that, in 1957, the anti-Rightist struggle was initiated, some famous philosophers such as Jin Yuelin (1896–1984), Feng Yu-lan (1895–1990), He Lin (1902–1992) and Tscha Hung (1909–1992) were to be criticized, and their suggestions could not be carried out.

Therefore, from 1958 to 1965, the philosophical debates in mainland China developed mainly among Marxist philosophers. In 1958, the Marxist philosopher and former vice president of the Central Party School, Yang Xianzhen's (1896–1992) viewpoint that "the identity of ideology and being is an idealist statement" was openly criticized, and condemned as opposing the general line for socialist construction proposed by Mao Zedong, namely: "Go all out, aim high and achieve greater, faster, better, and more economical results in building socialism"; this was because Yang was thought to negate the subjective activity of people, negating the theme that people's mental force can be transformed into material force.

In 1964, Yang's philosophical statement about the law of the unity of opposites, "two combines to one", was condemned as the doctrine of a fusion of contradictions which was thought to be working in concert with Khrushchev's revisionist line inside the international Communist movement, namely, "peaceful coexistence, peaceful competition, and peaceful transition" (Yang, 1981; Wu, 1978). Partly owing to this philosophical 'crime', Yang was discharged from his post as Vice-President of the Central Party School and was put in jail during the cultural revolution. He was rehabilitated and reinstated in 1980.

At the end of 1963, guided by Mao Zedong and the Propaganda Department under the Central Committee of the CCP, the Philosophy-Social Sciences Division of the Chinese Academy of Sciences convened its 4th enlarged Conference. The deputy director of the Propaganda Department under the CCCCP Zhou Yang (1908–1989) delivered the report entitled 'The Fighting Task of Chinese Philosophers and Social Scientists'. In the report, Zhou Yang called upon the philosophers and social scientists to criticize contemporary revisionism, and to learn and propagandize Marxism again; to criticize bourgeois humanism and its theory of human nature and replace it by the Marxist doctrine of class struggle and proletarian dictatorship. Zhou opposed describing Marx as a humanist, and proposed to substitute the idea of alienation for all attempts to propagate humanism. (Zhou, 1963)

III

It was ironic that, during the great proletarian cultural revolution which started in 1966, the directors of the Central Propaganda Department such as Lu Dingyi and Zhou Yang and others, were all overthrown as capitalist roaders and revisionists within the Party.

In August, 1966, at the 11th plenary session of the 8th Central Committee of the CCP, the 'Resolution on the Great Proletarian Cultural Revolution' was passed. This resolution stipulated that:

At present, our goal is to fight down capitalist roaders, to criticize reactionary bourgeois authorities.

To organize criticism of the capitalist representatives sneaked into the Party, of the reactionary capitalist 'authorities', including criticism of various reactionary viewpoints in philosophy, history, political economy, pedagogy, literature and art, theories of the natural sciences.

The Proletariat has to exercise overall dictatorship over the bourgeoisie in the superstructure, including in all the cultural domains.

In accord with this resolution, during the cultural revolution many philosophers in mainland China were criticized and cruelly persecuted; philosophical courses in colleges were stopped for a long time; research in philosophy in the research institutes was stopped; philosophical journals and books were stopped from being published or circulated; foreign philosophical publications were stopped from being imported. Chinese philosophical circles were in a situation of stagnation, retrogression and isolation.

During the cultural revolution, Lin Biao and the Gang of Four vigorously advocated the personality cult of Mao Zedong, emphasized the class struggle and proletarian dictatorship on the one hand, severely criticized humanism and the theory of human nature on the other hand, thereby whipping up public opinion for their fascist dictatorship. In that disastrous period, the value of a human being was lighter than a feather, and persons lost the dignity they deserved.

In the later period of the cultural revolution, from 1973 to 1976, under the direction of Mao Zedong, there developed a movement of 'criticism of Lin Biao and Confucius' in order to 'reevaluate legalists and criticize Confucians'. The Gang of Four incited a handful of hack philosophers to assign the scholars in Chinese history to the legalist school or the Confucian school, holding that legalists were all materialists who adhered to reform, and Confucians were all idealists who opposed reform and were conservative and reactionary: they thereby totally distorted the true history of Chinese philosophy.

IV

Since 1976, after the end of the cultural revolution, mainland China entered into a new historical period. In 1978, with the support of Deng Xiaoping (1904–), the criticism of 'two whatever' (namely, "Whatever chairman Mao decides, we should uphold resolutely; and whatever chairman Mao teaches, we should follow forever") was launched and the consideration that "practice is the only criterion of truth" was developed. In the meantime, Deng reaffirmed the Double Hundred Policy and called for the emancipation of mind. (Deng, 1984) Then, the philosophical circles in mainland China which had been as still as a pool of stagnant water began to be active.

Responding to the call of emancipation of the mind, the Marxist philosopher Yu Guangyuan (1915–) said: "Religion as religion is to be worshipped and followed abjectly; Law as law is to be obeyed and observed; Marxism is scientific, and as science, it should be discussed and criticized, should be developed in pace with practice."

Zhou Yang and the Marxist philosopher Wang Ruoshui (1926–), after reflection on the bitter lesson of the cultural revolution, diametrically changed their attitude to the criticism of Humanism and the theory of alienation of the early 60's, and began to advocate humanistic Marxism.

From 1980 to 1983, Wang Ruoshui published some articles such as 'The Human Being is the Starting Point of Marxism' (Wang, 1986, 200–216), 'On Alienation' (Ibid., 186–199), and 'A Defence of Humanism' (Ibid., 217–233) etc. He wrote:

An important inspiration drawn from Marx's *Economic-Philosophic Manuscripts (1844)* is to think highly of the human being.
  We should pay more attention to "the value, alienation, liberation of the human being." (*Ibid.*, 201–203)
  Socialism needs humanism.
  There is no essential distinction between the young, humanistic Marx and the mature Marx. (*Ibid.*, 217–233)

Wang believed that "there are alienations in socialist society, not only ideological alienation, but also political and economic alienation". (*Ibid.*, 186–199). In 1981–82 there was an important debate about humanism.

In the spring of 1983, at the centennial meeting of the birth of Karl Marx, Zhou Yang gave an influential speech. He said:

For a long period, we always criticized humanism as revisionism, thought that humanism was absolutely incompatible with Marxism. In this kind of criticism, there was a great deal of one-sidedness, even mistakes. . . .
  During the cultural revolution, Lin Biao and the Gang of Four pushed the criticism of humanism and the theory of human nature to the extreme, and whipped up public opinion in order to carry out inhuman, extremely tragic, feudal fascism. The wrong criticism of humanism and the theory of human nature in the past, theoretically and practically, brought about a very serious disastrous effect. This lesson must be borne in mind. (Zhou, 1983)

Zhou Yang did not propose to bring Marxism into the system of humanism; nor did he agree to subordinate Marxism completely to humanism. But he admitted that: "Marxism contains humanism. This is Marxist humanism". (Zhou, 1983)

From 1978 to 1983, humanism and alienation had become the focus of discussion inside philosophical circles in mainland China: 600 to 700 articles on this topic were published.

This situation was of great concern to the dogmatic Marxist Hu Qiaomu (1912–1992), the ideological authority in mainland China at that time. Hu and his colleagues thought there was "spiritual pollution" in the ideological and theoretical front. Early in 1984, Hu Qiaomu published a long article entitled "On Humanism and Alienation": he argued that taking the human being as the starting point of Marxism and "humanism as a world outlook and conception of history which can explain history and guide practice", is advocating humanistic historic idealism.

Hu admitted only "socialistic humanism as ethical theory or moral norms". He opposed "mixing up Marx's humanistic viewpoints in his early writings with the viewpoints of the mature Marx".

Hu Qiaomu resolutely denied that "there is alienation in socialist society" and thought that suggestion "seriously distorted socialist reality". (Hu, 1984)

When Hu Qiaomu's article was published, the National Minister of Education officially delivered a Notice which admonished the cadres, teachers and students in colleges to study this article and to clear away the "spiritual pollution". Meanwhile, Wang Ruoshui was discharged from his post as the Deputy Chief Editor of the *People's Daily* for his support of humanism. An academic debate was suppressed by political force and administrative measures. But few philosophers in mainland China could be persuaded in this way again and the mass of student youth was not concerned with Hu Qiaomu's official, dogmatic philosophy.

After the publication of Hu Qiaomu's article, Wang Ruoshui wrote an article entitled "My Viewpoints on Humanism – a reply and discussion" which was not allowed to be published until 1986. Wang's viewpoints were as follows:

Humanism is a kind of value, which is different from the explanation of the world and history, but its intension is broader than ethical theory and moral norms.

We should not only give scientific explanations of the world, but also make adequate value judgements: therefore, we need both materialism and humanism. Both are parts of a world outlook.

The assertion claiming that humanism as a world outlook and conception of nature is idealist, is illogical and groundless. (Wang, 1986, 239–274)

To Wang Ruoshui's reply, Hu Qiaomu made no response. But, in 1987, just after the anti-bourgeois liberalization struggle, Wang was expelled from the CCP.

Nevertheless, mainland China in the 1980's was quite different from what it was in the 1950's and the 60's. The reform and open door policy impelled Chinese philosophers to probe philosophical problems actively. Besides humanism and alienation, practical materialism, subjectivity in epis-temology, rationality and non-rationality, historical determinism and non-determinism were foci of the discussions. (Xing, 1991)

In this period, Western Marxism was extensively introduced. From Lukacs, Gramsci to the Frankfurt School, from humanistic Marxism, existentialist Marxism to analytical Marxism, all were introduced. Many representative works of various Marxist schools and famous scholars were translated into Chinese and published, although not deeply studied or objectively evaluated.

In this period, research into traditional Chinese philosophy and contemporary Western philosophy was vigorously developed. New Confucianism, the Taoism of Eastern philosophy, analytical philosophy (including logical empiricism, logical realism, holism, analytical philosophy of language, critical rationalism, the historico-sociological school), pragmatism, existentialism, structuralism, phenomenology, hermeneutics, . . . all were extensively introduced and studied and greatly interested numerous intellectuals, especially young philosophers, thereby widening the horizon of Chinese philosophy. It would be helpful to the Chinese philosophical circle to be able to keep pace with contemporary philosophical trends all over the world.

V

Since the reform and open-door at the end of 1970, Chinese and foreign philosophers began to visit each other frequently, to carry on academic communication directly and in wide-ranging areas. This is very advantageous to the development of philosophy in mainland China. In this aspect, Professor Robert Cohen has made outstanding contributions.

Before Cohen visited China in the autumn of 1985, the book series *Boston Studies in the Philosophy of Science* edited by him and his colleagues had already received a great reputation in Chinese philosophical circles. The items entitled 'Karl Marx' and 'Friedrich Engels' written by him for the *Dictionary of Scientific Biography* were translated into Chinese and published in the *Journal of Philosophical Problems in Natural Sciences* in 1983 and 1985. In these two items, Robert Cohen realistically appraised Marx and Engels as philosophers, pointed out the danger of simplification of the Marxist materialist conception of history and economic determinism, and the danger of a "vulgar" Marxism. It was very heuristic to the Chinese Marxists.

In 1985, at the invitation of the Association for the *Journal of Dialectics of Nature* of the Chinese Academy of Sciences, Cohen visited China for 22 days, and made a series of lectures entitled "The Development of Philosophy of Science in the Recent 40 Years", "Some Specific Problems in the Western Philosophy of Science", "Vienna Circle, Hegelian Leftists and the Frankfurt School in Contemporary Western Philosophy", "The Study of Science, Technology and Society in the West", "On Science, Religion and Ethics" and "On Marxism and Science", with fresh and abundant contents which were greatly welcomed by the audience of hundreds.

Cohen's paper entitled "Dialectical Materialism and Carnap's Logical Empiricism" was translated into Chinese by a colleague of mine and myself and was published as a booklet in 1988, 8,000 copies of which were soon sold out. In this paper, Cohen pointed out:

Marx and Feigl are united in two fundamental respects: they wish to reject Utopian fantasy as unrealistic in its appraisal of social possibilities; and they wish to preclude supernaturalism as seductively autistic in its cognitive claims. Both are scientific humanists." (Cohen, 1963, 99)

In mainland China, dogmatic Marxists persistently regarded logical empiricism as subjective idealism, as bourgeois ideology which was hostile to dialectical materialism. But Cohen believed that they were not "mutually hostile" but were "alike or supplementary", *(ibid.*, 101) because both of them believed in naturalism and humanism, both hoped "that science will be joined with a social movement to work toward a rational and cooperative society". *(Ibid.*, 102).

In this paper, Cohen also criticized Lenin's and the dogmatic Marxists' viewpoint that philosophy is partisan. He wrote:

Lenin was convinced that idealism in general, and subjectivism in particular, serve reactionary social ends, whereas materialism serves progressive goals. Unfortunately neither Lenin nor other Marxist scholars have furnished a comparative study of the history of the social relations of philosophy. The recent Marxist literature indicates that this problem of the social functions of philosophy has no simple solution. *(Ibid.*, 132)

Cohen also wrote: "Empiricism, rationalism and mysticism in different social settings may play different social roles (sometimes progressive, sometimes reactionary)." *(Ibid.*, 132–133)

As to Mach's philosophy, Cohen wrote:

Lenin thought that consistent positivism would corrupt the hypothetical character of advanced physical theory while encouraging the supernatural theological religion of established and reactionary churches.

But, according to Cohen's point of view:

In fact, Machism certainly helped to stimulate both Einstein's constructive critique of orthodox physics and the anti-theological perspectives of the Vienna Circle. (*Ibid.*, 133)

This paper, especially the above statements of Cohen, served as a forceful attack upon the Chinese dogmatic Marxists, and a strong support to the humanistic, liberal Marxists and non-Marxist logical empiricists. In spite of the fact that Cohen published his paper in the early 60's, it still had very practical and immediate significance in mainland China at the end of the 1980's. This was the reason why we translated and published it at that time.

In the autumn of 1988, Cohen visited China again. This time, the topic of his lectures was 'Marx, Marxism and Science'. In these lectures, Cohen introduced various explanations of contemporary Marxism: (1) analytical, (2) dialectical, (3) scientific sociological, (4) humanist, existentialist and phenomenological. When he talked about the relationship of Marxist philosophy and the special sciences, he emphasized that philosophy had to learn from the special sciences, and on the basis of the sciences, to attempt to unify the sciences. He also introduced contemporary, European and American schools of Marxism, and the teaching of Marxism in American colleges. All of these, were very helpful to the development of Marxist philosophy in China.

In the middle of June, 1992, the Beijing International Conference on Philosophy of Science was held. The topic of the Conference was 'Realism vs. Anti-realism in Science'. As a chairman of the Program Committee of the Conference, Cohen made excellent contributions to the preparation and progress of the Conference. In his speech at the closing session, Cohen encouraged Chinese Marxist colleagues to be open-minded, not to regard Marxism as dogma and to help to bring about positive evolution of mainland China in the 1990's.

As the director of the Center for Philosophy and History of Science, Boston University, Cohen received a number of Chinese philosophers as visiting scholars of the Center, and some were invited to give lectures at the Boston Colloquium for the Philosophy of Science. Cohen and

I co-edited a book entitled *History and Philosophy of Science in Contemporary China* as a volume of *Boston Studies*, which contains a number of selected papers from the *Journal of Dialectics of Nature* (1980–1985). Cohen really is a sincere friend of Chinese philosophers. On the occasion of his seventieth birthday, we wish him the best of health, and longevity; and hope that he will make further contributions to academic communication among international philosophical circles.

*Chinese Academy of Sciences*

### REFERENCES

Ai, Si-qi, 1979, *Mass Philosophy*. Beijing: Shanlian Bookstore.
Chen, Duxiu, 1923, 'Preface to Science and the Philosophy of Life', in *Science and the Philosophy of Life*. Shanghai: Oriental Book Co.
Robert Cohen, 1978a, 'Marx, Karl', *Dictionary of Scientific Biography* **15–16** (Supplement I): 403–417.
Robert Cohen, 1978b, *Ibid.*, 131–147.
Robert Cohen, 1963, 'Dialectical Materialism and Carnap's Logical Empiricism', in *The Philosophy of Rudolf Carnap*. LaSalle, Ill.: Open Court Pub.
Deng, Xiaoping, 1984, *Selected Works* (1975–1982). Beijing: Foreign Language Press.
Hu, Qiaomu, 1984, *On Humanism and Alienation*. Beijing: Renming Publisher.
Lu, Dingyi, 1956, 'Let a Hundred Flowers Blossom, Let a Hundred Schools of Thought Contend', *Xinghua Monthly*, no. 13, 74–80.
Mao, Tsetung, 1965, *Selected Works*, v. 1–4. Peking: Foreign Language Press.
Mao, Tsetung, 1977, *Selected Works*, v. 5. Peking: Foreign Language Press.
Wang, Ruoshui, 1986, *Defence of Humanism*. Shanlian Bookstore.
Wu, Jiang, 1978, 'On the Struggle between Two Philosophical Fronts', *Philosophical Research*, no. 1, no. 2.
Xing, Bensi, 1992, 'Some Issues in Question in Current Chinese Philosophy', *Philosophical Research*, no. 5, no. 6.
Yang, Xianzhen, 1981, *My Philosophical 'Crime'*, Renming Publisher.
Zhang, Dongsun (ed.), 1934, *A Polemic against Materialist Dialectics*, Mingyou Publisher.
Zhou, Yang, 1964, 'The Fighting Task of Chinese Philosophers and Social Scientists', *Xinghua Monthly*, no. 1, 74–79
Zhou, Yang, 1983, 'An Approach to Some Marxist Theoretical Problems', *People's Daily*, March 16.

BURTON DREBEN

## COHEN'S CARNAP, OR SUBJECTIVITY IS IN THE EYE OF THE BEHOLDER[1]

"Dialectical Materialism and Carnap's Logical Empiricism",[2] Cohen's long, learned, and remarkably interesting contribution to the Schilpp volume, *The Philosophy of Rudolf Carnap*, is an extraordinary *tour de force*. Written in 1955–57, during the height of the Cold War, it explored sympathetically and in detail Marxist, especially Leninist, criticisms of Logical Empiricism, free from the dogmatic rancor, self-righteous cant, and ideological zeal that characteristically mar such discussions. The point of so exploring was to see whether "[logical] empiricism and dialectic [i.e., Marxism]" could be viewed as "alike or supplementary", despite the long history of intense mutual hostility, the long history of "severely criticiz[ing] [each] other . . . [and of] be[ing] associated with contending forces in practical affairs."[3] At stake, for Cohen, is whether Scientific Humanism, Naturalistic Humanism – Cohen draws no distinction – is viable as a coherent and compelling world-view. He asserts that both Marx and the Logical Empiricists "are scientific humanists"[4]: both Marxism and Logical Empiricism seek "a unified scientific outlook . . . [and] attempt to preserve the role of human reasoning in a context of respect for observed facts and for scientific methods of inquiry, without recourse to supernatural explanation."[5] However, unless the sting of their mutually severe criticisms is drawn, Cohen fears for the viability of Scientific Humanism and for the viability of his fervent hope that "philosophy still has the social function of a persistent, historically concrete, and constructive analysis of human relations."[6] And what better way to draw the sting than to create a perspective from which the two antagonists are seen as complementary? This is the task Cohen, educated as a theoretical physicist, set for himself, "an observer who has participated in neither [Logical Empiricism nor Marxist dialectic] but who shares their repudiation of unconfirmable fancies and undisciplined speculations."[7] Not an easy task. It demands, at the least, instructive ways of construing Marxism and Logical Empiricism. To this end, in the course of his fifty-nine page essay, Cohen gave both a striking reconstruction of Lenin's *Materialism and Empirio-Critcism* as a "set of [27] cumulative theses,"[8] and a highly provocative reading of Carnap, a reading deeply informed by this reconstruction. The result is an important version of Naturalism linked to the form of Marxism taught by the Frankfurt School, particularly by Horkheimer and Marcuse. In this paper, my tribute to Bob Cohen, I shall briefly contrast this version of Naturalism with another that is also intimately connected with a highly provocative reading of Carnap occurring in the same Schilpp volume: Quine's enormously influential "Carnap and Logical Truth."[9] Part of my purpose is to suggest – for the most part implicitly – that just as much of the similarity between the two Naturalisms reflects Hegel's critique of Kant, so their essential difference reflects Hegel's critique of 18th century materialism. Both reject what they take to be Carnap's Archimedean, transcendent(al), stance, both insist that the philosopher cannot stand independently of his philosophy, outside of his philosophy, but they differ sharply on the uniqueness, the singularity, the special role, of his immanent stance. Cohen's hoped-for "naturalized . . . 'reason' "[10] is (or at least appears) alien to Quine's naturalized epistemology (see section 4). A further part of my purpose is to suggest – even more implicitly – how peculiar, how misleading, perhaps indeed how empty, have been philosophical debates about Subjectivity and its mirror image, Realism. And thus to vindicate what moved Carnap in the doing of philosophy, even if not all that he said – of course, under my reading of him (see section 3).

1

In raising as a question the ways in which Marxism and Logical Empiricism might be viewed as "supplementary" Cohen had implicitly rejected a major strain of Marxism. For many Marxists, Historical Materialism, the aspect of Marxism they call a science, and – even more – Dialectical Materialism, the aspect of Marxism they view as the replacement of philosophy, can only be (fully?) understood from a unique 'universal' perspective, the 'class-position of the proletariat'. To think otherwise, to think that one can adopt a 'neutral', an 'objective' standpoint from which one can assess and compare Marxism with Logical Empiricism or any other philosophical position is to completely miss the point, miss the essence, of one of Marx's major achievements: his discovery of the 'class-bound' nature, the 'ideological' nature of (almost all) philosophical discourse. For such Marxists, in the very setting of his task, Cohen had already gone profoundly astray, had already shown his inability to understand Marxism, had clearly disclosed (to use the technical jargon of the trade) his limited petit-bourgeois perspective. They would cite as conclusive evidence of his ideological blindness his feeling free to use the phrase "dialectical materialists" as a

rough label cover[ing] those independent dialectical philosophers who work mainly in social theory, and also adherents of several *different* political as well as philosophical tendencies (emphasis added).[11]

They would see nothing but total misunderstanding in his quoting two passages from the young Marx to justify his claim that Marxism is (a form of) Scientific Humanism, "a unified scientific outlook" on man and nature:

(1) communism as a complete naturalism is humanism, and as a complete humanism is naturalism;[12]

and from the same 1844 manuscript

(2) History itself is a real part of natural history, of the development of nature into man. Later natural science will include the science of man in the same way as the science of man will include natural science. There will be only one science.[13]

Against these passages – *prima facie* quite supportive of Cohen's claim – stands the creed of those to whom the young Marx is neither Marx nor Marxist, the creed summed up in 1968 by Louis Althusser, probably the leading postwar theorist of the French Communist Party: "The whole classical Marxist tradition has refused to say that Marxism is a *Humanism*."[14]

Logical Empiricism, too, is easily read as immediately condemning Cohen's question. Not, of course, by insisting that it can be understood only from its own (privileged) epistemological perspective. Nothing is more central to Logical Empiricism than the thesis that a statement is *cognitively meaningful* only if its confirmation turns on empirical observations available to (practically) everyone. As put by Feigl, in a paper cited by Cohen, a claim to *knowledge* is

in principle capable of test (confirmation or disconfirmation, at least indirectly and to some degree) on the part of any person properly equipped with intelligence and the technical devices [required for empirical] observation or experimentation.[15]

But then comes the rub. Under this criterion many of the most cherished claims of Marxism, let alone the purely normative claims from any quarter whatever, apparently cease to be claims to knowledge, indeed cease to be cognitively meaningful. Again, total incompatibility. Cohen's quest appears doomed from the outset. Or is it?

2

Cohen holds that Lenin's *Materialism and Empirio-Criticism*, "while . . . couched in tendentious language",[16] is primarily "an epistemological argument with historical references"[17] whose intent is that "of countering epistemological and ontological subjectivism."[18] Cohen further holds, in full agreement with Lenin, that such subjectivism must be countered, since it is both "*inadequate* as an account of scientific knowledge and *dangerous* in the restrictions [it] place[s] upon the use of reason in human affairs (emphases added)."[19] Yet it is most difficult to counter; it appears in many guises and is enormously seductive. Empiricism, just because it so emphasizes "the epistemological demand for a basis in experience,"[20] – a demand central to any adequate characterization of (natural) science – is especially susceptible. Examples of philosophical subjectivism, that is, "varieties of subjectivist thinking,[21] to which Empiricism has often been fatally drawn are "solipsism, pure conventionalism, monadic atomism, [and] phenomenalism."[22] Positivism, best seen as "the phenomenalist tendency within the empiricist tradition,[23] is the subject of Cohen's 27th (and final) thesis on Lenin.

Lenin might sum up his argument in this way: Positivism is a philosophy of untroubled, passive, isolated and fixed, atomic facts; it is false to the world of troubled, active, interrelated and developing natural entities. It is mechanically true to the stable aspects of society; it makes no sense of the unstable aspects.[24]

Cohen is not anachronistic; neither Lenin nor his reconstruction is blindly followed. Indeed, one of Cohen's major concerns is to demonstrate through an examination of Carnap's history "to how considerable an extent the Leninist criticisms of logical empiricism have been overcome by developments [within logical empiricism and independent of these criticisms] since the first decade of this century."[25] The developments particularly prized by Cohen are what he views as the turning from Phenomenalism to Physicalism, and the emphasizing of Semantics over Syntax, an emphasis that yielded both the distinguishing of truth and meaning from confirmation and the entertaining of intentional and modal notions in the analysis of science. Briefly put, one might say that Cohen sees latter-day Logical Empiricism and in particular Carnap as having abandoned Positivism and as adopting many of the basic tenets of what is often called (but not by Cohen) "Scientific Realism". Cohen does not attribute this shift to Lenin's attack on Mach, but he does think it shows much of the point and much of the value of that attack. Thus, he sees his examination of Carnap's history as strengthening, as sharpening, the underlying thrust of Lenin's philosophical intent. Cohen contends:

> if . . . subjectivism remains today as only a trap for the empiricist philosophy of science rather than an essential presupposition, it [is] nevertheless . . . illuminating to see how deep the trap has been and might yet be.[26]

### 3

Not unexpectedly, Carnap in his reply to Cohen denies the charge of "subjectivism" both for himself and for his fellow Logical Empiricists, i.e., fellow members of the Vienna Circle. Indeed, he goes further. Using language uncharacteristically strong – except perhaps when discussing Popper or Heidegger – the mild mannered Carnap asserts:

> I certainly agree with Cohen with respect to the dehumanizing effects of the contemporary forms of social and economic organization. But I see *no basis* for the view, frequently maintained by Marxists (since Lenin's criticism of Mach) and apparently also held to a certain extent by Cohen, that *there is a close, perhaps causal connection* between the harmful effects of the social and economic order surrounding us, on the one hand, and a way of thinking labelled *subjectivist* and allegedly represented by positivists, empiricists, and pragmatists (these terms understood here in a wide sense of these terms), on the other (emphases added).[27]

The gauntlet is thrown. Carnap sees "no basis" for Lenin's (and his numerous followers') condemnation of Logical Empiricism as socially "dangerous" nor for Cohen's concern to protect it form such condemnation. Why? Because:

> The theoretical theses of logical empiricism, based on *analyses* of procedures of knowledge and of the structure of languages and conceptual frameworks, are as such *neutral* with respect to possible forms of organization of society and economics (emphases added).[28]

Obviously, little could be more infuriating to any Leninist (probably to any Marxist) than the claim that a philosophical position is "neutral" with respect to different forms of society. The whole notion of superstructure, of ideology, is being challenged. But Carnap's interest is not to deny one of the core tenets of Historical Materialism. Rather, it is to emphasize two core tenets of Logical Empiricism early, middle, and late:

(1) Metaphysical theses are pseudo-theses, are meaningless.

(2) Meaningful philosophical theses, e.g., the "theoretical theses of logical empiricism", do *not* make genuine truth-claims; they are either proposals about linguistic frameworks, or the consequences of such proposals and hence analytic, empty of material content, like the theses of logic and pure mathematics.

Usually, a philosophical insight does not say anything about the world, but is merely a clearer recognition of meanings or of meaning relations. If an insight of this kind is expressed by a sentence, then this sentence is, although meaningful (as we would maintain in contrast to Wittgenstein's view), not factual but rather *analytic*. I would interpret, e.g., the principle of verifiability ( or of confirmability), or the empiricist principle that there is no synthetic *a priori*, as consisting of proposals for certain explications (often not stated explicitly) and of certain assertions which, on the basis of these explications, are analytic. Such philosophical principles or doctrines are sometimes called theories; however, it might be better not to use the term 'theory' in this context, in order to *avoid the misunderstanding that such doctrines are similar to scientific, empirical theories* (emphases added).[29]

Carnap never wavers in his insistence on the meaninglessness of the great classical questions of philosophy (or at best their transformation into questions of mere meaning analysis). The changes, the "developments", that Cohen charts in Carnap's history mark no retreat in this insistence. On the contrary, they afford – in Carnap's eyes – more effective ways of formulating, of clarifying, of explicating, of appreciating this meaninglessness. But they neither afford nor purport to afford (theoretical) *arguments* that *prove*, that *establish*, that *justify* the assertion of meaninglessness. *None can be given, and none are needed.* So Carnap has always held – at least since 1928. Nothing is more essential to him and for the understanding of him, and nothing more conclusive than his declaration from the late fifties:

My friends and I have maintained the following theses (see *Scheinprobleme in der Philosophie*, 1928):

(1) The statement asserting the reality of the external world (realism) as well as its negation in various forms, e.g., solipsism and several forms of idealism, in the traditional controversy are *pseudo-statements*, i.e., devoid of cognitive content.

(2) The same holds for the statements about the reality or irreality of *other minds*,

(3) and for the statements of the reality or irreality of *abstract entities* (realism of universals or Platonism, vs. nominalism).

At the time of the Vienna Circle, the views just stated were sometimes misunderstood. For example, our rejection of the thesis of realism was interpreted as indicating an idealistic position. This interpretation overlooks the important point that we rejected the thesis of realism not as false but as being without cognitive meaning ("meaningless", as we usually said at that time, following Wittgenstein). If the thesis is meaningless so is its negation. . . . *In later years, we assumed that our conception with respect to this question was generally known, although, of course, not generally accepted. Therefore, we did not restate it frequently in recent years, but referred to it incidentally* (emphasis added). To my surprise I see that some empiricists still misunderstand our view.[30]

Hence it should occasion no surprise in us that Carnap rejects any description of his history as a progression from the false position of Phenomenalism to the true position of Scientific Realism. He accuses Cohen of failing

like most critics of positivism and empiricism, to distinguish with sufficient clarity between two fundamentally different meanings of the term 'phenomenalism'. Sometimes, and perhaps in most instances, this term refers to a certain *ontological* thesis which asserts, roughly speaking, the primary reality (in the metaphysical sense) of phenomena, e.g., sense-data, in contrast to material objects. . . . Phenomenalism in the second, *methodological* or *linguistic* sense, may be understood as the *proposal* of a phenomenalistic language as the basis of the total language [of science] (emphases added).[31]

Carnap immediately goes on to claim that even the Phenomenalism in his book *Der logische Aufbau*, written before 1928, was of the second, the linguistic kind, and he reaffirms that under the influence of Wittgenstein's *Tractatus* he and the rest of Vienna came to declare the first or metaphysical kind to be meaningless.

Even before I came to Vienna, I emphasized in my book *Der logische Aufbau* that, although I constructed the language on a phenomenalistic basis, taking sense-data or experiences as starting points, this construction did not imply an acceptance of the metaphysical thesis of phenomenalism. In the Vienna Circle, under the influence of Wittgenstein, we made the distinction in an even more radical way by declaring the metaphysical theses of phenomenalism, solipsism, and idealism, together with the counter-thesis, viz., metaphysical materialism, as *pseudo-theses*, i.e., as devoid of cognitive content (emphasis added).[32]

Carnap's autobiographical remarks show how extraordinarily consistent he has been in *refusing to engage* in "traditional philosophical ways of thinking."[33] They bear quoting at length.

Since my student years, I have liked to talk with friends about general problems in science and in practical life, and these discussions often led to philosophical questions. ... Only much later [1922–25], when I was working on the *Logische Aufbau*, did I become aware that in talks with my various friends I had used different philosophical languages, adapting myself to their ways of thinking and speaking. With one friend I might talk in a language that could be characterized as realistic or even as materialistic; here we looked at the world as consisting of bodies, ... [W]ith another ... I might adapt myself to his idealistic kind of language. ... [and] consider the question of how things are to be constituted on the basis of the given. With some I talked a language which might be labelled nominalistic, with others ... Frege's language of abstract entities ... which some contemporary authors [e.g., Quine] call Platonic. I was surprised to find that this variety in my way of speaking appeared to some as objectionable and even inconsistent. ... When asked which philosophical positions I myself held, I was unable to answer. ... Only gradually, in the course of years, did I recognize clearly that *my way of thinking was neutral with respect to the traditional controversies*. ... When I developed the system of the *Aufbau*, it actually did not matter to me which of the various forms of philosophical language I used, because to me they were merely modes of speech, and not formulations of positions. ... With respect to the problem of the basis, my attitude was again ontologically neutral. For me it was simply a methodological question of choosing the most suitable basis for the system to be constructed, either a phenomenalistic or a physicalistic basis. *The ontological theses of the traditional doctrines of either phenomenalism or materialism remained for me entirely out of consideration*. This neutral attitude toward the various philosophical forms of language, based on the *principle that everyone is free to use the language most suited to his purpose* has remained the same throughout my life. It was formulated as 'principle of tolerance' in *Logical Syntax* and I still hold it today, e.g., with respect to the contemporary controversy about a nominalist or Platonic language (emphases added).[34]

Not that Carnap denies that what look like arguments can be given for and against methodological phenomenalism, and for and against physicalism, i.e, "methodological materialism."[35] But if these are arguments, they are of a special kind. They are "practical"[36] arguments, "practical deliberations and decisions"[37] – truth and falsity are at best instrumental, never primary – "concerning the choice of certain language forms"[38] for "purposes of science."[39] Carnap writes – and not without irony:

I prefer those arguments against methodological phenomenalism which show that a language constructed on a physicalistic basis is more suitable for the purposes of science than a language constructed on a phenomenalistic basis. These arguments against a phenomenalistic method seem to me much clearer and more convincing than those brought forward by dialecticians, be it Hegel or Lenin or contemporary Marxists; but this may be a *subjective* bias on my part (emphasis added).[40]

4

Would Cohen be quelled by Carnap's rebuttal? Would he fall silent? Hardly. It takes little effort to quickly compose a reply totally consonant with Cohen's text, and then to immediately envisage Cohen politely but firmly – Lenin would have been savage – turning the irony thus. "Yes my dear Carnap, I regret to say that you *are* exhibiting subjective bias. My reading of *your* history is more objective, more accurate, than your own. The apparent distinction you draw between ontological and methodological phenomenalism is but a dodge. Your entire apparently neutral, Olympian, transcendental stance, summed up in your Principle of Tolerance, is but a dodge. A dodge that hides, that masks, the harmful social role, the pernicious 'ideological' role, that a refusal to engage in (traditional) philosophical questions plays wittingly or unwittingly."

What is most striking in this (imagined but textually inspired) reply is what is not present. In a long essay brimming with more than one hundred and thirty references Quine is neither used nor mentioned. Cohen never attacks the Carnapian distinctions with the famed – to some the notorious – weapons so carefully forged by Quine for that very purpose in 'Two Dogmas of Empiricism' and 'Carnap and Logical Truth'. This is no slip, no oversight. Cohen's quarrel with Carnap is not Quine's. Cohen has no quarrel with a sharp conventional/factual distinction or a sharp analytic/synthetic distinction as such: "from every philosophical viewpoint, the logical clarification of scientific knowledge demands delineation of factual and definitional components, of objective from arbitrary."[41] Cohen, of course, does question strongly and persistently Carnap's claim that (meaningful) philosophical theses are merely analytic, but such questioning just reflects the deeper ground for Cohen's quarrel with Carnap, a ground totally foreign to Quine: the absence in Logical Empiricism generally and in Carnap particularly of what Cohen views as a sufficient awareness of the social price that a claim of philosophical neutrality, which for Cohen can only be a pose, exacts. It "incapacitates philosophy for its role as critic . . . undermine[s] the possibility of a rational critique of any given state of man or society."[42] To Cohen this absence is but part of a larger absence in Logical Empiricism, namely, the absence of "a dialectical social theory"[43] in which the *role of the philosopher* is itself a central topic of inquiry. And that larger absence is the immediate result, so the physicist Cohen argues, of Logical Empiricism taking physics as the touchstone, the paradigm of knowledge. For physics is not "reflexive",[44] is not self-reflective.

(1) Physics cannot *criticize* physical nature; . . . physics does not discuss nature at large, for its concepts do not *include the physicist* or his *socio-historical* sources and behavior (emphases added).[45]

Logical Empiricism – even in its post-Positivist incarnation (see section 2 above) – is hobbled from the start by a far too restricted, a far too limited vision of *"the knowing activity"*.[46] A vision that

> (2) ignores the very . . . social facts which provide the intersubjective agreement with which physicalism begins; it is not an agreement of *behavioristically* defined men, but rather of *(introspectively) conscious* observers, not objects of study but subjects who engage in inquiry (emphases added).[47]

A vision that leads directly to "a distrust of what was felt to be unverifiable historical theory and hence the *divorce of history from social science* (emphasis added)".[48] But, the argument continues, such a divorce precludes any adequate social theory, any adequate theory of man as a knower and doer, as a social being. For only "a general *historical sociology* . . . a general science of society . . . [can provide a] . . . unified science of knowledge [that] is prerequisite to an adequate philosophy of knowledge (emphasis added)."[49] Because only a general historical sociology is sufficiently reflexive to "account for the social theorist"[50] as well as the social facts, that is, to be indeed unified. And only a general historical sociology can provide

> (3) an *appraisal* of the . . . social universe of observed facts . . . with respect to their genesis, development, epistemic relations, and potentialities. Such [critical] appraisal . . . is foreign to classical science as much as to positivist philosophy (emphasis in original).[51]

Hegel *redivivus* or, at least, Hegel as rewritten first by Marx, and then by the Frankfurt school whose works are Cohen's prime examples of "dialectical social theory". (No reader need be reminded that it was Hegel who asserted the centrality of self-consciousness for history and the centrality of history for philosophy.) We have come to the nub of Cohen's critique, the very heart of his difference with Carnap. It is here, in his insistence on history as a fundamental source of understanding and of knowledge, and in his construal of the nature of that historical understanding, that Cohen shows his true intellectual allegiance and how truly distant he is from Logical Empiricism. And how even more distant he is from Quine. For Quine's critique of Logical Empiricism and of Carnap's transcendental stance, i.e., of Carnap's use of the analytic/synthetic distinction to explicate his conception of philosophy (section 3 above), is linked to a conception of materialism, of physicalism, that both challenges Cohen's strictures and would seem – at least to Carnap

– to be a far more obvious target for those strictures than Carnap's "methodological" physicalism. A few quotations from Quine should suffice[52] to highlight the nature of this challenge. Each should be contrasted with the quotations from Cohen in the preceding paragraph. In particular quotations (1'), (2') and (3') should be contrasted with quotation (1), quotation (4') with quotation (2), and (1') with (3).

(1') I am a physical object sitting in a physical world. Some of the forces of this physical world impinge on my surface. Light rays strike my retinas; molecules bombard my eardrums and fingertips. I strike back, emanating concentric air waves. These waves take the form of a torrent of *discourse* about tables, people, molecules, light rays, retinas, air waves, prime numbers, infinite classes, joy and sorrow, good and evil. My ability to strike back in this elaborate way consists in my having assimilated a good part of the culture of my community, and perhaps modified and elaborated it a bit on my own account. All this training consisted in turn of an impinging of physical forces, largely other people's *utterances*, upon my surface, and of gradual changes in my own constitution consequent upon these physical forces. *All I am or ever hope to be* is due to irritations of my surface, together with such latent tendencies to response as may have been present in my original germ plasm. And all the *lore of the ages* is due to irritation of the surfaces of a succession of persons, together, again, with the internal initial conditions of the several individuals. (emphases added.)[53]

(2') I hold that knowledge, mind, and meaning are part of the same world that they have to do with, and that they are to be studied in the same empirical spirit that animates natural science.[54]

(3') In a general way, therefore, I propose . . . to ponder our talk of physical phenomena as a *physical phenomenon*, and our scientific imaginings as activities within the world that we imagine. [emphasis added.][55]

(4') When a naturalistic philosopher addresses himself to the philosophy of mind, he is apt to talk of language. Meanings are, first and foremost, meanings of language. Language is a social art which we all acquire on the *evidence* solely of other people's overt *behavior* under publicly recognizable circumstances. Meanings, therefore, those very models of mental entities, end up as grist for the *behaviorist's* mill . . . (emphases added.)[56]

De la Mettrie *redivivus*? The words are the words of Quine, the spirit perhaps that of Julien Offray de la Mettrie whose *L'Homme Machine* was criticized as "one-sided" by Hegel in his *Lectures on the History of Philosophy*, and who is a perfect example of the "mechanical", non-dialectical, materialists so criticized by Marxists since Engels's *Anti-Dühring*, the precursor of *Materialism and Empirio-Criticism*.

Further insight into what is fundamental for Cohen is gained from two remarkable footnotes – the 56th and the 60th – whose force is but heightened when we recall that Neurath, Carnap's "close associate,"[57] was in practice far more to the Left and thought himself far more a Marxist than Cohen ever was or thought himself. First footnote 56:

> The [logical] empiricists, upon reflection, recognized that [the demand for reflexiveness] penetrated their own attitude toward social science also. Much might have been made of Otto Neurath's remark that "The general revolution of our age is the ground of a scientific sociology," in *Empirische Soziologie* (Vienna, 1931), 146, if he had not already repudiated any rational assessment of prospective *fundamental* social change; "We must wait for the new phenomena in order that we might then discover lawlike regularities for them in their turn," *ibid.*, 1106.[58]

The key words are, of course, "rational assessment of prospective ... change". They echo "appraisal of ... potentialities" in quotation (3) two paragraphs back, where the contrast between historical sociology and natural science is sharply drawn. Both phrases flag what to Cohen is the core of historical understanding, the point of historical knowledge, and what he takes, not incorrectly, as the basic claim of Historical Materialism: it provides a mode of critically discerning, of rationally estimating, underlying social trends and tendencies. (Another nod to the (purist) reader learned in Marxism. Yes. The title of the essay with "Historical" in place of "Dialectical" would have been less misleading, even granted Cohen's "rough labels" – see section 1 above.)

Now to (the first fourth of) footnote 60:

> It is positivism and conventionalist relativism, rather than dialectical materialism, which support the public philosophy of Big Brother in George Orwell's political novel *1984*. Whenever a political regime undertakes to provide a theoretical defense of its manipulative distortions we see what subjectivist weakness in theory may legitimate in practical life, e.g. rewriting history. Thus, when the editors of the *Large Soviet Encyclopedia* instructed their subscribers to remove the old page about the late L. Beria in order to substitute not an objectively corrected page about Beria but a new discourse about the Bering Sea, they abandoned, *in social practice*, the Marxist idea of objective truth, and embraced relativism. The inability of positivist thinkers to cope with relativism in social theory is linked with their subjectivism; Max Horkheimer pointed to the "naive" espousal of relativism by such distinguished positivist thinkers as the sociologist Otto Neurath and the legal theorist Hans Kelsen, (whose political commitments were genuinely democratic), and he contrasted Mussolini's realistic appraisal of the ideologically subversive role of relativism.[59]

Carnap's reaction to footnote 60 documents the divide. Cohen would see Carnap as avoiding the issue, being much too defensive. Carnap sees himself as exercising practical reason, showing political wisdom.

> Cohen believes that it is subjectivist positivism rather than dialectical materialism which would be the totalitarian philosophy of Big Brother (in Orwell's sense). Let us examine Cohen's view with regard to two illustrations which he himself uses to characterize this

dangerous positivism which is supposed to lead to totalitarianism. The first is a certain change made in a new edition of the *Large Soviet Encyclopedia*, the second is Hans Kelsen's legal theory. The latter theory is correctly classified by Cohen as positivistic (in a wide sense). But to what kind of totalitarianism did Kelsen's thinking lead? Cohen says of him only that he adhered to relativism; therefore he mentions him in the same breath with Mussolini, because the latter also maintained relativism in his essay, 'Relativismo e fascismo'. Leaving aside the ambiguity of the term 'relativism', I think this association of Kelsen with Mussolini will appear amazing to those who remember Kelsen as one of the main creators of the constitution of the Austrian Republic in 1919, a constitution which is regarded by many throughout the world as the model of a democratic constitution. And that the Soviet regime is here taken as representative of positivism in contrast to dialectical materialism will surprise most readers, even if they do not know that, since Lenin's book against Mach's philosophy, no author in the Soviet Union has dared to discuss neopositivism with sympathy or even only with an attitude of objective, unbiased criticism.[60]

Carnap closes, first, by quoting one of Cohen's most passionate sentences:

Without a concept of Reason which will permit realistic *criticism* of the world as revealed by experience, positivistic empiricism constructs a world of empty and inhuman mechanisms, a cyclic flux, divorcing the *human spirit from natural process* (emphases added)[61]

And, then, by displaying his own passion in the guise of displaying bewilderment:

I find it difficult to reconcile this picture with what Cohen himself has said at other places. . . . He has repeatedly explained that since Vienna, especially Neurath and I, have *criticized* the existing order of society as unreasonable and have demanded that it should be reformed on the basis of scientific insights and careful planning in such a way that the needs and aspirations of all would be satisfied as far as possible; this attitude is the core of our *scientific humanism*. . . . [he] has explained our physicalistic and naturalistic conceptions which are diametrically opposed to any dualism that divorces *human spirit from natural process*, and which emphasizes that man with all his experiences in various spheres of life, not excluding those of philosophical thinking, is just part of one all-embracing nature (emphases added).[62]

There is a genuine divide between the two. But is there a genuine debate between the two? Or do we find here grounds for Rawls's claim that fruitful political philosophy need not (should not?) rest on any metaphysical or comprehensive philosophical position? And would that claim not add some weight to Carnap's fundamental attitude to the doing of philosophy?

Thus ends my dialectical tribute to Bob Cohen, but not the dialogue.

*Boston University*
*Harvard University*

## NOTES

[1] This paper grew out of conversations with John McNees about Hegel and Quine, and with Michael Shepanski about Carnap and Quine. I am most grateful to them. I am also grateful to Robert Cohen, Juliet Floyd, Charles Griswold, and John Rawls for advice and comments on earlier drafts. Richard Creath, Michael Friedman, Warren Goldfarb, W. D. Hart, Jaakko Hintikka, Peter Hylton, Daniel Isaacson, Edward Minar, Hilary Putnam, and Thomas Ricketts in their different ways of differing with me about Carnap on philosophy have taught me much that is reflected, often obscurely, in this paper. I thank them.
[2] Schilpp, pp. 99–158.
[3] Schilpp, p. 101.
[4] Schilpp, p. 99.
[5] Schilpp, p. 101.
[6] Schilpp, p. 158.
[7] Schilpp, p. 101.
[8] Schilpp, p. 134.
[9] Schilpp, pp. 385–406. Much of this paper had appeared in 1956 in Sidney Hook's *American Philosophers at Work*.
[10] Schilpp, p. 154.
[11] Schilpp, p. 101, ft. 5
[12] Schilpp, p. 99.
[13] Schilpp, p. 101.
[14] Althusser, *Lenin and Philosophy*, p. 22.
[15] Feigl, "The Scientific Outlook: Naturalism and Humanism", *Readings in the Philosophy of Science*, p. 11.
[16] Schilpp, p. 133.
[17] Schilpp, p. 133.
[18] Schilpp, p. 134.
[19] Schilpp, p. 105.
[20] Schilpp, p. 105.
[21] Schilpp, p. 105.
[22] Schilpp, p. 105.
[23] Schilpp, p. 105.
[24] Schilpp, p. 140.
[25] Schilpp, p. 134.
[26] Schilpp, p. 134.
[27] Schilpp, p. 865.
[28] Schilpp, p. 865.
[29] Schilpp, p. 917.
[30] Schilpp, p. 868
[31] Schilpp, p. 863.
[32] Schilpp, p. 863.
[33] Schilpp, p. 17.
[34] Schilpp, pp. 17–18.

[35] Schilpp, p. 864.
[36] Schilpp, p. 869.
[37] Schilpp, p. 862.
[38] Schilpp, p. 869.
[39] Schilpp, p. 864.
[40] Schilpp, p. 864.
[41] Schilpp, p. 113.
[42] Schilpp, p. 106.
[43] Schilpp, p. 127.
[44] Schilpp, p. 128.
[45] Schilpp, p. 127.
[46] Schilpp, p. 128.
[47] Schilpp, pp. 147–148.
[48] Schilpp, p. 127.
[49] Schilpp, p. 128.
[50] Schilpp, p. 127.
[51] Schilpp, pp. 127–128.
[52] For a discussion of these quotations from Quine, see my paper 'Putnam, Quine – and the Facts' and Putnam's "Reply to Dreben" in *Philosophical Topics* **20**, No. 1, Spring 1992. See also my paper 'Quine' in *Perspectives on Quine*. (Only the first of these quotations was printed before 1957.)
[53] Quine, 'The Scope and Language of Science' (1957), reprinted in *Ways of Paradox*, 2d ed., 228.
[54] Quine, 'Ontological Relativity', *Ontological Relativity and Other Essays*, p. 26.
[55] Quine, *Word and Object*, p. 5.
[56] Quine, 'Ontological Relativity', *Ontological Relativity and Other Essays*, p. 26.
[57] Schilpp, p. 157, ft. 128.
[58] Schilpp, p. 127, ft. 56.
[59] Schilpp, pp. 128–129, ft. 60.
[60] Schilpp, p. 866.
[61] Schilpp, pp. 154–155.
[62] Schilpp, p. 867.

## BIBLIOGRAPHY

Louis Althusser, 1971, *Lenin And Philosophy, And Other Essays*. New York: Monthly Review Press.

Robert Barrett and Roger Gibson (eds.), 1990, *Perspectives On Quine*. England: Blackwell, paperback 1993.

Herbert Feigl and May Brodbeck (eds.), 1953, *Readings In The Philosophy Of Science*. New York: Appleton-Century-Crofts.

Sidney Hook, 1956, *American Philosophers At Work*. New York: Criterion.

'*Philosophical Topics*' **20**(1), Spring 1992. Fayetteville, Ark.: University of Arkansas Press.

W.V. Quine, 1960, *Word And Object*. Cambridge, Mass.: MIT Press.
W.V. Quine, 1969, *Ontological Relativity And Other Essays*. New York: Columbia University Press.
W.V. Quine, 1976, *The Ways Of Paradox And Other Essays*, 2nd edition. Cambridge, Mass.: Harvard University Press.
Paul Arthur Schilpp, 1963, *The Philosophy Of Rudolf Carnap*. La Salle, Ill.: Open Court.

WŁADYSŁAW KRAJEWSKI

# STALINISM AND SOVIET SCIENCE

The impact of Stalinism on Soviet science, especially biology, is one of the paradigmatic examples (besides the Inquisition and Nazism) of the negative influence of external factors upon the growth of science. In this paper we present the harmful impact of Stalin's policy on Soviet science and some cases of resistance to it.

We shall speak of science in the narrow sense of this term, i.e. about the natural sciences. However, at the beginning – few words about the situation of humanities in Russia after the October Revolution in 1917.

In the first years of Soviet rule a deep purge took place in departments of humanities in all Russian universities. All professors of philosophy were removed from universities on the charge of idealism. They were indeed adherents of various kinds of spiritualism. Many professors of other branches of humanities – history, economy, law, etc. – were removed as well, although in no branch was the purge so total as in philosophy. Arrests of professors were at that time rare. By the end of 1922, three hundred professors of humanities – among them famous philosophers like Berdiayev, Lossky and others – were expelled from Russia. A special train organized by Lenin took them; they became professors of the universities in Germany. France and other countries. Professors of humanities who preserved their jobs in the USSR were under an increasing ideological pressure. They had to deliver their lectures in the spirit of Marxism and socialism.

The situation of the natural sciences was quite different. The Soviet government attached great importance to science and technology. It was aware that the active participation of scientists is necessary for the realization of great plans of industrialization, electrification, etc. It appealed to the Academy of Sciences which remained for many years without change from the pre-revolutionary time for participation in the plan of electrification, in investigation of geological resources, especially the so-called Kursk Magnetic Anomaly, etc. Almost all members of the Academy were hostile towards Communism; therefore the response was not immediate. However, the task was in agreement with the vocation of science and no ideological declarations were at that time demanded from the natural scientists. Hence, they took up the job, some of them after lingering, some others, with enthusiasm, since during the civil war the Communist government started to modernize Russia.

People ruling the country – the role of Lunacharsky was here important – understood that it is impossible to develop the applied sciences in the long run without the growth of the basic, theoretical sciences. Therefore, they took care on the conditions of scientific investigation in all branches, not only in those which gave immediate practical effects. Scientists were privileged in the rationing of food in the time of hunger. The equipment of laboratories was also considered an urgent task. The most striking example was the building of a special scientific station in the vicinity of Leningrad for the greatest Russian scientist, Nobel prize winner, Ivan Pavlov, although he did not hide his hostility to Communism. He even demanded the building of a chapel in the station saying that his wife is very religious, and the chapel was built! Later on Pavlov praised the Soviet government for its generosity. At the same time he protested against arrests of some his students. Another example: A professor of physics at the Leningrad university O. Khvolson was a spiritualist in philosophy. He had been sharply criticized earlier by Lenin in *Materialism and Empirio-Criticism* for the publication of a "Black-Hundreds reactionary" pamphlet against Ernst Haeckel. Nevertheless, Khvolson kept his university job till his retirement (without any self-criticism).

Of course, a strong ideological campaign in the 20's and later concerned natural sciences, too. This campaign was led by the members of the Communist Party, mainly philosophers. Their papers and books combatted idealism – genuine or spurious – in all sciences, including physics, chemistry, biology and even mathematics. In these papers "false" opinions of Western and Soviet scientists were criticized. However, the criticism was directed only against philosophical commentaries, not against the scientific theories themselves. All theories in natural sciences were allowed, all new conceptions were presented in books and journals, e.g. Relativity Theory of Einstein, Quantum Mechanics of Bohr, Heisenberg and Schrödinger, new ideas in chemistry, astronomy, genetics, ecology, even psychology (Freud) and pedagogics (Dewey). Soviet scientists participated in their elaboration.

The situation began to change for the worse at the turn of the 20's and 30's. Ideological declarations were demanded from scientists: the approval of "diamat" (dialectical materialism) and the criticism of idealism and of mechanism. In the introductions to handbooks and even monographs the mention of diamat and citations of the classics of Marxism (mainly, Engels and Lenin) became obligatory. Mostly scientists did it, sometimes with conviction, more often as a duty. They preserved freedom inside science and this was the most important thing for them.

We may notice an analogy with the situation of science in Nazi Germany. As Heisenberg and others wrote later, German physicists agreed to such ritual gestures as raising of the hand in fascist greeting when entering the lecture room or ending each official letter with the words "Heil Hitler!". In the USSR such gestures were not used but others were. Beside the above mentioned remarks in the introductions, we may notice such duties as participation in political meetings, standing up and applauding when Stalin's name was uttered.

In the second half of 30's it turned out that such gestures were not sufficient. The period of the "great terror" came. Millions of people were arrested and sentenced to concentration camps, many of them to death. Scientists were, of course, among them. True, among the natural scientists, especially mathematicians, physicists, chemists the number of the victims was essentially lower than among the professors of humanities, not to mention engineers, the military, government officials, etc. Nevertheless, there were arrests and every arrested person was proclaimed to be an "enemy of the people". They were condemned at political meetings, and not joining in the condemnation was dangerous, so not many dared avoid this.

The best-known victim of Stalin's terror among scientists was Nicholai Vavilov, the greatest Russian biologist, one of the pioneers of genetics. At the beginning of 30's Trofim Lysenko, a member of the Academy of the Agricultural sciences, led by N. Vavilov, began to attack the "formal genetics". He propagated "Michurinian biology", also called "Soviet creative Darwinism", although it had little to do with genuine Darwinism. (Darwin was praised by Marx and Engels, hence his name was useful.) According to Lysenko, the whole of living matter is the bearer of heredity and not just the chromosomes and the mere genes ("nobody has seen them"). The Lamarckian idea of the inheritance of acquired characteristics was proclaimed, hence the possibility of changing the nature of species by changing the condition of their life. Vavilov and other geneticists argued against these unjustified ideas. Soviet philosophers were mostly neutral in this controversy at that time. Only a few of them backed the Michurinian biology.

Lysenko succeeded in removing Vavilov from his post as President of the Academy of Agricultural Sciences and soon acquired this post himself. In 1939 N. Vavilov was arrested (under some false fantastic accusation) and soon died in prison. His public defence became impossible but the defence of genetics was still possible – till 1948. In this year the famous session of the Academy of the Agricultural Sciences took place where Lysenko made a militant speech against genetics and announced that his speech was approved by the Central Committee of the Communist Party. Then the discussion was terminated. Soon adherents and flatterers of Lysenko seized all important posts in Soviet biology. Some of Lysenko's adversaries in biology (not only in genetics) were forced to engage in self-criticism but the most eminent ones like Dubinin or Shmalhausen refused such "self-criticism" and then were removed to the margins of scientific work. There were several cases of suicide.

Lysenko's adherents in biology, together with philosophers who joined them, began a campaign against genetics, called also "Weissmanism" or "Mendelism-Morganism". It was branded as idealist and metaphysical because it considered the "hereditary substance" to be autonomous and "separated from the body" (like the soul . . .). Besides, it was held to be irrelevant for agriculture because it dealt with flies (drosophila) and not with cultivated plants. The drosophila became a symbol of evil, its breedings were annihilated. Finally, genetics as a science was charged with being associated with eugenics and racism.

At the same time in neurophysiology and in medical science, in general, a parallel ideological campaign started. A dogmatic version of the theory of the late Ivan Pavlov was propagated. Not only were Pavlov's adversaries like Beritashvili blamed, but also his followers like L. Orbelli, who did not agree with the officially approved version of Pavlov's theory elaborated by Bikov and Ivanov-Smolensky.

In agriculture, the two-field system of Williams (backed by Lysenko) was propagated and other conceptions were condemned and some of their adherents were arrested. In the 50's Lysenko proclaimed the possibility of developing new species of corn by changing the environment. He approved pseudo-scientific experiments performed by Boshyan, Lepeshinskaya and others. All of them presupposed the possibility of deep hereditary changes by simple outer action.

In general, Lamarckian ideas of the decisive role of the environment were popular among Communists because they permitted the moulding of people's characters according to the plans of educators. Therefore, in pedagogics, all conceptions which admitted the essential role of heredity were condemned. They were associated with racism and petrification of social inequalities. These associations were not entirely groundless but the assumption of equality of people (in theory, not in practice!) assumed an extreme form. The measuring of IQ was refuted as alleged justification of the unsurmounted intellectual inequality of students and the helplessness of educational efforts. Education was held to be omnipotent.

We see that the official ideology had a harmful influence upon Soviet biology and other sciences. However, it also had a positive impact. Materialism fostered the investigation of such problems as the origin of life on the Earth, without religious inhibitions. As is known Alexander Oparin was a pioneer in this domain. His merit was widely proclaimed in the USSR and later acknowledged by world science.

In the more exact sciences like mathematics, physics, chemistry, the situation was much better than in biology. However, here also the criticism of opinions held to be idealistic or mechanistic increased. It was, in general, still the criticism of the philosphical interpretation of scientific theories and not of these theories themselves.

The Copenhagen interpretation of Quantum Machanics, elaborated by Bohr and Heisenberg, was condemned as idealistic and indeterministic, although it was approved by the vast majority the world's physicists. At the same time the opposite interpretation of QM by the Parisian school (de Broglie, Vigier) was held to be mechanistic. A vivid discussion took place among Soviet physicists. The majority of them in fact admitted the Copenhagen interpretation although rather abstained from the direct expression of this view, avoiding "philosophical reasoning". Only some of them actively engaged in philosophical discussions. W. Fock, in general, approved the Copenhagen interpretation although he criticized some philosophical remarks of Bohr. Another leading physicist D. Blokhintsev elaborated his own interpretation of QM, the so-called "quantum ensemble" conception opposing it to Copenhagen and to the Paris interpretations. The third philosophizing physicist, Y. Terletsky, was closer to the Paris school but he was rather isolated. The main philosophical discussion took place between Fock and Blokhintsev. The former ascribed probability to single quantum events, the latter, only to ensembles of events. The discussion lasted many decades, in fact until the death of both adversaries. Fortunately, the Communist Party did not intervene in this discussion and other discussions in the domain of physics, although this threat was near.

Soon after the biological attack in 1948, preparation for an analogous attack in the domain of physics was initiated. Some most dogmatic philosophers claimed that both Quantum Mechanics and Relativity Theory in their received shape are idealistic, they blamed some leading Soviet physicists for the neglect of diamat, etc. However, soon after it began, this campaign suddenly stopped. It was evident that its interruption (like its beginning) was decided by Stalin. Why? Probably this interruption was indebted to the construction of nuclear weapons. According to information stemming from I. Kurchatov, the head of the nuclear project, he had a conversation with L. Beria who was the supervisor of this project on behalf of the government. Beria asked: are QM and RT indeed idealistic? Kurchatov answered: without them the construction of the atomic bomb will be impossible. Beria rejoined: the most important thing is the bomb. Probably, Beria reported this to Stalin who decided to stop the campaign. As a Russian journalist wrote not long ago, "the bomb saved physicists".

The situation with relativity was more quiet till 1952. Both special and general relativity were acknowledged. Sometimes Einstein's philosophical views were criticized but rather rarely. It is worth noticing that in 1922 Lenin called Einstein "a great reformer of science" and criticized only the idealistic interpretations of his theory by some philosophers. This muted the criticism of Einstein. In the 40's only A. Maximov, the most dogmatic philosopher of science, criticized the Relativity Theory. However, in 1952 a collection of papers on philosophy of physics was published, the so-called "green book" (the cover was green). In it a violent attack on RT was contained. An expression "reactionary Einsteinism", was used several times. It described the alleged idealistic interpretation of RT by Einstein. But some months later Stalin died and the situation had changed. Leading physicists condemned the authors of the "green book" as ignoramuses in physics. The authors of the book kept silent. The expression "reactionary Einsteinism" was never repeated. It lived only half a year.

A long time after this Soviet philosophers still rejected the relativistic cosmology. It was held to be idealistic because it admitted the finiteness of the universe. (In fact it admitted only such a possibility). The infinity of the universe both in time and space was a dogma and it took some time before Soviet philosophers got free of this dogma. The situation was paradoxical: general relativity was acknowledged and its application to cosmology was condemned. This situation came to an end only in the 60's.

At the turn of the 40's and 50's another ideological campaign started: against cybernetics. It was proclaimed to be mechanistic because of the alleged neglect of the essential difference between human beings and machines. In the press, articles appeared with the following titles: *Cybernetics – a pseudo-science, Whom does cybernetics serve*? The hostility to cybernetics was, probably, caused by the fact that N. Wiener initiated it in the USA in the period when the cold war began and America was held to be a nest of evil. This campaign lasted not long until the end of the 50's.

The beginning of the cold war was a cause of the most mournful ideological campaign: a battle against "cosmopolitanism". It was not quite connected with Marxism. It was rather a revival of Russian nationalism. This campaign had two sides: the anti-Semitic one (an ethnic purge began) and the anti-Western one. In science, a campaign of the "defence" of the priority of Russian scientists took place. When a Russian had some role in a discovery, even a small one, he was proclaimed as *the* discoverer. We may adduce some examples: In the XVIIth Century I. Polzunov constructed a primitive steam engine without any practical use (a similar one was made by Heron of Alexandria) – he was proclaimed the discoverer of the engine before Watt. In the XVIIIth Century M. Lomonosov formulated the law of conservation of matter in a very vague way (an analogous formulation can be found in Lucretius' poem). He was proclaimed the discoverer of this law before Lavoisier. In the XIXth Century A. Butlerov investigated the structure of molecules simultaneously with Kekulé. The Russian chemist was proclaimed the discoverer of molecular structure, although Kekulé's work was more important. In general, it is the case that several scientists participate in each scientific discovery. It is a "many-subject process" – to use the expression coined by my friend Elzbieta Pietruska. When only one of the participants is mentioned, and this is not even the main one, then the picture of the history of science is distorted. At the beginning of the 50's, the nationalistic campaign assumed a grotesque form: the portraits of Western scientists were removed from the walls in the schoolrooms and replaced by the portraits of Russian scientists. After Stalin's death the wave of Russian nationalism gradually died down, but not completely.

As we see, Stalinism led to many serious damages to Soviet science. However, history is rarely simple and one-directional. Under Stalin some positive changes in the ideological situation of science also took place. In the 20's dialectical materialism was proclaimed in the USSR to be a general method of all sciences, including the applied ones. At the beginning of the 30's, this was regarded as a dogmatic exaggeration. In the Communist Party newspaper *Pravda* an article appeared under the title: Don't use dialectics in the smithy! A popular writer A. Korneychuk in a theatre play presented a grotesque person, a woman in the department of health, who organized a "scientific" conference with papers: *The dialectical method of the treatment of tuberculosis, Idealism in the surgery*, etc. Since that time there were no more ideological requirements in the applied sciences.

I want also to note, though it goes beyond our theme, that in the middle 30's some positive changes took place in school-teaching, especially in the humanities. (I remember them from my student-experience). Lessons in history were no longer reduced to social-economic changes and class-struggle, and the policy of governments was also considered as was the role of eminent persons, including kings, etc. Lessons in literature began to consider aesthetic values beside the social content of novels, etc. However, the level of teaching of history and other humanities was still low, in contradistinction to the level of mathematics, physics and chemistry which was always high in Soviet schools.

Let us go back to the natural sciences. The situation of biology after Stalin's death was complicated. Geneticists got the opportunity to work with drosophila and similar material, but Lysenko retained a strong position. These two currents ran parallel. There were two institutes in the same domain of biology without any communication between them. Lysenko did not give up because he was still supported by Khrushchev who knew him from the time they both worked in the Ukraine. Khrushchev believed that Lysenko would revolutionize agriculture whereas the geneticists cultivated flies. This situation changed immediately after the removal of Khrushchev in 1964. The Central Committee of the Communist Party appealed to leading biologists to write articles denouncing Lysenko who was soon removed from all his posts. It was a paradox: as is known, Khrushchev criticized many of Stalin's deeds and Brezhnev restrained this criticism. In the biology we have to do with the opposite: Khrushchev backed Lysenko and Brezhnev did not.

We may say that in 1964 Stalinism in Soviet natural sciences eventually came to an end. Of course, in the social sciences it lasted till Gorbachev, but this is not our theme.

* * *

In the period of Stalinism not all Soviet scientists submitted to the pressure. Some of the leading physicists not only ignored ideological pressure but often actively resisted it. Landau, Tamm, Kapitsa and many others defended not only QM and RT but also their founders against ideological attacks. They tried to avoid philosophy and even more, politics, but sometimes they expressed unorthodox views. e.g., Kapitsa mentioned Freud and Keynes positively in some articles while they were condemned by the official ideology. They protested against arrests of some his students, etc. The mathematician A. Lapunov led a seminar in Novosibirsk on cybernetics in the time when it was condemned as pseudo-science.

A little more about the striking case of the geneticist Nicholai Timofeyev-Ressovsky. His story was told in 1987 in a documentary novel by the well-known Russian writer Daniil Granin, under the title *Zubr*.

*Zubr* (aurochs) is a European variety of bison. It lives in a large forest at the border of Poland and Belorussia. Timofeyev-Ressovsky was called by friends "zubr" for both his physical and (especially) his moral strength. In 1925 he was delegated to work at the Brain Institute in Berlin. There the young Russian scientist found good conditions for work, he led a big laboratory and got a high position in science. In the 30's he refused to return and remained in Germany. During the period of Nazism he managed to save some colleagues of Jewish origin by doctoring their documents. In 1945 he declined a suggestion to move to the USA, waited for the Red Army and was, of course, arrested as a "refusenik". He was soon sent to a special camp in the Urals with the task of investigating biological effects of nuclear radiation. He had as colleagues both arrested Russian scientists and German war prisoners, including his former collaborators in Berlin. He managed to continue some of his genetics research with drosophila – besides the investigation which he was obliged to do. His direct chief, an NKVD officer Uralets turned out to be an extra-ordinary person. He took interest in the genetics research of Timofeyev-Ressovsky and not only allowed it but hid it in his reports to the superior authorities. Again, a paradox: in the time when genetics was banned and even the Drosophila cultures were annihilated, there remained an isolated camp in the Urals where genetics research was carried on by arrested people and war prisoners!

A known economist and publicist Gavril Popov (who later became the Mayor of Moscow), inspired by Granin's novel, published in 1988 an article *The System and the Zubrs*. He noticed that Timofeyev-Ressovsky was not alone. There were also other "zubrs" who resisted the Stalinist system: N. Vavilov, A. Lapunov, P. Kapitsa and even Uralets from NKVD who risked his head.

Again, we may see an analogy with the behavior of great German scientists in the Hitler time. Max Planck, Werner Heisenberg, Max von Laue after 1933 did not manage to defend their colleagues of Jewish origin. (These latter were removed from their posts and mostly emigrated); but they never renounced relativity and were contemptuous towards the "German physics" of Lenard and Starck. They were, to use the terminology of Granin and Popov, "zubrs" in Nazi Germany. Both Soviet and German "zubrs" made some "ritual gestures" of compliance (some of them to a very little degree, Timofeyev-Ressovsky not at all) but never gave up the crucial causes of science.

*University of Warsaw*

PETER MARCUSE

# HERBERT MARCUSE ON REAL EXISTING SOCIALISM: A HINDSIGHT LOOK AT *SOVIET MARXISM*

Herbert Marcuse's *Soviet Marxism*, published in 1958,[1] has, I believe, a great deal to contribute to an understanding of current developments in the Soviet Union and eastern Europe. I admit that I am not unbiased in the matter,[2] but I hope the argument presented here will be examined on its own merits. I do not know whether my father discussed any of the issues in the book with Bob Cohen. The book was certainly written before they knew each other, but I know that they discussed many issues together, and that my father valued Bob's opinions and advice highly both on political and on theoretical matters. My father's critical stance vis-a-vis the Soviet Union was, I believe, consistent throughout his life, and if *Soviet Marxism* was written without the benefit of discussions with Bob, his other major piece on the topic was not: the last (and, to my knowledge, the only other) piece he wrote directly on Eastern Europe, the text of the talk given at the Bahro Congress in Berlin in 1978, "Protosocialism and Late Capitalism: Toward a Theoretical Synthesis Based on Bahro's Analysis."[3]

The recent events in the Soviet Union and its successor states has had a major impact on Marxism, both in theory and in practice, in the First World and in the Third World. It is easy to see in the Third World; in many countries national struggles took place in the space opened by the tensions between the great powers, and in others liberation movements had become largely dependent on the Soviet Union for their strength: Cuba, for example. In the First World, the impact has been stranger: while the left continues to adhere to the position that Soviet-style systems had nothing to do with their own goals, the collapse of the Soviet regimes has, perhaps unconsciously, been internalized as a defeat for the principles of socialism and the political parties that had espoused socialism. It was certainly seen as a defeat for socialism by the right, and certainly successfully sold as such in the popular media.

Marxist theoretical analysis has likewise had a strange aspect, strange in that it has largely shied away from analysis of what actually happened within the Soviet Union and its kindred states, what in the past produced the present, and speaks only of the present and the future, predicting the disasters attendant on the forced introduction of a market system in a hitherto centrally planned state, etc. At best Marxist theory has been applied to the present class structure of the successor states. The occasion has not as yet been used to reflect on whether the present events require a reexamination of the past, and whether that reexamination would lead to a different analysis of the long-term processes of transformation and the "transition to socialism" with which Marxism has always been concerned. And yet, it seems to me, a reexamination is indicated, for, with very few exceptions, the course of recent history was hardly predicted by even the acutest analyses of the past. Predictions of ultimate collapse, indeed, abounded, from both left and right; but it can hardly be said that the development of the reform movement within the Soviet Union, symbolized by glasnost and perestroika, was widely foreseen, nor that the almost spontaneous melting away both of entrenched Stalinist regimes and of the efforts at their reform would occur as they did.

I would like to argue that Marxist theory would, indeed, suggest the likelihood of the trends that have led to the present results. I would further like to argue that both *Soviet Marxism* and the Bahro review, "Protosocialism . . . ," can contribute significantly to an understanding of those trends and of their consequences today.

Both pieces deduced the presence of trends towards change, towards reform, in the Soviet system. In *Soviet Marxism* the analysis suggests an internal necessity of liberalization in the Soviet Union, although it is skeptical as to whether such "liberalization" will change the essentially non-socialist character of the system. In the text on Bahro, Marcuse calls Bahro's book, *The Alternative: A Contribution to the Critique of Actually Existing Socialism*, "the most important contribution to Marxist theory and practice to appear in several decades," and Bahro in turn sees in internal developments in the German Democratic Republic the hope of a fundamental change in the character of the Soviet-style society from within.

I want here to focus on the discussion of these liberalizing trends in the Soviet Union in these two works, although in each that discussion is really secondary to the main analysis. In "Protosocialism," Marcuse's main attention is devoted to the implications of Bahro's analysis of east European societies for the possibilities of change in the west, in particular the relationship of "base" and "superstructure" and between the change to individuals and the change of societal structures. Bahro finds agents of change present in Soviet-style systems; Marcuse explores the existence of analogous agents in the West. In the process, he accepts rather than explicitly critiquing Bahro's analysis of the situation in the G.D.R. Silence suggests consent, although perhaps criticism of Bahro's views of G.D.R. society was muted because of Bahro's immediate position, then in jail for precisely those views.

The formulations of *Soviet Marxism* also were undoubtedly influenced by the circumstances in which it was written. Alone among Marcuse's works written under contract, it was the product of stays at the Columbia and Harvard Russian Research Centers during the period of McCarthyism and at the height of the agitation justifying the Cold War. Outright defense of the Soviet Union was not in the cards at either institution, nor would Marcuse have wished to undertake such a defense; but an explicitly Marxist approach based on the validity of Marxist conceptions would not have been widely understood either. On the other hand, a wholesale attack on all aspects of Soviet society would have been easily misconstrued as an attack on socialism and a rejection of Marxism as a whole. Thus the formulation of an "immanent critique" of Soviet Marxism, with which Marcuse begins the book and which he is at pains to justify theoretically in language that, today, seems forced and unnecessary to the main task;[4] but it served the purpose of permitting a Marxist critique of a pseudo-Marxist theory and a pseudo-socialist (later "protosocialist") reality, which was the real aim of the book. The "immanent critique" is productive; its discussions of the transformation of Marxian theory as it "ceases to be the organon of revolutionary consciousness and practice and enters the superstructure of an established system of domination" is fascinating; the detailed discussion of the dialectic as it is transformed from a critical tool of social analysis to an all-embracing philosophical system, for instance, is a model of clarification of the history of ideas.[5]

Marcuse's critique of Soviet Marxism as theory and ideology is not, however, that on which I wish to focus here. My concern here is rather to examine the discussion of Soviet-style societies to gain a better understanding of the forces for change in that society, and what implications that analysis has for an assessment of the forces of change within our own society. A better understanding of the "base-superstructure" relationship is important for this process, for the question of what changes, or changes to what extent, in the "base," can produce fundamental changes in to societal system as a whole, and/or to what extent "superstructural" changes play in societal transformations, remains a central question both for theoretical analysis and for political practice.

If the Soviet Union was not capitalist, it was not socialist either, "in the sense envisaged by Marx and Engels."[6] In classic Marxist terms, the difference lay in the ownership of the means of production: they were nationalized, but not socialized, not put in the control of the "immediate producers." This was seen as an intermediate stage in the transition to socialism in early Soviet Marxism.[7] The beginning question then is, what forces of change, and in what direction, might be foreseen from the early and "transitional" revolutionary period? How might one, given a Marxist analysis, expect things to develop?

Understanding the early changes poses no particular problem; Marcuse spends little time on them, and presents nothing that is radically different from previous accounts. The Bolshevik revolution took place – "it is assumed that the initial intention and objective of the Bolshevik Revolution was to build a socialist society . . ."[8] – not in an industrially advanced country, but in a backwards one. Without outside help – historically, it was the success of the German revolution that never came that Lenin had counted on – socialism was on weak footing. Add to the lack of positive help the presence of capitalist hostility and encirclement, and no normal development towards socialism, no smooth transition, could be expected. Marcuse places the turning point early, as far back as 1923, when it became clear that there would be no immediate revolution either in Germany or in any other advanced capitalist country. No "choice" was presented to the Soviet leadership under the circumstances; all energy had to be directed towards the building of the industrial base, leading to an ever growing "priority of the Soviet state over Soviet workers."[9] Whether the development was a result of internal weakness or the international context remains unclear; on the one hand, international events "defined" Soviet Marxism,[10] on the other hand, "there are no 'extraneous' causes . . . for all apparently outside factors and events will affect the social structure only if the ground is prepared to meet them, . . . if they 'meet' corresponding developments within . . ."[11] However that may be (and one may argue that the distinction between internal and external is inapplicable in this situation in any event because "the class struggle is international by its very nature"[12]),

if the dialectical law of the turn from quantity into quality was ever applicable, it was in the transition from Leninism (after the October Revolution) to Stalinism. The "retardation" of the revolution in the West and the stabilization of capitalism made for qualitative changes in the structure of Soviet society.[13]

But where do we go from there? Is the result, neither capitalist nor socialist, stable? Or does

> Soviet nationalization, under the historical condition of its progress, . . . possess an inner dynamic which may counteract the repressive tendencies and transform the structure of Soviet society . . . ?

Marcuse, writing in the first years of the Khrushchev regime, but thirty years before Gorbachev, gives a clear "yes" answer. Why?

A number of threads come together to supply the answer. The first gives primacy to the external situation. It is worth quoting at length the key passage:

> The 'class interest' of the bureaucracy (that is, the common denominator of the special interests of the various branches of the bureaucracy) is linked to the intensified development of the productive forces, and administrative progress into a 'higher stage of socialism' would most effectively secure the cohesion of Soviet society. On the other hand, the Soviet state has consistently diverted a very large sector of the productive forces (human and material) to the business of external and internal militarization. Does this policy forestall the transition to the 'second phase?' The compatibility of an armament economy with a rising standard of living is more than a technical economic problem. The maintenance of a vast military establishment (armed forces and secret police) with its educational, political and psychological controls perpetuates authoritarian institutions, attitudes, and behaviour patterns which counteract a qualitative change in the repressive production relations. Inasmuch as the bureaucracy is a separate class with special privileges and powers, it has an interest in self-perpetuation and, consequently, in perpetuating repressive production (and political) relations. However, the question is whether the repressive economic and political relations on which this bureaucracy was founded are not increasingly contradicting the more fundamental and general interests and objectives in the development of the Soviet state.
>
> If our analysis of Soviet Marxism is correct, the answer must be affirmative. The fundamental Soviet objective in the present period is the breaking of the consolidation of the Western world which neutralizes the 'interimperialist conflicts' on whose effectiveness the final victory of socialism depends . . . In the Soviet Marxist analysis, Western consolidation is based on a 'permanent war economy,' which . . . sustains the rapid development of productivity in the capitalist countries and the integration of the majority of organized labor within the capitalist system . . . The capitalist war economy is in turn sustained by the 'hard' Soviet policy, which also stands in the way of Soviet progress to the second phase where it can effectively compete with capitalist capabilities. Consequently, the first step must be the relaxation of the 'hard' policy. This, however, is a matter of internal as well as foreign reorientation, of shifting the emphasis from military and political to more effective economic competition, and of liberalizing the Stalinist bureaucracy.

One might fantasize that Gorbachev had read these words, were it not that Marcuse was banned reading in the Soviet Union. To my knowledge, no translation of *Soviet Marxism* was ever made there. It is certainly a quite precise description of the direction of the Soviet leadership's foreign and internal policy after 1985.

But there were more purely internal reasons to anticipate a liberalization in the Soviet Union also. One goes back to the question of class structure. The bureaucracy dominates the decisions of the state, but its class base is uncertain. It does not "own" the means of production, it merely controls them. The distinction is important.[14] At bottom it means that the appropriation of the profits of production by the bureaucracy is not legitimized; its political and legal foundations are weak. If the bureaucracy is to consolidate its position, even in the short run, the increase in production that can give rise to an increase in the standard of living, a general sense of progress, is indispensable.[15] Given such progress, the continuance of overt repression becomes not only unnecessary but counterproductive.

Technological progress itself requires liberalization, goes another strand in the argument supporting its likelihood. Technological rationality is inconsistent with rigid, repressive, and command-centralized organization of economic activity; fully developed, it even "contains an element of playfulness."[16] While technological rationality and human freedom are hardly identical, the former is a means to the increased productivity which, today, any form of social organization must have if it is be stable. The technological rationality that Marcuse foresaw as a necessity for the Soviet state to survive did not of itself promise human freedom, and Marcuse is clear in his view that liberalization is not identical with socialism, that the necessity of technologically rational development does not imply the necessity of socialization of the means of production, of their control by their immediate producers. It is not socialism that Marcuse sees as the result of the internal dynamics of Soviet development, but a relaxation of overt repression.

An element of determinism creeps into the logic that links technological rationality with political liberalization. We know little of the social dynamics involved in the production of Sputnik and the Soviet Union's space program. Whatever it was, it produced highly advanced technology in a very repressive over-all environment. And in just what sense can one speak, as Marcuse does at the end of the first long paragraph in the above quotation, of a "contradiction" between the interests of a bureaucracy in control of the state apparatus and the "more fundamental and general interests and objectives in the development of the Soviet state?" The movement towards increased productivity sometimes seems to take on a life of its own, a "law of history" governing the actions of men and women.[17] But the grounds for believing in a strong pressure for improved production in Soviet society are strong even without appeal to such laws.

Marxist theory itself provides a further impetus for liberalization in a state that historically takes such theory seriously, however it may distort, codify, or subvert its context. ". . . The continued promulgation and indoctrination in Marxism may still turn out to be a dangerous weapon for the Soviet rulers."[18] For Marxist theory holds out the prospects of the free play of human faculties, the expression of creativity, liberation from repressive relationships in productive work and in play – concepts which can be tested against immediate experience and raise problems if the gap is too large and too visible.[19]

Given the strength of these arguments, liberalization becomes merely a matter of time. Krushchev's policies seemed to bear out the predictions of theory as Marcuse was writing; Gorbachev's policies, after Marcuse's death, seem incontrovertible confirmation. But then the question arises, why did liberalization fail, and the entire Soviet system, not merely its most repressive elements, fall?

Here we must go beyond *Soviet Marxism* for an answer, and look at some of the more far-reaching implications of its analysis.

Marcuse clearly expected the Soviet economy to continue to grow, to increase both in production more and more and consumption at the same time as basic production advanced. He quotes Krushchev's claim that the Soviet Union had, already in 1953, "the means for high-speed, simultaneous development of heavy industry, agriculture, and light industry,"[20] and he believes, "given conditions under which the growing production . . . is not . . . utilized for wasteful and destructive purposes, production is likely to generate the material and cultural wealth that would permit . . . the second phase."[21] "Permit," not "produce;" this is not technological determination, and Marcuse insists that radical social change must accompany technological progress for technology to be liberating;[22] but that technological progress would occur, he essentially had no doubt.

But of course events did not progress smoothly or in linear fashion in the Soviet Union, or anywhere in Eastern Europe. The arms race in fact intensified, partially because of direct pressures from conservative

United States administrations. The initial response in the Soviet Union was not liberalization, but its opposite; Khrushchev's hold on power was broken by the mid-1960s. Thus when liberalization came, it may have come too late.

The arms race not only undercut the full use of the resources of the Soviet Union in developing its economy; under the particular conditions of political repression, it also undermined the advance of technical knowledge and the technical foundations for advancing productivity. Technological rationality, in Marcuse's exposition, is both a necessary ingredient of advances in productivity and is itself a function of, made increasingly possible by, such advances. In a technologically underdeveloped society, a level of production must first be reached that makes technologically rational behaviour both necessary and effective. But that rationality is itself a prerequisite for the productivity level to be reached. Thus a vicious circle exists. It cannot be broken all at once; the areas in which technological thinking is most advanced slowly spill over into other areas, so that advance is made in different areas at different times and places. But precisely this process was aborted in the Soviet Union, in the concentration of the best of technical work in the space and armaments program and the creation of mammoth blocks to the interchange of advances there with those in other sectors of the economy. Thus to the inherent chicken-and-egg dilemma were added blockages to the "normal" ways around. The failure to develop balanced and wide-ranging technological progress was in part the foreseeable result of developments Marcuse's analysis did in fact explore.

Coupled with these externally-rooted explanations for the decimated state of Soviet, and indeed most East European, economies by the late 1980's were purely internal factors, economic problems inherent in any socialistically-organized and Marxist-grounded society. Some economic problems of course had little to do with socialism, but were simply decision, that could have gone either way, under the control of the leadership: excessive centralization, distortions of investment policy, lack of responsiveness to technological changes and to changes in consumption desires, too rigid education policies, an inflexible command structure, failure to utilize markets at least as sources of information and the clogging of other information flows, etc.[23] A repressive political system and the absence of market indicators made errors in economic decisions more difficult to correct, but indications enough of problems existed; a "wiser" leadership might indeed have done much better, even within structures inherently required by Marxist theory in a socialist economy.

But other aspects of the retarded progress of the Soviet economy have more to do with its unambiguously socialist characteristics.

The first has to do with the role of Marxist theory. No system that relies for its legitimacy, if not its direction, on Marxism, even in the form of Soviet Marxism, can afford to engage in activities absolutely counter to the fundamental principles of that theory. Thus, for instance, the exploitation of workers needs to ameliorated, not exacerbated, over time. Except in conditions of wartime or other dire emergency the living and working conditions of industrial workers must be improved. In the Third World, likewise, a political system basing itself on the concepts of Soviet Marxism cannot exploit workers as imperialist countries would. Unemployment cannot be tolerated on any wide basic; thus the simple lay-off of workers whose jobs become obsolete is a difficult matter. In these and other ways, a Soviet Marxist system suffers from competitive disadvantages compared to a system without such inhibitions. It might be expected, therefore, that a Soviet-style system would lag behind in the competition with advanced capitalist economies, even under the best of circumstances.

The second socialist-grounded factor involved in the impeded progress of the Soviet economy has to do with the role of the bureaucracy and the intelligentsia. Technical advance comes from a technical intelligentsia. However recruited, however organized, whatever their ideology, a level of education and training and ability is necessary to produce innovation, and those possessing these levels are among the critical components of the intelligentsia. Bahro speaks of them as developing a level of "surplus consciousness" under real existing socialism: "free human capacity that is no longer absorbed by the struggle for existence."

> The industrial, technological-scientific mode of production, in which intellectual labor becomes an essential factor, engenders in the producers . . . qualities, skills, forms of imagination . . . that are stifled or perverted in capitalist and repressive noncapitalist societies.[24]

Both Bahro and Marcuse saw such surplus consciousness as a factor, perhaps the factor, that would permit a break in the "chains of domination, the subjugation of human beings to labor."[25] Marcuse, in his discussion in 1958, was not concerned to look at the impact of the increase in such surplus consciousness on technological progress, but the exploration is potentially fruitful. For the specific forms by which "skills, forms of imagination . . . are perverted" is quite different in capitalist and repressive noncapitalist societies. Capitalism provides rewards for their application to inherently unrewarding tasks, real existing socialism did not. Marcuse (and Bahro) saw surplus consciousness as leading to the ending of the domination of compensatory interests[26] over emancipatory interests. Real existing socialism did not permit the full expression of emancipatory interests, and therein lay the potential for an explosive rupture of its system of domination. This was the main point of Bahro's analysis,[27] and Marcuse applied it, *mutatis mutandis*, to capitalism.

But within capitalism compensatory interests are much more fully addressed than within real existing socialist societies. That was not an issue that either Bahro or Marcuse, in his essay on Bahro, addressed, although clearly, from extensive discussion beginning with *One-Dimensional Man* and onwards, it was an essential part of Marcuse's over-all assessment of the strengths and weaknesses of capitalism. The consequence is that surplus consciousness can be better harnessed to the interests of technological rationality in a capitalist than in a repressive socialist society. Even at the level of observation of every-day life, the result is evident. The dynamism, the energy, the search for innovation, that are found in the leading advanced industrial societies, however distorted and unproductive it may be in a human, seem altogether absent under real existing socialism, appearing, if at all, in artistic work, but certainly not in industrial production or the commercial service sector. Put crudely, the Soviet-style systems gave up one set of incentives for technological progress and increasing productivity without substituting any equally effective alternative.

Thus, Marcuse's analysis would suggest that, even apart from "external" pressures and even apart from the particular mistakes of particular leaders or particular organizational strategies, it is unlikely that repressive Soviet-style socialism could make good on its promise for increased productivity and a steadily increasing quality of life for its populations. Whether, under the best of circumstances – no "external" problems and "wise" leadership – a socialist economy might be expected to perform as or more efficiently than a capitalist one we cannot tell from the historical record, but there is certainly some theoretical reason to doubt it. At least in the short or intermediate range, heightened productivity is not that which can give the goal of socialism its appeal, its promise for the future. That conclusion, which is implicit in all of critical theory, emerges concretely from Marcuse's analysis of Soviet Marxism.

The bureaucracy itself could have, however, attempted to overcome these inherent difficulties in advancing productivity, either through measures designed to increase the allocation of resources to consumption, and thus to enhance the satisfaction of compensatory interests for the intelligentsia, or through reforms in the organizational hindrances to progress: the over-centralization, etc. There is some evidence that it tried to do so, e.g., in the G.D.R. Honecker's shift of industrial capacity to the production of consumer goods in 1971. But the reaction to the difficulties encountered at the beginnings of the reform process in the Soviet Union was not a rallying of the bureaucracy around Gorbachev for the implementation of further reforms within the system based on their own self-interest in the maintenance of that system, as Marcuse would have expected based on the analysis in *Soviet Marxism*. It was rather an abandonment of the system as a whole, a snow-balling and quickly complete surrender of the existing bases for their power and prestige to forces of change hostile to it. Why did this unexpected surrender of the system by the bureaucracy it had produced take place? If the "class interest" of the bureaucracy lay in the reform of the system, as Marcuse states at the beginning of the long quotation above, why did it so quickly abandon that reform?

Because the bureaucracy was not, indeed, a "class" whose ownership of the means of production was the basis of its power and privileges. It did control the productive processes, but control and ownership are not the same thing, as I have argued above. Its class base was uncertain.

Bureaucracy by itself, no matter how huge it is, does not generate self-perpetuating power unless it has an economic base of its own from which its position is derived, or unless it is allied with other social groups which possess such a power base.[28]

When "free market" pressures appeared and received powerful support from the outside, when internal political division gave the upper hand to market-oriented forces of change, the bureaucracy quickly realized it could as easily exercise its power in the new system as in the old. Not being dependent on relations of ownership, it had little to lose by a change in those relations, and possibly even something to gain. History, with hindsight, vindicates the "non-class" analysis.

Marcuse recognized, from the outset, that internal reform was only a possibility in the Soviet Union, and that the likelihood that such reform would break through the bounds of real existing socialism to some form more akin to what Marx and Engels had envisaged was an even slimmer possibility. History remains inconclusive as to whether that possibility ever existed. If it did, it probably came closest to manifesting itself in the G.D.R. during the brief period of the *Wende* or in Czechoslovakia at the very beginning of the velvet revolution.[29] In the first case, German unification quickly wiped out whatever possibilities existed; in the latter case, neither the ideological nor the practical political support for a real reform of socialism was ever substantial. In both cases, the external context sealed the fate of whatever possibility for a reform socialism might have developed.

Thus the Soviet Marxist chapter of the narrative of socialism seems closed. Does Marcuse's analysis lead us to any insights into the future?

Marcuse's Bahro review, although it abjures any convergence theory,[30] points out strong parallels in real existing socialist and real existing capitalist societies. He finds a drive towards technological rationality in each, although with different motors and different effects. He finds disparate forces for change in each, but none that comes close to an assurance of progress towards a radically different social order – no "revolutionary subject" on either side. Rather, he finds an internalization of subordination, a "transformation of freedom into security,"[31] in both social orders. But he also finds, in both, serious sources of instability, principal among them the existence of a surplus consciousness, of unsatisfied human drives, aspirations, desires, hitherto incapable of fulfillment, but now visibly within the range of the possible.[32]

What inhibits the realization of that possibility, what prevents instability from maturing to fundamental change?

Here the answer is quite different in the two systems. In the one, the Soviet, it was the combination of internal repression and the external "threat" which justified that repression. Liberalization, he foresaw, might be one step in the direction of stability, whether or not of further change. In the capitalist world the situation is rather the opposite. The external threat serves both to justify an internal economic policy, sometimes called the "permanent war economy," and to provide legitimacy to a liberal regime although that regime falls far short of fulfilling the potential of the technical progress it has made possible. Externally, Soviet Marxist theory had always counted on conflicts among the capitalist powers to provide it with a respite within which to solidify its position internationally; the reality had proved otherwise, ironically, in that the very existence of the Soviet Union and its allies had provided a basis on which the Western powers has been able to come together and bury their own conflicts. That analysis is not one with which Marcuse significantly disagreed,[33] although he saw many more forces for stability and potentials for progress in the West than the theorists of Soviet Marxism ever saw or acknowledged.

The disappearance of the Soviet Union changes this picture dramatically in the capitalist countries. The threat from outside, which so long justified massive military expenditures and investment in wasteful technology, is harder and harder to find. It is harder and harder to explain the reasons for the continued existence of poverty alongside wealth, repression in some countries while others more powerful profess commitments to freedom, rising unemployment while public investment decays, injustice, racism, xenophobia – in a world in which the possibility of plenty for all is more and more apparent and its postponement less and less able to be justified by the threat of an outside menace. If world-wide competitiveness increases to the point where economic crises follow each other in an accelerating tempo, the original "anomalous" position in which the Soviet Union found itself at its birth might not confront another protest from below. If, on the other hand, that competitiveness is brought under control, and progress does indeed continue more or less smoothly, the means for capturing surplus consciousness within the confines of compensatory interests may become slimmer and slimmer. Environmental constraints and their human meaning, to which Marcuse was increasingly turning his attention at the time of his death, suggest other limits on the extent to which compensatory interests can forever be at the same time stimulated and satisfied. So the "surplus consciousness" of those doing well, coupled with the discontent of excluded, under the constraints of a finite natural environment, may yet open the door to a form of liberation which neither Soviet style socialism nor anti-Soviet style capitalism have yet made possible.

It was in the theoretical analysis of these issues, and in their engagement in efforts to find real solutions, that the friendship between Marcuse and Cohen had its impersonal rationale.

*Columbia University*

## NOTES

[1] *Soviet Marxism: A Critical Analysis.* New York: Columbia University Press, 1958. A Vintage Press edition was published in 1961 with a new preface, and the subsequent French edition likewise has a new preface. See Douglas Kellner, *Herbert Marcuse and the Crisis of Marxism.* Berkeley: University of California Press, 1984, pp. 197–228, both for bibliographic information and a substantive critique.

[2] Apart from filial affection, those of my own experiences that may color this discussion are described in *Missing Marx. A Personal and Political Journal of a year in East Germany, 1989–1990.* New York: Monthly Review Press, 1991.

[3] It is reprinted in Ulf Wollter (ed.), *Rudolf Bahro: Critical Responses*, White Plains, New York: M. E. Sharpe, Inc., 1980.

[4] 'This study attempts to evaluate some main trends of Soviet Marxism in terms of an "immanent critique," that is to say, it starts from the theoretical premises of Soviet Marxism . . . The critique employs the conceptual instruments of its object, namely, Marxism, in order to clarify the actual function of Marxism in Soviet society . . . [It assumes] that Soviet Marxism (i.e. Leninism, Stalinism, and post-Stalin trends) is not merely an ideology promulgated by the Kremlin in order to rationalize and justify its policies but expresses in various forms the realities of Soviet developments.' p. 1. If one accepts that 'the theoretical premises of Soviet Marxism' are indeed the theory developed by Marx and Engels, and that theory, in Marxist understanding, plays a historical role going beyond ideology, one might as easily have said, 'this study is a Marxist critique of Soviet Marxism.'

[5] Chapter 7: 'Dialectic and its Vicissitudes.'

[6] P. 8, fn. 1. All page references, unless otherwise noted, are to the 1958 edition of *Soviet Marxism.*

[7] Marcuse, agreeing, says 'the abolition of private property in the means of production does not, by itself, constitute an essential distinction as long as production is centralized and controlled over and above the population.' p. 81. In 'Protosocialism,' he speaks of 'the abolition of private ownership of the means of production' as the 'indispensable precondition of socialism . . . [the real difference lies] in the way in which the material and intellectual forces of production are used.' p. 26.

[8] P. 8, fn. 1.

[9] P. 74.

[10] P. 6.

[11] P. 3.

[12] P. 96.

[13] P. 74.

[14] I have explored it in legal terms in 'Privatization and its Discontents,' in Michael Harloe *et al.* (eds.) *The Post-Socialist City.* London: Blackwell, forthcoming.

[15] P. 118.

[16] P. 257. See, in general, chapter 12, 'Ethics and Productivity.'

[17] In a brief discussion of the concept of historical laws in Marxism, Marcuse refers to the 'irreversibility' of historical processes determined by the 'basic form of societal reproduction,' but the example he gives is of the emergence of the feudal system out of the agricultural economy of the late Roman empire! pp. 3–4.

[18] P. 265.

[19] P. 267.
[20] Quoted at page 177 from *Current Digest of the Soviet Press*, p. 11. no. 39 (November 7, 1953).
[21] P. 170.
[22] '... the truly liberating effects of technology are not implied in technological progress per se; they presuppose social change, involving the basic economic institutions and relationships.' p. 256.
[23] Innumerable Western texts expand on these and other issues. One of the less ideological is Kornai, Janos. 1980. *Economics of Shortage*. 2 vols. Amsterdam and New York: North- Holland Publishing Company.
[24] 'Protosocialism,' p. 27.
[25] 'Protosocialism,' p. 28.
[26] '... not through a policy of reducing consumption but through a "genuine equalization in the distribution of those consumer goods which determine the standard of living".' Marcuse, at p. 35, quoting Bahro.
[27] Based on it, Bahro believed an overthrow of the existing regime in the G.D.R. from the inside was possible. Marcuse did not take a position on Bahro's belief. 'Protosocialism ...', p. 36.
[28] P. 109
[29] For more detailed discussion, see my *Missing Marx*, supra.
[30] In *Soviet Marxism* he already spoke of the 'fundamental difference ... paralleled by a strong trend toward assimilation,' p. 81. Elsewhere, he speaks of 'an essential link between the two conflicting systems ... in the technical-economic basis common to both systems, i.e., mechanized ... industry as the mainspring of societal organization in all spheres of life,' p. 6.
[31] P. 191.
[32] Marcuse emphasized that the 'turn to subjectivity,' in Bahro's formulations, was ambivalent, 'Protosocialism,' p. 46. While Marcuse was also much concerned with what Bahro called the 'essentially aesthetic motivation' of socialism, I would suspect Marcuse would not have followed Bahro on the road Bahro subsequently took on the relationship of the subjective to the political.
[33] See for instance p. 99.

JOHN STACHEL

# MARX ON SCIENCE AND CAPITALISM[1]

There are many "Marxisms" today, most of which have as little to do with Marx's views as most current Christian creeds have to do with the outlook of another Jewish revolutionary who is reputed to have been crucified by the Roman state about 2000 years ago for subversive activities. After over fifty years of engagement with Marx's thought and various movements purporting to act in his name, the approach to Marx that I have found most fruitful is based on the central position in his work of his critique of capitalist society. If Marxism cannot help us to understand the contemporary dynamics of this social system, cannot contribute to our efforts to transcend it, then it amounts to little more than a page in the history of the nineteenth and early twentieth century.

So, in spite of my interest in current discussions of Marxism and philosophy,[2] I must agree with a comment of Lucio Colletti: "If Marxists continue to remain arrested in epistemology and gnoseology, Marxism has effectively perished."[3]

A French Communist (Jean Fallot) once complained that Marx "linked the theory of science to that of surplus value, while one would have preferred it to follow from the materialist theory of knowledge."[4] Jean-Jacques Lenz retorted quite justly: "The naiveté of the Communist academic eager for theories of knowledge is striking in this proposition, which would be correct . . . if Marx had the same preoccupations as Engels and the founders of Diamat."[5]

Indeed, I maintain that the contemporary relevance of Marx's ideas on the natural sciences lies precisely in the link they establish between "the theory of science" and "that of surplus value." Here, I shall try to explain the nature of that link and why its understanding is important for attempts to work out a new strategy of opposition to contemporary capitalism. Space requires brevity, and brevity tends to be the mother of dogmatism, so I apologize in advance for the unintendedly assertive tone of some of my remarks, as well as for the neglect and/or compression here of many links in what should be a more complete chain of argument.

The starting point for this work is Marx's analysis of capitalism, the combined exposition of the laws of development and critique of the capitalist mode of production that he produced in the course of his work on *Capital* (I remind you that its subtitle is *A Critique of Political Economy*). This work includes the many drafts of *Capital*, starting in 1857 and going up to almost his last year,[6] as well as the published texts of Vol. One, which Marx saw through two editions, and Vols. Two and Three, which were edited after his death by Friedrich Engels.[7] I can already hear an objection: are not his views on political economy among the most outmoded elements in Marx's work? What modern economist – even so-called Marxist ones – takes seriously his account of the labor theory of value, for example?[8] What can such views tell us about contemporary capitalism? I can only respond: a lot. I suggest that skeptics take a look, for example, at Alain Lipietz's work to see how fertile Marx's approach can be when used as the basis for dealing with many of the economic mechanisms operating – or failing to operate – in contemporary capitalism.[9] Of course I cannot enter here into any detailed discussion of these mechanisms.

## THE RELEVANCE OF MARX

If, without entering into details, we take a global look at Marx's analysis of capitalism, we do not find the set of "iron laws" so often present in mechanistic-deterministic interpretations of his theories.[10] Rather, we find an account of a structured, hierarchical network of laws that operate as tendencies,[11] often generating counter-tendencies,[12] which are in constant conflict with each other for dominance in determining the actual motion of the system. These tendencies are constantly generated and regenerated in ever new forms by the two basic conflicts of the system: between wage-labor and capital and between individual capitals. We find the portrait of a system that is unable to confine these basic conflicts and their manifold manifestations within any fixed, permanent limits. Rather, it is a system that is continually driven by these ever-changing manifestations of its basic conflicts to improvise, to develop quasi-stationary solutions to them, temporarily effective means of containing them. The nature of these solutions is by no means pre-determined by the nature of the conflicts, but results from the relative strength of the various tendencies and the ingenuity of human agents in either imposing a solution by force or guile, or in finding an acceptable compromise. Often the solution to the same basic set of conflicts will differ markedly from one capitalist social formation to another (Japanese capitalism differs from American capitalism in many non-trivial ways, for example[13]).

But whatever their nature, as long as they remain within the basic parameters of the capitalist system, these solutions cannot solve the basic conflicts of the system. So these very solutions form the starting point of a process that generates new conflicting tendencies, new manifestations of the basic conflicts, which in turn impel the agents of the system to develop new solutions. Of course, each critical period when the old solutions are failing represents an opportunity for those agents seeking to transcend the system – but no more than an opportunity – to develop an effective strategy of opposition to the system as a whole, and the possibility of radically transforming its nature. From the point of view of the interpretation I am advocating, there is no such thing as a "final crisis," in the sense of a set of circumstances from which no escape is possible that stays within the basic parameters set by capitalism. There are only critical moments for the system, which offer unusually favorable potentialities for the kind of effective opposition that can lead to opportunities to break out of the system. Whether we recognize these opportunities, and what we make of them is our responsibility.

In contrast to this viewpoint, Marx's theory has been interpreted – with good warrant in some of his own words, it must be admitted – as implying the inevitable advent of a post-capitalist social formation, variously called socialism or communism, which would resolve the manifold conflicts of capitalism by abolishing the system itself. Yet, if we look carefully at Marx's critique of capitalism, what we find there are quite convincing arguments for the constant renaissance of conflicts within the capitalist system and for the impossibility of ever resolving them once and for all within the system. We do not find any real argument for the inevitability of their successful resolution by a transcendence of the system. Nor does the evidence of the last century of struggles offer any examples of such a successful resolution – or revolution, if you prefer – in any of the advanced capitalist countries. And it is only success in such countries that has any direct relevance to Marx's vision of socialism as the transcendence of the conflicts of capitalism.

What are we left with then, those of us who are not happy with the prospect of capitalism for the indefinite future? If we are not given a theory of the inevitability of socialism, we are left with a critique that emphasizes – in Marcello Cini's phrase – the non-inevitability of capitalism.[14] Contrary to those economists who claim to find capitalism among the cave-people, Marx's analysis stresses the historical genesis of the capitalist system, and here we *can* call history to witness: There have been many fundamentally non-capitalist social systems, very few of which have spontaneously – and then only under certain specific conditions – given rise to capitalism. Even more, as I noted earlier, Marx's critical analysis shows why the capitalist system can never reach the kind of steady state – so to speak – that many non-capitalist systems were able to reach. We are left with an understanding of how and why the system will continue to be driven onward by attempts to resolve ever new – and every more menacing – manifestations of its basic conflicts. We are left with an understanding that any socialist movement can only hope to succeed by inserting itself effectively into the struggle over just how – and at whose expense – these various manifestations of the basic conflicts are to be resolved. Therein lies the only hope of transcending capitalism.

This will not be enough for some. Many people have found in Marxism a substitute for a lost religious faith, and many more feel the need for the sense of security provided by *some* totalizing dogma – a historical theodicy with a guaranteed happy outcome for mankind. Such persons, who feel that a Marxism with an open-ended vision of potentialities is not enough, will have to look elsewhere in their quest for certainty. Indeed, many East European former Marxists seem to have found a quick-fix substitute faith in the fetichism of the "free market."

However, in this time of capitalist ideological triumphalism, replete with assertions of "the end of history" and deifications of the market one the one hand; and growing evidence on the other hand that world capitalism is passing through one of its periods of severe structural crisis; at a time when American capitalism, in particular, cannot provide many of its people with the basic items needed to sustain existence at a minimum survival level – to say nothing of a level of human dignity – let alone provide them with the ever-expanding standard of living that its post-World War II ideology promised and to some extent did provide for several decades – at such a time, I find great need for a well-founded analysis reaffirming the non-inevitability of capitalism and providing some indication of the tendencies within the system that might lead to movements radically breaking with capitalism, and analysis based on critical thought and not blind faith.

This introduction may appear to have been a long digression from my topic; but I believe it is important in this time of disillusionment for many socialists to at least sketch why someone can still be interested in Marxism as more than an exercise in the history of ideas. Now

I shall turn to my main topic. Before I can turn to the role of science, however, I must sketch Marx's approach to capitalism.

## THE NATURE OF CAPITALISM

Like any economic system, the capitalist system can be subdivided – at least conceptually – into the elements of production (who makes what), exchange (who trades what), and distribution and consumption (who gets what and does what with it). Marx's analysis is based on the centrality of the element of production, and in particular on the social relations of production that characterize the system, and determine its relations of distribution and consumption.[15] Under capitalism, the most important social relation of production is that between wage-labor and capital. Marx emphasizes that, for example, a machine as such is no more capital than one human being as such is a capitalist and another a wage-worker.[16] This requires a social setting in which the machines are owned by an individual or group (a capitalist, for short) that is able to hire for wages (exchange) other human beings (workers, for short) and put them to work at the machines (production), take the entire product of their labor as the property of the capitalist (distribution), sell them on some market (exchange), and pocket the profits – if any – as his own – to eat caviar and reinvest in new machinery while the workers eat beans (consumption). It is only when they are enmeshed in this web of social relations that the machine functions as capital and the human being functions as a worker.

Early, I spoke repeatedly of the basic conflicts of the system and their constantly-changing manifestations. This distinction between the underlying dynamics of the system and their surface manifestations is characteristic of the capitalist system. In contrast to all pre-capitalist societies, the nature of capitalist social relations of production is peculiarly opaque: the essence of the system is often masked by its appearances. Under slavery, for example, it is pretty clear that a field slave works part of the day to support him/herself – the part of the total labor time that Marx calls necessary labor time – and the rest of the day to support the master and mistress and their retinue of flunkies – what Marx calls surplus labor. It is clear that, in this sense, the slave is economically exploited; the rate of this exploitation can be defined quantitatively by the ratio of surplus to necessary labor time. Under capitalism, it is not so obvious that the wage-worker is similarly exploited: wages *appear* to be the pay for the *entire* day's work; the never-ending struggle over wages and hours appears to be a struggle for "a fair day's wages for a fair day's work," rather than a struggle over the division of the day's work into necessary and surplus labor times.[17] Similarly, profit does not appear as that portion of the product of surplus labor over his costs that the capitalist is able to retain; rather, it appears to be the result of a magical "productivity of capital," where capital is identified with the physical being of the machine, rather than with the machine's role in the process of wage-labor, that is with capital as a social relation.

These are just some important examples of the systematic tendency under capitalism for the underlying social relations of production between people to appear as relations between people and things, or even as relations between these things themselves. It appears that *things* get taken to market and exchanged; what is not so apparent is that *people* thereby establish certain social relations between themselves. They are mediated by the commodities, of course, but if we hope to understand capitalism, we must focus on these social relations of production and distribution.

Marx refers to this systematic tendency of the capitalist system to present social relations between people as relations between things as fetishism, by analogy with religious fetishism, in which people relate to fetishes or idols – the products of their own brains and hands – as if these idols possessed inherent creative powers.[18] These idols even appear to enter into relations with each other, when it is actually the people who made them who are doing so. More generally, Marx constantly emphasizes the distinction between the surface manifestations of the capitalist system – the exoteric economy – and the underlying social relations of production – the esoteric economy – that generates and ultimately governs this realm of appearances.[19] But it must be emphasized that the surface appearance of the system under discussion is not a swindle, consciously perpetrated by the capitalists – although there is certainly no shortage of conscious swindling under capitalism. The surface manifestations are an inherent element of the reality of the capitalist system; they are the way it presents itself to all its agents, capitalists as well as workers, who make their day-to-day decisions based on these surface appearances – until they are caught short by some particularly unpleasant manifestation of the underlying conflicts. For example, money appears to have an inherent power to command all goods and services until runaway inflation destroys this power overnight. Perhaps an analogy from astronomy is helpful to an understanding of the difference between a deep structure and the apparent phenomena that it explains but does not replace: the sun and stars appear to turn around the earth, even though we regard this appearance as a manifestation of a deeper structure: The laws of dynamics, showing we arrive at a more profound understanding of the solar system by using a frame of reference in which the earth is really rotating once each day, while the stars and sun are at rest (to a good approximation) in this frame. Yet, when we look at the sun each day or the stars each night, this knowledge does not affect the appearances.

Marx shows that it is in the very nature of the capitalist system to constantly generate this opacity of its basic social relations with the help of ever-changing surface manifestations. It could not function otherwise, and it takes a profound scientific analysis to penetrate this opacity in theory to reach an understanding of the esoteric economic relation, an understanding which – like the knowledge of the earth's rotation – does nothing as such to dissipate these appearances, of course.

## TECHNOLOGICAL DETERMINISM?

An important example of this opacity brings me closer to the role of science. This is the tendency of capital not only to appropriate and utilize all the skills of labor in its own interests, but to present these skills of labor as if they were powers of capital – capital identified, remember, not as a social relation but with its material embodiment in technologies such as machines. How often do we read or hear these days about the marvels of technology and how much they have done for mankind; or more to the point these days, about technological imperatives that force reluctant capitalists to cut wages and benefits?[20] Marx's approach to technology continues to be the source of a great deal of misunderstanding. Even some "Marxists," to say nothing of many of his critics, regard him as a technological determinist (admittedly with some warrant from his own ill-chosen words in some of his more concise statements of historical materialism, such as "The hand-mill will give you society with the feudal lord, the steam-mill society with industrial capitalism"[21]). They base themselves on the supposed causal chain, enshrined in Stalin's notorious article on "Dialectical and Historical Materialism": the forces of production, identified with technology and taken as an independent variable, determine the relations of production, taken as the dependent variable. Whatever rhetorical excesses Marx may have committed in some abstract formulations, the thrust of his concrete analysis of the origins of capitalism argues against such a relation between forces and relations of production. Let me cite a particularly explicit passage:

> In fact, it happened historically that capital, at the beginning of its formation, not only takes the labor process as a whole under its control (subsumes it), but also the particular real labor processes as it finds them technologically ready, as they have been developed on the basis of non-capitalist relations of production. It encounters the real process of production – the determinate mode of production, and initially only *formally* subsumes it under itself without altering anything in its technological character. Only in the course of its development does capital subsume the labor process under itself not only formally, but also transforms it, shapes anew the mode of production itself, and thus first creates the mode of production peculiar to itself.[22]

It would be hard to conceive a less ambiguous statement of Marx's position: Initially, the capitalist relations of production utilize the existing technologies, just as they have been developed before the advent of capitalism (he calls this the formal subsumption process); and only subsequently, once this take-over process has been completed, do the capitalist relations of production begin to reshape the forces of production in order to create a mode of production characteristic of capitalism (elsewhere he calls this the real subsumption process). In a word, the relations of production shape forces of production adequate to realize the tendencies inherent in these relations.

## THE ROLE OF THE NATURAL SCIENCES

And here we come at last to the natural sciences: They play a vital role in this process of real subsumption, the shaping of a mode of production peculiar to capitalism. But before examining their relation to capitalism, let us look at Marx's intrinsic characterization of the natural sciences. True to his paradigm of the primacy of production, Marx applies this category not only to material but also to intellectual processes. It is sometimes assumed that he identified all intellectual production under capitalism with ideology (which I shall briefly characterize here as intellectual production in the service of the ruling classes), leading to such misconceptions as the class nature of all knowledge. Actually, Marx made a clear distinction between what he called "the ideological components of the ruling class" and "the free intellectual production of a particular social formation."[23] He clearly identifies the natural sciences as one component of the latter, speaking of "advances in the area of intellectual production, i.e. the natural sciences and their application."[24] He characterizes science as "the product of the general development of history in its abstract quintessence,"[25] "the most solid form of wealth, both its product and its producer." He refers to "the development of science," "this ideal and at the same time practical wealth," as "only one aspect, one form in which the development of human productive powers, that is, of wealth, appears."[26]

Like any other form of production, intellectual production requires labor: Marx speaks of "the development of intellectual labor, in particular of the natural sciences."[27] The relation of this form of intellectual labor to the wage labor exploited by capital changes in the course of development of capitalism. In its early stages, the ties between the two are not very strong. Remember Marx divides the development of capitalism, broadly speaking, into the period of the formal subsumption of labor by capital and the period of its real subsumption. During the first period, roughly the period of manufacture (literally, "made by hand") based on the old handicraft technologies, the major method of increasing the rate of exploitation was by increasing surplus labor time. This could be done by lengthening the working day and better organizing the application of existing techniques within the (manu)factory to cut down time "wasted" by the workers. During this period, scientific labor was also organized and executed as a handicraft, by and large independently of any direct (even formal) control by capital. Links certainly existed between science and production, technology sometimes inspiring science, and science sometimes being applied to manufacture. But these links were by and large sporadic and unorganized. The following comment by Marx applies essentially to this handicraft period:

The product of intellectual labor – science – always stands far below its value, because the labor time needed to reproduce it has no relation at all to the labor time required for its original production. for example, a schoolboy can learn the binomial theorem [discovered by Newton] in an hour.[28]

But there are natural limits to the possibility of extending the workday. These limits on labor time, together with the increased struggles by the working class first to resist its extension and then to reduce its length, propelled capital onto the path of intensification and rationalization of the work process itself, the search for methods of increasing surplus labor by reducing necessary labor, inaugurating the period that Marx calls the real subsumption of labor by capital.

The contrast was drawn by Marx: In the earlier period:

> Handicraft remains the basis, a technically narrow basis that excludes a really scientific division of the production process into its component parts, since every partial process undergone by the product must be capable of being done by hand, and of forming a separate handicraft.[29]

On the other hand:

> Modern industry never views or treats the existing form of a production process as the definitive one. Its technical basis is therefore revolutionary, whereas all earlier modes of production were essentially conservative. By means of machinery, chemical processes and other methods, it is continually transforming not only the technical basis of production but also the functions of the worker and the social combinations of the labor process.[30]

This transformation of the labor process results in the development of what Marx refers to as cooperative or collective labor; that is, the integration of the labors of many individuals into a collective process, forming what Marx calls the collective laborer.[31]

This process can only proceed systematically if scientific labor is integrated systematically with the collective labor process under the more-or-less direct control of capital. Marx refers to scientific labor so integrated as universal labor:

> We must distinguish here ... between universal labor and cooperative labor. They both play their part in the production process, and merge into one another, but they are each different as well. Universal labor is all scientific work, all discovery and invention. It is brought about in part by cooperation of the living, in part by utilizing the labor of their predecessors. Cooperative labor, however, simply involves the immediate cooperation of individuals.[32]

As we see currently, there is an ever growing interdependence between these two forms of labor, the universal and the cooperative. Manual workers today, instead of being the possessors of individual skills that would enable them to independently create a usable product – or at least a recognizable, distinct part of a product – as in the handicraft era, now must intertwine their labors with those of many others – often at the other end of the globe – in order cooperatively to create such a product. Those other labors will almost invariably include the universal labor of many scientists, inventors, technicians, etc., whose work is increasingly organized under the direct control of capital (or its intellectual agents and flunkies) in research institutes and development laboratories. In contrast to the situation when science was organized on an independent handicraft basis, the costs of this vast research and development establishment now constitute a non-negligible part of the value of many commodities, in some cases – notably the pharmaceutical industry – the major portion.[33]

The manifold fruits of this vast global social process of production are thus the result of the creative efforts of a collective laborer consisting of the hands and brains of literally hundreds of millions of manual and intellectual workers around the world. But here the opacity of the system enters the story. This is not how things appear to the participants in the capitalist system, even to the individual workers whose creativity makes up the collective laborer. The fetishism of capital prevents this.

> The transposition of the social productivity of labor into the material attributes of capital is so firmly entrenched in people's minds that the advantages of machinery, the use of science, invention, etc. are *necessarily* conceived in this *alienated* form, so that all these things are deemed to be the *attributes of capital*. The basis for this is (1) the form in which objects appear in the framework of capitalist production and hence in the minds of those caught up in that mode of production; (2) the historical fact that this development first occurs in capitalism, in contrast to earlier modes of production, and so its *contradictory character* appears to be an integral part of it.[34]

By "its contradictory character," Marx is referring primarily to the antagonistic character of the relations between labor and capital, that (again contrary to appearances during some periods of expansion) in the long run, a gain for one constitutes a loss for the other (what we might today call "a zero sum game").

If the trend toward the real subsumption of labor, both collective and universal, by capital continues unchecked, the picture of its future that Marx draw is anything but rosy:

> With the development of machinery there is a sense in which the conditions of labor come to dominate labor even technologically and, at the same time, they replace it, suppress it, and render it superfluous in its independent forms. In this process, then, the *social* characteristics of their labor come to confront the workers so to speak in *capitalized* form: thus machinery is an instance of the way in which the visible products of labor take on the appearance of its masters. The same transformation may be observed in the forces of nature and in science, the product of the general development of history in its abstract quintessence. They too confront the workers as the *powers* of capital. They become effectively separated from the skill and knowledge of the individual worker; and even though they ultimately are themselves the products of labor, they appear as an *integral* part of capital wherever they intervene in the labor process. The capitalist who puts machinery to work does not need to understand it . . .But the science realized *in the machine* becomes manifest to the workers in the form of *capital*. And in fact, all these applications of science, of the forces of nature, and of the products of labor on a grand scale, all based upon socialized labor, indeed even appear as only *means for the exploita-*

*tion of labor*, as means of appropriating surplus labor, and powers belonging to capital as opposed to labor. . . .And so the development of the *social* productive powers of labor, and the conditions of its development, appear as the *achievement of capital*, an achievement to which the individual worker only relates passively, and which progresses at his expense.[35]

In the light of the experience of the hundred years since these lines were penned, it is hard to fault Marx's analysis. I hope my selection of quotations has given you a glimpse of the crucial role that the natural sciences play in Marx's vision of the internal logic of capitalism, a role that is emphasized in an extraordinarily prescient passage in which Marx summarized:

Three cardinal facts about capitalist production:

(1) The concentration of the means of production in a few hands, which means that they cease to appear as the property of the immediate workers and are transformed on the contrary into social powers of production. Even if this is at first as the private property of capitalists. The latter are trustees of bourgeois society, though they pocket all the fruits of this trusteeship.

(2) The organization of labour itself as social labour: through cooperation, division of labor, and the association of labour with natural science.

(3) Establishment of the world market.[36]

As I look around our contemporary world, I see nothing to contradict and much to confirm his prognosis for capitalist development. In short, neither history nor what Marx called "the hired pugilists of the capitalist class" have delivered the so-often-claimed knock-out punch to his picture of capitalism.

## SOME STRATEGIC IMPLICATIONS

Let me close with few thoughts about the implications of this picture for the development of an adequate contemporary strategy of struggle for a post-capitalist society. These thoughts, like Marx's analysis of capital on which they are based, perforce remain at a high level of abstraction, based as they are on the most universal – and basic – elements common to any contemporary capitalist society, however much these societies may and do differ in the manifestations of these basic conflicts. In addition to an insight into these common elements, any long-range strategy for the labor movement of any country – let alone its day-to-day tactics – must also be based on a concrete analysis of the specific strategies being used by the capitalist class, both locally and globally, to contain and control the concrete manifestations of these basic conflicts.

Nevertheless, it seems to me that certain broad strategic conclusions can be drawn, perhaps of even more universal import than heretofore. For awareness of the trends noted by Marx helps us to understand why the problems of the capitalist system tend more and more nowadays to lose their local character and increasingly assume global form. The internationalization of capital in response to the formation of a more and more truly world market certainly is one important factor in this global process. Unless and until the response by labor to the international character of this process and the resulting crises also assumes more active global forms – as opposed to the current primarily local and therefore primarily reactive responses – it is hard to see how an effective defensive strategy can be worked out, let alone one that enables the working class to take the offensive and raise the question of power.[37]

But "global" should not be taken here just in the geographical sense. It also involves a recognition of the growing intertwining of manual and mental labor in the collective labor process, including the important and growing role played by universal – that is, scientific and technical – labor in this process. Segmentation of the working class – by color, by gender, by nationality, by age, etc. – has always been one of the most effective tools of the capitalist class strategy, and ways to more effectively combat these perennial – but continually renovated – methods of segmentation must continually be developed. The analysis presented here suggests the crucial contemporary role of another traditional form of segmentation, that between mental and manual labor.[38] Many who have been drawn into the capitalist labor process as brain workers fail even to recognize their new roles as members of the working class; while many members of the more traditional strata, well aware of their position as workers, fail to recognize potential allies in such newly-enlisted members of their class. Among these brain workers is an ever-growing – in both numbers and in importance – segment of scientific and technical workers, increasingly subject to hierarchical forms of subordination that leave little room for creative endeavors but plenty for overt forms of exploitation. Ways of successfully coping with the current scission between mental and manual workers will be a vital component of a successful contemporary strategy for labor.

In particular, such an alliance would contribute a great deal to overcoming the feelings of powerlessness so effectively inculcated into many workers, both manual and mental, feelings that inhibit many sectors of the labor movement from waging effective defensive actions against the current onslaught of capital, an onslaught not only on wages but on conditions of labor taken for granted in the advanced capitalist world since the Second World War. Moreover, they powerfully inhibit the working class from raising issues of power: the power of workers within the factory to control their labor process; and the power of the entire class to help shape the answers to the urgent social issues facing our society today.

Whether my outlook should be called optimistic or pessimistic I cannot say. I would call it realistic, in the sense that, until answers to some of these strategic questions are found – not just at the level of theory, but in action – the question of a working class offensive to take power at the head of a broader alliance for a socialist reconstruction of society cannot realistically be posed. Certainly, one sees few signs today of the emergence of such a concrete strategy to challenge capitalist hegemony, and many signs of the obstacles to its creation. But one must have faith in capitalism – faith in its inability to lay to rest the many conflicts it continually creates, conflicts and crises that pose ever graver dangers to the future of humanity; but also faith in its ability to constantly create and recreate its own potential gravediggers, men and women able and ready to take their destiny in their own hands.

*Boston University*

## NOTES

[1] This paper, affectionately dedicated to my dear friend, colleague and mentor, Bob Cohen, is based on a talk first given at the Boston Colloquium symposium on Marxism and the Natural Sciences: Critical Reflections, April 21, 1992. Other versions of the talk have been given at meetings of the Boston and San Francisco Bay chapters of Solidarity. An earlier version was published in Greek translation in *Outopia*, no. 3, Sept.–Oct. 1992, pp. 93–105.

[2] I cannot resist just a few words on my outlook on the problem of a Marxist epistemology. The question of how the natural sciences attain knowledge of their objects if they are embodiments of certain social relations is just a particular case of the general question of how any human activity (praxis) unites both intellectual and material aspects. Like all such questions, it should not be discussed abstractly, but in a concrete context. Marx emphasized: 'In order to examine the connection between intellectual production and material production it is above all necessary to grasp the latter itself not as a general category but in *definite historical* forms' (*Theories of Surplus Value* 1: 276. Moscow: Progress Publishers, 1963).

[3] 'A Political and Philosophical Interview', *New Left Review*, no. 86 (1974). Since then Colletti has completely abandoned his Marxist outlook (see *Le Déclin du Marxisme*, Paris: Presses Universitaires de France, 1984).
[4] *Marx et le machinisme*, Paris: Editions Cujas, 1968.
[5] *De l'Amerique et de la Russie*, Paris, Le Seuil, 1972.
[6] The most important published drafts include the *Grundrisse*, the first draft of *Capital*, volume one, published as *Zur Kritik der Politschen Ökonomie (Manuskript 1861–1863)*, a sixth section of volume one (*Resultate der unmittelbaren Produktionsprozesses*), and the historical-critical manuscripts published as *Theorien über den Mehrwert*.
[7] I shall cite *Capital*, vol. 1, transl. by Ben Fowkes, Hammondsworth: Penguin Books in association with New Left Review, 1976, *Capital*, vol. 3, transl. by David Fernbach, ibid., 1981. The manuscripts of volumes two and three have not been published, but a comparison of selected portions with the text published by Engels suggests that a fuller comparison might reveal significant differences (see Norman Levine, *Dialogues within the Dialectic*, London: George Allen and Unwin, 1984, chap. 4).
[8] John Elster's influential 'analytic' reconstruction of Marx's work speaks of his theory of value as 'conclusively shown to be invalid' (*Making Sense of Marx*, London: Cambridge University Press, 1985). Paradoxically, as Alain Lipietz notes: 'On the one hand, from listening to politicians of all parties, or from reading the economic press, it would appear that the [labor] theory, in its most vulgar form, had become universally accepted. It is taken for granted that prices and competitiveness are determined by the labour time necessary to produce the commodities, their "labour value" (or its mathematical converse, productivity); that the way out of the crisis lies in the reduction of this labour time; and that the distribution of added value (and more especially the rate of profit) depends upon the length and intensity of labour, the purchasing power of wages, and, in short, the "rate of exploitation". Yet in the small world of economists or associated intellectuals who have a smattering of Marxism, even those with the least to contribute feel qualified to criticize the mistakes, logical fallacies, and insoluble contradictions in Marx's economic writings' (*The Enchanted World: Inflation, Credit and the World Crisis*, pp. 139–140. London: Verso/New Left Books, 1985).
[9] See especially *The Enchanted World: Inflation, Credit and the World Crisis*. This remarkable work shows the crucial importance of Marx's distinction between the 'esoteric' relations of production and their 'exoteric' manifestations for an understanding of many features of contemporary capitalism. Marx's views on such currently crucial issues as money, credit, and crises, for example, cannot be understood if this distinction is neglected, as it usually is even by self-styled Marxists. Lipietz goes beyond Marx on certain questions, notably money and credit, but he does so on the basis of a firm grounding in Marx's theory. For Lipietz's analysis of more recent economic developments, see *Towards a New Economic Order*, New York: Oxford University Press, 1992.
[10] Perhaps the most influential such recent interpretation is found in G.A. Cohen, *Karl Marx's Theory of History: A Defense*, Princeton: Princeton University Press, 1980. As pointed out by John Foster Bellamy, 'Cohen finds himself compelled by the very nature of his argument to jettison all logical connections between historical materialism, on the one hand, and (1) Marx's understanding of capital, (2) the labor theory of value, and (3) the ontological/anthropological basis of human beings in concrete human praxis – all of which Cohen himself explicitly rejects – on the other' (p. 32, 'Introduction' to Joseph Ferraro, *Freedom and Determination in History According to Marx and Engels*,

New York: Monthly Review Press, 1992, which includes a detailed critique of Cohen's views).

[11] Marx states explicitly that 'a general rate of surplus value' is 'a tendency, like all economic laws' (*Capital* **3**: 275, transl. by David Fernbach, Hammondsworth: Penguin, 1981), and notes that 'it is only in a very intricate and approximate way, as an average of perpetual fluctuations which can never be firmly fixed, that the general law prevails as the dominant tendency' (*ibid.*, p. 261). Gérard Duménil's valuable *Le Concept de Loi Économique dans 'le Capital'*, Paris: Francois Maspero, 1978, cites all uses of the word 'law' in *Capital* ("Annexe," pp. 401–429).

[12] Marx's discussion of 'The Law of the Tendential Fall in the Rate of Profit' is immediately followed by a discussion of 'Counteracting Factors' and of the 'Development of the Law's Internal Contradictions' (*Capital* **3**: 317–375 'Part Three').

[13] For a recent Marxist analysis, see Makoto Itoh, *The World Economic Crisis and Japanese Capitalism*, New York: St. Martin's Press, 1990.

[14] Cini used this phrase in a talk at a Boston Colloquium symposium on 'Marx and Science' held to celebrate Bob Cohen's sixtieth birthday.

[15] Perhaps the clearest statement of his methodology, with its emphasis on the primacy of production is in the 'Introduction' to the *Grundrisse*. While a number of translations of this exist, the most useful commentary is found the edition edited by Terrell Carver, *Karl Marx: Texts on Method*, New York: Barnes and Noble, 1975.

[16] 'A cotton-spinning machine is a machine for spinning cotton. Only in definite [social] relations does it become *capital*. Torn from these relations, it is no more capital than gold in and of itself is *money* or sugar the *price* of sugar' ('Wage-Labour and Capital', in Karl Marx, *Selected Works in Two Volumes* **1**: 263–264. New York: International Publishers [n.d.], translation modified).

[17] See Friedrich Engels's analysis of the slogan in his article with this title, originally printed as the first of a series in *The Labour Standard*, reprinted under the title *The British Labour Movement*, New York: International Publishers, 1940.

[18] Marx first introduces the term in a section on "The Fetishism of the Commodity and Its Secret," *Capital* **1**: 163–177.

[19] '[T]he vulgar economist thinks he has made a great discovery, when, as against the disclosures of the inner connection, he proudly proclaims that in appearance things look different. In fact, he is boasting that he holds fast to the appearance, and takes it for the last word. Why, then, any science at all?' (Karl Marx, *Letters to Kugelmann*, p. 74. New York: International Publishers, 1934).

[20] For 'a general reference guide for college students and working people who want to understand the forces behind technological change,' see Chris de Bresson with Jim Petersen, *Understanding Technological Change*, Montréal/New York: Black Rose Books, 1987.

[21] *The Poverty of Philosophy*, Moscow: Foreign Languages Publishing House, 1931, p. 92, translation modified. This statement is quoted with approval in a recent book that purports to 'describe when, how, and why technology became a particularly dangerous enemy' (Neil Postman, *Technopoly/ The Surrender of Culture to Technology*, p. xii. New York: Vintage Books, 1993; Marx is quoted on p. 21). To understand what Marx had in mind, his remark should be read together with another on p. 112: 'The hand-mill presupposes a different division of labor from the steam-mill.'

22 *Zur Kritik der Politischen Ökonomie (Manuskript 1861–1863) Text – Teil I*, Berlin: Dietz, 1976, p. 83, my translation. An English translation of this text is now available in Karl Marx and Frederick Engels, *Collected Works*, vol. 30, *Karl Marx: 1861–1863*, New York: International Publishers, 1988.
23 *Theories of Surplus Value* 1: 277, op. cit., translation modified.
24 *Capital* 3: 174.
25 "Results of the Immediate Process of Production," in *Capital* 1: 941–1084.
26 *Grundrisse*, p. 540. Hammondsworth: Penguin, 1973.
27 *Capital* 3: 175.
28 *Theories of Surplus Value* 1: 347, op. cit.
29 *Capital* 1: 458, op. cit.
30 *Capital* 1: 617, op. cit.
31 For a review of recent discussions of the labor process by Marxists and others, with an extensive bibliography, see Paul Thompson, *The Nature of Work/ An Introduction to Debates on the Labour Process*, 2nd ed., London: MacMillan, 1989.
32 *Capital* 3: 198, op. cit., translation modified.
33 'One of the most important differences between the capitalism of Marx's time and modern capitalism is the increasing importance and commercialization of knowledge. ... [V]ast numbers of agents are now engaged in the production of science and techniques and that production is increasingly commercialized, that is, either carried out by business as "in house" research, as a business in itself, or directly or indirectly influenced by business,' Guglielmo Carchedi, *Frontiers of Political Economy*, p. 18. London/New York: Verso, 1991.
34 'Results of the Immediate Process of Production', op. cit., p. 1088.
35 'Results of the Immediate Process of Production', op. cit., p. 1055, transl. modified.
36 *Capital* 3: 375.
37 Even if it has not yet led to the development of an effective strategy, a consciousness of the need for transnational cooperation by labor to resist the transnational offensive of capital seems to be developing among a growing number of trade unionists. Witness, for example, the growing cooperation between United States, Canadian and Mexican trade unionists in the attempt to resist NAFTA (The North American Free Trade Agreement); and the call for international cooperation by a delegation of leaders of the French CGT (General Confederation of Labor) touring the United States (remarks by Louis Viannet, General Secretary of the CGT, and other French and American trade unionists at a meeting in Boston, 14 April 1994).
38 The use of these traditional terms is not meant to suggest that there is not a large intellectual component in 'manual labor', nor that strenuous physical effort is not involved in much 'mental labor'. Like gender appelations, these terms are social constructs – based on a biological substratum, no doubt – but serving to define social relations in the production process rather than biological functions. For an attempt to define mental and manual (or material, as the author calls it) labor within a Marxist framework, see Carchedi, *Frontiers of Political Economy*, op. cit., Sec. 2.3, 'Elements of a Theory of Material and Mental Labour,' pp. 18–27.

REN-ZONG QIU

# KARL POPPER AND KARL MARX

One of the most important events of the twentieth century is the worldwide spread of socialist practice and its subsequent retreat. Of course, it may be said that strictly speaking, the retreat is not that of socialist practice in general, but rather of a particular pattern, namely the Leninist version of socialism, because it is Lenin who greatly revised Marxism and attempted to adapt it to the Russian socio-cultural context. However, it is reasonable to reexamine the principles and method of Marxism in general on the basis of seventy years' experience. This is an arduous task. Among other things, it requires a review of some criticism of Marxism by other schools. In this paper I shall review Karl Popper's criticism and the alternative he provides.

In his intellectual autobiography [1, p. 36] Popper recounts the quite special role Marxism had played in his life. His encounter with Marxism was one of the main events in his intellectual development. As he says, it made him a fallibilist, and highly conscious of the difference between dogmatic and critical thinking. In conjunction with his encounters with the work of Alfred Adler and Sigmund Freud, Marxism also bears credit for his answer to the 'demarcation-problem': what is distinctive about scientific statements is that they are open to empirical refutation and critical assessment, i.e., they should be falsifiable. Since then Popper became a relentless critic of Marxism. However, he cannot help regarding Karl Marx's personality with reverence, when he writes:

There can be no doubt of the humanitarian impulse of Marxism . . . Marx made an honest attempt to apply rational methods to the most urgent problems of social life . . . although he erred in his main doctrines, he did not try in vain . . . [2, v. 2, pp. 81–82]

A return to pre-Marxian social science is inconceivable. All modern writers are indebted to Marx, even if they do not know it. . . . even his mistaken theories are proof of his invincible humanitarianism and sense of justice. [2, v. 2, p. 121]

Marx's faith . . . was fundamentally a faith in the open society. [2, v. 2, p. 200]

For Popper, Marx's social theory is reducible to two main doctrines, namely historicism and economism. In his opinion, Marxist and other social theoreticians are committed to historicist views which prevent their working out any falsifiable social theories. Now what is historicism?

## HISTORICISM

For Popper, historicism involves the claim that:

> it is ... the task of social sciences to furnish us with long-term historical prophecies based on laws of history. [2, v. 1, p. 3]

Or, historicism is:

> An approach to the social sciences which assumes that historical prediction is their principal aim, and which assumes that this aim is attainable by discovering the 'rhythms' or 'patterns', the 'law' or the 'trends' that underlie the evolution of history. [3, p. 3]

On the basic of these and other passages [2, v. 2, p. 82f, 86, 106, 136, 319] historicism can be characterized by the following theses:
(1) The method of social sciences is historical, a matter of the study of history.
(2) The study of history can discover the inherent, necessary, inexorable laws of social development.
(3) These laws form the basis of the 'prophecies', that is, the unconditional predictions about the future course of social development.

For Popper, this characterization of historicism implies that there is a fundamental difference between natural sciences and social sciences in the following senses:
(1) The laws of natural sciences are universal, like Newton's laws of motion and gravitation, but the laws of social sciences are not, and apply only to specific societies. The logical consequence is that there are no general laws of history to provide predictions for any given society.
(2) A society is not reducible to sets of individual human beings and their relationships. As it showed in Marx's thesis "Society does not consist of individuals, but expresses the sum of relations, connections within which these individuals stand" [4]: that societal structure has an impact on individual behavior that cannot be understood by the individual engaged in the behavior.

The distinction of social sciences from natural sciences makes statements about societies and histories immune to falsification. Evidences from one society cannot be used to refute any statement about another society. It also enables people to describe theoretically a perfect society as an ideal to pursue at all costs. Because of this, historicists typically have strong ties to some form of utopianism. And since there is no universal law in social sciences, historicists attempt only to determine what direction the society will in fact take. They often hold that the direction of the development of societal history is a necessary consequence of the structure of the society, as Marxists have held that capitalist societies must destroy themselves due to their own internal contradictions. For historicists, long-term forecasts of the future of a society are the staple of a study of a society.

Popper's account of historicism, at least in the case of Marxism, is not very fair. Marx and many of his proponents have argued the unity of natural science and social science; and they did claim that there are universal laws in social sciences, such as the law that production relations correlate with productive forces in every society, and that all human societies pass through primitive communist, slave, feudal, capitalist, socialist, and then communist stages, in that order. One can argue as to whether such laws do exist in societies, but one cannot deny that Marxists do claim that there are universal laws in the social sciences, like those in the natural sciences. On the other hand, Popper himself has made some statements which would lead people to think that in fact he also claims that there is distinction between the social sciences and the natural sciences. For instance, Popper strongly argues against reductionism in [5]. However, when he argues for methodological individualism, which is the view that facts about a society can always be explained by a series of individual human actions, he seems to be a social reductionist, given that we put the problem of whether a society can be totally explained by individual actions aside. So is Popper's charge of Marxism as unfalsifiable. This will be discussed later. But Popper's criticism of historicism, including Marxism, as having affinity with utopianism and his subsequent remarks on "historical trends" or "long-term forecasts" are plausible.

Now let us look at Popper's arguments against historicism.

## UNFALSIFIABILITY

Popper's fundamental thesis is that there can be no prediction of human history by scientific or any other rational means. It may mean that all such predictions are unfalsifiable. However, the Marxist hypothesis that all human societies pass through primitive communist, slave, feudal, capitalist, socialist and then communist stages can be turned into a highly falsifiable and strict universal generalization. Later when Popper argues against the possibility of making scientific forecasts of historical events, he relies on the fact that the predictions or forecasts may be wrong. The reason is that the evidence on which historicist predictions are based does not validly guarantee their truth. Because if we produce a hypothesis based on the observation of similarities in the evolution of various societies, we have no valid reason to expect, of any apparent repetition of a historical development, that it will continue to run parallel to its prototype. And if we engender a conjecture that a particular pattern or trend is a general feature of a society after the pattern or trend has been observed over a long period of time, nevertheless, as Popper observes, a trend which has persisted for hundreds or even thousands of years may change within a decade, or even more rapidly than that. Popper's argument for the fallibility of historical predictions is plausible, but it turns out to imply that historicist predictions are as easily falsifiable and fallible as any predictions made in science, and therefore his accusation of historicist unfalsifiability collapses.

Popper criticizes historicism on the ground that its 'laws of development' are not really laws at all but 'trends' extrapolated into the future in a way which overlooks the dependence of trends on initial conditions. [3, p. 118] Popper holds that practically every evolutionary sequence in the physical or social world has its precise character determined by a variety of initial conditions; however, if the conditions had changed, then evolution would have taken a completely different path. The ebb and flow of the communist movement since the beginning of this century thus corroborates Popper's claim of trend depending on initial conditions.

The argument Popper raised against the possibility of making forecasts of historical events based on the fact that they may be wrong is self-defeating. If any of these predictions may be wrong, it means that the theory from which these predictions are derived is falsifiable. And the failure of predictions does not always provide a strong reason to reject the theory from which they are derived. It is perfectly acceptable for scientific theories to be revised in the light of failures of predictions. Failure of predictions is a refutation only if proponents of the theory insist that mistakes are not possible in the correct application of the theory. But only dogmatic Marxists, not all Marxists, insist on this. In many cases, Marxist predictions are vague. They outline a development but do not or cannot provide a timetable for its development in real time. Thus there is no time at which failure of predictions constitute refutation, and the theory is being treated as unfalsifiable. But Marxists could argue that the development of history does not require totally accurate temporal prediction, and that only general goals to be achieved in a society can be set in advance, the details being established in the course of realizing them.

## UTOPIANISM

However, the most decisive argument against historicism Popper has raised is the one which appears in the Preface to [3]. This argument demonstrates that

for strictly logical reasons, it is impossible to predict the future course of history

because:

(i) The course of human history is strongly influenced by the growth of human knowledge.
(ii) We cannot predict, by rational or scientific methods, the future growth of our scientific knowledge.
(iii) We cannot, therefore, predict [by rational or scientific methods] the future course of human history. [3, pp. vi–vii]

Popper is right in the passage cited above: We can only make a falsifiable and testable prediction concerning later members of a sequence based on a generalization about the members of that sequence, only in the case of numerical or letter sequences, but not in the sequence of events in human history, because of the impact of knowledge on human history and the unpredictability of the growth of knowledge.

Many honest Marxists fight for an ideal to build a society without exploitation and oppression between human beings, but they confuse the noble ideal with scientific law. Their belief in the necessary attainment of communism implies that both science and human beings are omnipotent without any uncertainty and finitude: human beings can certainly predict the future of human history on the basis of scientific laws. However, no scientific law can help us to predict the future of mankind, nor to predict the future of science itself. Marx prescribes that the principle of distribution in communism should be 'From each according to his ability, to each according to his needs'. But the implementation of this principle presupposes that

(a) Products are plentiful enough to meet the needs of all members of the society;

(b) All members of the society are willing to do and actually do all they can to bring their ability into full play.

However, it is doubtful that any of these presuppositions can be met at any time. Because, first, needs, wants or desires are ever expanding, and become inflated by the progress of technology in particular. The advances of technology can help meet human needs, but at the same time they enormously stimulate people's needs, wants or desires. As soon as China opened its doors, and eagerly assimilated advanced technology, people were no longer content with radio. They needed TV sets, from black-white to colour, from small-size to large size, screens from old style to new style. Any new product, when it begins to develop, is always scarce, not plentiful. Now the question is: who should be among the first to get it? In this case the second part of the principle 'to each according his needs' does not apply.

Secondly, human beings are not manufactured products, they are not the same and cannot be made the same. Each person has 23 pairs of chromosomes on which are attached 50 to 100 thousand genes and he/she is a product of interaction between these genes and a complex physical and socio-cultural environment. It is inconceivable to make everybody willing to do, and actually to make them do what they can to bring their ability into full play.

The conclusion is: the communist principle of distribution is an ideal, but at the same time it is a utopia.

Besides, Marxists in almost all ex-socialist and socialist countries made many wrong predictions on the role of the market and of democracy in social, economic and political life, a role which they ignored; and on the class struggle, which they exaggerated; and on the conflicts between social or ethnic groups, which they underestimated; and they carried out a series of mistaken polices based on these wrong predictions which led to the retreat or even collapse of the socialist system.

## HOLISM

As for what concerns the relationship between individuals and society as a whole, I think both the Marxist and Popper each grasp one half of a dichotomy: Popper argues that historicist conceptions cannot explain what happens when a societal institution is changed, because individuals must change institutions by their actions; institutions cannot change themselves. In addition, the historicist view that institutions control human behavior cannot explain what will happen when two institutions or two societies come into conflict. Popper argues that not only does the construction of institutions involve important personal decisions, but the functioning of even the best institutions will always depend, to a considerable degree, on the persons involved. He analogizes institutions with fortresses which must be well designed and manned. That is right. However, once institutions or fortresses are established, they must have an impact on persons who stay within or without them. So it may be plausible to assume that there is interaction between institutions and individual actions: Institutions may cause individual acts which in turn change institutions which in turn cause individual actions once again.

Moreover, there is a problem for Popper here, related to his "World 3" which, as he claims, is independent of individual decisions and not reducible to human actions, and once established, will have coercive consequences for human thoughts. It would seem that Popper's argument for the autonomy of World 3 applies with equal force to the social world – 'World 4,' – especially in the light of Popper's emphasis on the situation that individuals confront and the unintended consequences of their actions in those situations. Indeed, such human knowledge as written constitutions and laws form a part of the social situation. Recognizing the impact of social institutions or situations on individual actions would be consistent with his arguments about downward causation. [5] So Popper's reductive individualism is inconsistent with his doctrine of World 3, and also inconsistent with his anti-reductionism in the physical and mental world: rejecting the reduction of chemistry to physics and the reduction of mental states to physical states. Social holism is wrong when its proponents, no matter whether they are Marxists or other historicists as Popper labelled them, deny the upward causation from individual actions to social institutions, take individuals as mere means for achieving the mission of a given privileged social group, race or class, and ignore the individual good. As the practices in socialist countries show, politicians who usually pretend to be the representative of the whole society exploit individuals as mere means to achieve their ambitions and suppress dissidents as heretics. However, reductive individualism seems to be much stronger a position than is needed to combat social holism. After all, there is something in a society which is more than the sum of individuals. Individualism and holism both hold for a half of the dichotomy; there is some truth in them, but not the whole truth. There is a perpetual tension between individual good and social good. When they conflict, we have to take a pragmatic approach to resolve the conflict case by case.

The inconsistency between Popper's reductionism in the social world and anti-reductionism in the natural world might be explained by Popper's emphasis on the creativity of the human mind. A material object cannot go beyond downward causation, but a human mind can, in some degree. Even so, it should not be neglected that individual actions including individual creativity are conditioned by their socio-cultural context. Thus Popper's social theory is perhaps best revised by agreeing that social institutions are in many cases not reducible to individual actions and admitting that institutions as well as World 3 are basic for an explanation of human society. It could then be argued that changes in social institutions are typically indeterministic and that individual actions may influence such changes. There is some interaction between them.

## TOTALITARIANISM

Popper is right when he finds a tendency towards totalitarianism in Marxist thought. If the ideal society is a necessary product of the inexorable laws of social development as historicists assert, the plan of the ideal society is designed in advance based on the knowledge of these inexorable laws. But it never can be realized as Utopians envision it, and therefore, individuals should be subordinated to the public good, and it will be reasonable and ethical to use the knowledge along with power to coerce others into a path they might not choose to follow, even though totalitarianism has to be employed as a means of achieving such an ideal society. The practices of Marxist socialist countries has proved that what Popper said is true. The affinity of socialist systems with totalitarianism does not come from the Russian or Chinese version of Marxism, but from Marxism itself. No matter whether Marx expressed his own preference for totalitarianism or not, it is an objective consequence of Marxian theory. But this totalitarianism stifles people's creative and critical power and the open criticism which must underlie decision making in the open society which Popper describes as the best form of social structure.

For Popper, a society in general can be viewed as a problem-solving organization. In his remarks on the open society it is shown how his scientific method of conjectures and refutations can be thoroughly applied in the social field. In society, as in science, the best conjectures may come from anywhere, not only from the top. So the flow of information should not only pass from the top to the bottom, but also from the bottom to the top. For this, free discussion at every level and free communication among all of its members are required, and hence some form of liberalism and democracy is indispensable. So Popper's open society is essentially a liberal democracy in which the free discussion characteristic of a scientific community should be a model of the social community. Although I think Popper's normative picture of an open society is preferable to a closed one, there are some inadequacies in it.

First, it is not known how relevant information can be distributed to all members of an economically and educationally backward developing country. First and foremost is the requirement that members of the society have time to listen to information provided and are competent to understand the information. If making a living occupies their full time, or they are illiterate, they are in a difficult position to take part in free discussion and make any rational criticism or decision. When some Western people eagerly export their democracy to developing countries, they always forget this simple truth. Where they did this, it resulted in chaos, but no democracy at all. The same is true of free discussion. Even in the Western developed countries the poor and disadvantaged groups always find it difficult to take part in sharing of information and free discussion because the relevant economic freedom is circumscribed, and exploitation is highly developed there.

It has been argued that in large societies, various social, ethnic and cultural groups have divergent ends which need to be evaluated and ranked if the society is to make coherent decisions, and groups operating at cross-purposes may not be able to rationally resolve their agreements by debate as Popper suggested. There is a half truth in this counterargument. Popper's fault is his strong belief in human reason. However, human reason has its limits as does also science or authority. The disagreements or conflicts between different social, ethnic and cultural groups should be and can only be resolved by dialogue, negotiation and consultation after debate. Debate is necessary, but not sufficient. What is needed is compromise. Each group should be prepared to compromise with others at any time. I agree with Winston Churchill that democracy is the best of the worst. We have to remember that democracy is better than totalitarianism and authoritarianism, but it is not a living heaven.

## ECONOMISM

Now let us examine Popper's criticism of Marxism as economism. For Popper, economism is

> the claim that the economic organization of society, the organization of our exchange of matter with nature, is fundamental for all social institutions and especially for their historical development. [2, v. 2, p. 106]

In Popper's opinion, Marx uses economism to supplement the thesis of historicism by specifying that the laws of development are laws of economic development. Consciousness is epiphenomenal, or even reducible to matter involved in production. A consequence is that since

> all politics, all legal and political institutions as well as all political struggles are just a way in which the economic or material reality and relations between classes which correspond to it make their appearance in the world of ideologies and ideas, they can never be of primary importance. Politics is impotent. [2, v. 2, pp. 118–118]

We can find some passages in Marx's texts to confirm Popper's accusation of economism like following:

> Capitalist production begets, with the inexorability of a natural process, its own negation. [6, p. 791]
>
> With the change of the economic foundation the entire immense superstructure is more or less rapidly transformed. [7, p. 9]

But other passages cannot be said to confirm Popper's accusation:

> The first step of revolution by the working class is to raise the proletariat to the position of the ruling class . . . The proletariat will use its political supremacy to wrest . . . all capital from the bourgeoisie. [8, p. 481]

Popper's argument against economism is that this doctrine, by taking history to be determined by class structure based on economic factors alone, relies on a biased and misleading conception of the motivation for human actions, and the non-economic factors are no more and no less material than the economic factors.

In my opinion, the concept of economism or economic determinism is not well-defined. If it means that in the human history only economic factors matter, all others do not, and all changes in human history can be explained only by economic changes, this sort of economism or economic determinism is oversimplified and untenable. However, Marx and his close comrade Friedrich Engels cannot be said to be proponents of this sort of economism or economic determinism. Engels once said:

> According to the materialist conception of history, the ultimately determining element in history is the production and reproduction of real life. More than this neither Marx nor I have ever asserted. Marx and I are ourselves partly to blame for the fact that the younger people sometimes lay more stress on the economic side than is due to it. We had to emphasize the main principle vis-a-vis our adversaries, who denied it, and we had not always the time, the place or the opportunity to give their due to the other elements involved in the interaction. [9, p. 493]

In the interaction between social, economic, political, cultural and ideological factors, the economic ones have played a special role which all pre-Marxian historians or social scientists ignored. The special role lies in the fact that the existence and development of human beings depends on the production of material means which forms the basis of a society. However it does not exclude the role played by other factors which are involved in even the change of the mode of production. So Popper's picture of Marx's theory is oversimplified. That Marx sees the invisible, namely the special role of economic factors in human history, more clearly than other scholars is precisely his merit, not his fault. Contemporary human beings on this planet have vivid experience of the importance of economic factors. On the contrary, it is Lenin or Mao Zedong's fault that primary importance was given to political factors. When the Western politicians tenaciously impose their political system on developing countries, but do not concern themselves with people's existence and development, they are making the same mistake.

All attempts to repudiate Marx's theory as nonsense or pseudoscience based on the retreat or failure of socialist practice overlooks the importance of the Marxian reorientation of our thinking about human beings and societies. The analysis introduced by Marx and the Marxists has proved valuable even to non-Marxists as basis for more sophisticated understanding of the social and economic dynamics of modern societies.

*Department of Philosophy of Science and Technology*
*Institute of Philosophy*
*Chinese Academy of Social Sciences*

## REFERENCES

[1] Karl Popper, 1978, *Unending Quest: An Intellectual Autobiography*, Fontana.
[2] Karl Popper, 1952, *The Open Society and its Enemies*. London: Routledge and Kegan Paul.
[3] Karl Popper, 1957, *The Poverty of Historicism*. London: Routledge and Kegan Paul.
[4] Karl Marx, 1969, 'Thesen über Feuerbach', in *Marx Engels Werke*, Band 3, p. 535. Berlin: Dietz Verlag.
[5] Karl Popper and John Eccles, 1977, *The Self and its Brain*, pp. 14–21. Berlin: Springer.
[6] Karl Marx, 1972, *Capital*, in *Marx Engels Werke*, Band 23, Berlin: Dietz Verlag.
[7] Karl Marx, 1972, 'A Contribution to the Critique of Political Economy', in *Marx Engels Werke*, Band 13, Berlin: Dietz Verlag.
[8] Karl Marx and Friedrich Engels, 1972, 'Manifesto of the Communist Party', in *Marx Engels Werke*, Band 4, Berlin: Dietz Verlag.
[9] Friedrich Engels, 1967, 'A Letter to Joseph Bloch, September 21, 1890', in *Marx Engels Werke*, Band 37.

PAUL T. DURBIN

# TECHNOLOGICAL PRAXIS: REFLECTIONS

> [I] characterize some of the objective conditions of the fourth revolution [in the history of technology], ... namely, those conditions which politicize technology as a central question of national policy, the national economy, international competition, rivalry, or war, and governmental or global regulation of massive hazards for species life. All this is new [though ...] this does not mean that aspects of such problems did not already show themselves much earlier. ...
>
> The fourth revolution, by contrast to the first three, introduces a terrifying option; it makes technological or maker's truth hostage to political power, in a decision-procedure that tests policy against the lives of millions, against the planet's future. ... However loose the fit between intentions and outcomes in policy matters, good faith requires some reading of the relevant facts, in their best determination, upon which the policy decision is crucially based. The willful distortion or suppression of facts, or even of reasonable conjectures and arguments about the facts, in the interests of some favored policy goal, or of some exercise of power, is the most dangerous corruption that the politicization of technology makes possible in the context of the fourth revolution.
>
> – Marx W. Wartofsky[1]

Wartofsky's emphasis on the *willful* distortion of the facts the public needs to know, in making good democratic decisions where a decision "tests policy against the lives of millions," or "against the planet's future," suggests another question. Suppose that distortions are not willful but ideologically blindered; and suppose that the ideological blinders affect not only leaders but the entire populace. Is not that situation even more terrifying than the one Wartofsky describes?

I believe we do face terrifying decisions in our technological society, and I believe we need to face squarely the possibility of ideological influences on the social praxis of technological societies. What I offer here is an examination of the concept of *praxis* as it has been used at various stages in the history of Western philosophy.[2]

First, however, some texts to reflect on. I begin with John Dewey's claim that philosophy both ought to be and always has been praxical:

> It has been stated [here] that philosophy grows out of, and in intention is connected with, human affairs. . . . [This] means more than that philosophy ought in the future to be connected with the crises and the tensions in the conduct of human affairs. For it is held [here] that in effect, if not in profession, the great systems of Western philosophy all have been thus motivated and occupied.[3]

More recently, George Allan has devoted an entire book – *The Realizations of the Future: An Inquiry into the Authority of Praxis*[4] – to an analysis of the meaning of praxis. Allan begins with an exceedingly general definition:

> I would suggest . . . that we begin this inquiry, this search for a worthy practice, by first roughing in a working sense of what any practice involves. . . . The answer seems obvious enough. If human being is its becoming, then the shape of that becoming, the contour of human practice, should itself exhibit the necessary conditions within which the substantive character of the ethical ends we seek are to be found. . . . Let me propose the term "praxis" as a name for such a putative general framework of human action.[5]

Armed with this broad definition, Allan explores concepts of praxis from Aristotle to Kant and Marx, and on to such recent deconstructionists as Lyotard.

## SOME DISCUSSIONS OF PRAXIS IN THE HISTORY OF WESTERN PHILOSOPHY

While Allan begins with Aristotle, I want to begin with Plato. Paul Friedländer, an otherwise respected commentator on Plato's dialogues, goes slightly against the general grain of scholarship in accepting Plato's Seventh Letter as both authentic and autobiographical. And on that basis he concludes: "Plato did not set out in quest of this world [of eternal forms]. He set out in quest of the best state, and on this quest he discovered the world of forms."[6]

Friedländer discovers the occasion for this quest in Plato's encounter with Socrates:

> The moment of history into which Socrates was born confronted him with the task of searching for Dike, who had all but disappeared from the sight of men. . . . [H]e searched, through the *Logos*, through continuous dialogical or dialectical inquiry, for the true

meaning behind words, for "what is," for the meaning of "justice." . . . He even died for this "justice."[7]

On this basis, Friedländer cites the Seventh Letter as authoritative:

> The generations of mankind . . . would have no cessation from evils until either the class of those who are true and genuine philosophers came to political power or else the men in political power, by some divine dispensation, became true philosophers.[8]

Friedländer admits that this bold claim may have been made ironically, and of course not everyone agrees that the political efforts of Plato and Socrates were benign[9] – or that philosopher-kings are a good idea.[10] But at least some people believe, with Dewey, that Plato's dialectic was intimately linked to social and political movements of his time.

Nor is this far removed from fairly standard interpretations at least of Plato's early dialogues. We need only think of the *Euthyphro* as unmasking religious pretense, of the *Apology* as confronting Athenian citizens with the challenge that "the unexamined life is not worth living," and of both as leading, by dialectical examination, to the search for Justice as triumphant over factional differences in the *Republic* – where the same claim about the benefits of philosopher-kings is made. So, at least in this respect, Friedländer's acceptance of the authenticity of the praxical claims of the Seventh Letter is not too far out of line.

Jumping forward quickly over a thousand years to the Middle Ages, it is not generally thought that the unmasking of religious and other traditional authorities was a principal aim of the practice of medieval philosophy. There may be reasons for qualifying this commonly-accepted claim, but to dwell on that issue would distract me from my aims here.

This brings us to Descartes and the beginnings of modern philosophy, where the first thing to note is how often praxical philosophers of recent decades have taken Descartes to be the very paradigm of an anti-praxical philosopher. But even as relentlessly intellectual, even analytical, an interpreter of Descartes as Bernard Williams[11] cannot ignore a link to at least the intellectual world of Descartes' time:

> This seemingly rather odd combination of attitudes is more than an accident of Descartes's temperament. The early seventeenth century was only just beginning to develop the apparatus of scientific communication, the foundation of an international scientific community, which is familiar today. The tireless Abbé Mersenne acted as a post office between the many scientists, mathematicians and others that he knew: Fermat at Toulouse, Debeaune at Blois, Desargues occasionally at Lyon, Descartes in Holland.[12]

Williams sees Descartes as contributing to this rising new community a defense against the skepticism engendered not only by the "new science" but also by the Reformation – even while Descartes strove to maintain his good standing with, or at least to avoid condemnation by, the authorities of the Roman Catholic Church. Others more attuned to the social turmoil of Descartes's time than Williams are even more inclined to see Descartes's efforts as related to the cultural events of his time – even to the revolutionary impact of modern science on traditional religion and traditional authority.[13]

When we move on from the beginnings of modern philosophy, alongside modern science, to Enlightenment philosophy as represented in Kant, the same ambivalence – but also the same culture-linked aim – is clear. Lewis White Beck, certainly a traditional interpreter of Kant, makes this praxical point:

> Kant was not just an absolute metaphysician and epistemologist, or a moral philosopher distant from the urgent problems of his times. He was a learned student of history, and intellectually and emotionally very caught up in the stirring political issues of the late eighteenth century, especially the American and French Revolutions.[14]

This would not necessarily assign a priority to praxis in Kant's thought, but elsewhere Beck makes explicit reference to the predilections of Kant that have made him a hero of the Enlightenment:

> For six years [in Prussia] there was harsh dispute between the forces of religious reaction and the more freethinking professors, publicists, and enlightened clerics. Kant was a leader in the campaign to warn of the consequences of both religious intolerance and the intervention of the state in matters of scholarship and private belief.[15]

The fact that Kant, like Descartes, dissimulated in deference to the authorities in no way eliminates the close connection between his philosophizing and the broader cultural movement of the Enlightenment.

If we make another long historical leap, from Kant to the early twentieth century, we find further evidence of the truth of Dewey's claim – even in the school of thought he opposed so strenuously, Logical Positivism, which one might think is farthest removed, in the academic isolation of the positivism-based philosophy of science community, from political and social affairs.[16] Contrary to this expectation, the Vienna Circle positivists have, quite properly, been linked closely with a broader "modernist" movement throughout Europe:

> Since the heyday of Vienna Circle positivism in the 1920s and 1930s, the philosophy of science has been thoroughly intertwined with what it is now fashionable to call the politics

of modernity. In the case of the Vienna Circle itself, the parallels to modernist movements in other domains of culture are very strong. Peter Galison [*Science in Context* 2, 1988] has recently noted that the militant internationalism and antitraditionalism of the Vienna Circle's manifestoes for unified science echoed the contemporary pronouncements of the Italian Futurists and the Bauhaus. But the parallel was more than just rhetorical. Logical positivism was a sweeping program for the critique of culture, whose basic motivation was formalist.[17]

The fact that the positivists' rigid fact-value dichotomy led to the almost total invisibility of their cultural and political persuasions in no way removes the connection of their philosophy to praxis.

## SOME PROBLEMS WITH A BROAD DEFINITION OF PRAXIS

None of this proves that Dewey is right in saying that all philosophical systems are or ought to be linked to cultural, social, and political affairs, but the evidence is at least suggestive. Plato, Descartes, Kant, the Logical Positivists – formalists or rationalists all, in some sense – were all also deeply involved in tradition-critical social movements. A case could much more easily be made for utilitarians such as Bentham and the Mills. And similar cases have been made recently for phenomenologists and existentialists and all sorts of other Continental philosophers.

However, the wide application of Dewey's seemingly-paradoxical claim suggests caution. If all philosophers are in some sense praxical, it must be important to make distinctions with respect to the meaning of the term "praxis."

This shows up already in George Allan, with whose broad definition of praxis we began. Allan ends the key chapter in his book this way:

This is the metaphysical understanding to which we have been led in our long inquiry regarding human praxis and its meaning . . . [namely] the kingdom of ends, the beloved community. . . . [W]e affirm with Whitehead: "The concept of 'God' is the way in which we understand this incredible fact – that what cannot be, yet is."[18]

The chapter which ends with this enigmatic conclusion begins with a discussion of Hegel. It progresses in a dialectical mode, correcting perceived deficiencies in Hegel with analyses of Errol Harris; correcting Harris with the help of Pierre Teilhard de Chardin; and so on. Allan's conclusion, though it cites Whitehead, is based on a reinterpretation of Whitehead by Charles Hartshorne:

> Hartshorne agreed that concrete achievement is what is fundamental [to praxis], but argued that therefore it is not sufficient that the abstract ideal of the realized perfection of the cosmos be all that remains after the deaths of all the attempts to actualize it. Despite what seems in *Adventures in Ideas* to be his agreement with [concrete achievement as fundamental] . . . , Whitehead here [at the end of *Process and Reality*] provides what Hartshorne wants: a way to affirm that multiple temporal perishing achievements are primary but that their significance is everlasting.[19]

In his concluding chapter, Allan plays off Arnold Toynbee – echoing the Hegelian and Whiteheadian view – against Jean-Francois Lyotard. He is self-critical enough to acknowledge that "all Toynbee can come up with is a version of the old, anemic idea of the Kingdom of God realized in the hearts of everyone" as the guarantee of effective praxis that Whitehead and Hartshorne were seeking.[20] Leaning on Lyotard's deconstructionism, Allan concludes that aiming praxis at achieving "the final reconciliation of all its instances in an ultimate realized totality" has "something fabulous" about it. In the end, "Our conversation is all there is. . . . First me, now you. Strutting out there all alone, each of us conversing only with ourself."[21] Nonetheless, Allan's reading of the nature of praxis – in which "the voice of the activist [cannot] be slighted, nor favored, over that of the reflective interpreter"[22] – would be singularly unsatisfying to most praxical philosophers today.

Moreover, for my purposes here – that is, in the context of addressing the possibility of ideological distortion of technological decisions affecting millions of lives and the fate of the natural world – either of Allan's interpretations of praxis, as needing quasi-religious guarantees or as the speculative musings of isolated individuals, seems to offer no protection against ideological distortion. Indeed, it could be said to invite it.

Allan is not naive, and he explicitly refers, at the very beginning of his book, to more activist and social interpretations of praxis – notably those of Dewey and Marx. With respect to Dewey, he even admires the scientific resources he had "for resisting both a dualistic ontology and a monism based either on subjects or on objects."[23]

In spite of this – and despite numerous other positive uses, throughout the book, not only of Dewey's ideas but of his approach – Allan seems much less worried than Dewey was about religious and metaphysical interventions blocking progressive social praxis. In *A Common Faith*,[24] Dewey does make the well-known distinction between anti-progressive religious institutions and the religious attitude that can motivate even the most progressive social action – and it is conceivable that Allan means for his Hartshorne-based Whiteheadianism to be virtually indistinguishable from this aspect of Dewey's thought[25] – but nearly everywhere else in his writings[26] Dewey shows himself to be the implacable enemy of religious and metaphysical and ideological (though he does not use that term) impediments to the progressive social problem solving needed in a world he saw as increasingly technology-dominated.[27]

The best positive statement of the demands of progressive social praxis in Dewey's work is to be found in *Liberalism and Social Action*.[28] There Dewey insists:

Vital and courageous democratic liberalism is the one force that can surely avoid . . . a disastrous narrowing of the issue[s]. . . . The question[s] cannot be answered by argument. Experimental method means experiment, and the question[s] can be answered only by trying, by organized effort.[29]

Dewey adds immediately why such social action is urgent:

The reasons for making the trial are not abstract or recondite. They are found in the confusion, uncertainty and conflict that mark the modern world.

In my view, Dewey would agree with Wartofsky that today, with technology's even greater power and scope than when Dewey was writing, the urgency is all the greater. So is the need to unmask ideologies that can unwittingly distort major technological decisions with their enormous consequences.

But perhaps, radical critics will say, even progressive liberal social action is not enough. It too can be subject to ideological distortion. Even Allan recognizes the special nature of Marx's interpretation of radical social praxis:

Marx was concerned with consequences, with altering the material conditions of life, with making the world a better place. The better place was for him one where human activity is at last everywhere fully human. The goal of revolutionary praxis is to liberate human activity from the forces that debase it.[30]

Though Allan does not say so, the goal of revolutionary praxis is also to liberate human activity from the ideological blinders that accompany economic and political exploitation and domination of the working class.[31] Allan's mode of liberation is through personal self-criticism and reflection; Marx's mode of liberation is through revolutionary social action.

Later Marxists – and notably the members of the Frankfurt School[32]

– have been even more emphatic that revolutionary praxis demands resistance to the use of technological means – not just overt propaganda but more subtle means such as mass media advertising, and control of the film and entertainment industry, along with the schools – to keep the potentially revolutionary masses pacified. This need not be thought of in conspiratorial terms,[33] though no doubt astute members of the managerial classes have recognized the potential for obfuscation that the media provide.[34]

Here at last we come to the most terrifying prospect of all with respect to technological decisions affecting the lives of millions and the fate of the earth: they may be (are being?) made by ideologically-blindered anti-democratic leaders consciously or unconsciously (more often the latter) interested in nothing more than the maintenance of their own positions of power, and without resistance from the masses who have been brainwashed, in hidden as well as overt ways, to think that the way things are – whether that means wars of mass destruction or the endless consumption of trivial goods leading to the destruction of the environment – is the way things should be.[35]

## IS A RADICAL CRITIQUE OUR ONLY SALVATION IN A TECHNOLOGICAL WORLD?

This radical critique might lead us to think that the only way out is to heed the radical critique and act accordingly. Unless the late-capitalist ideological blinders, of leaders and the masses, are removed, there is no way of avoiding technological catastrophes affecting millions of people – or even techno-blunders that might destroy life on earth.

The problem with this kind of revolutionary rhetoric today is the end of the Cold War and the demise of Communism in Eastern Europe. Almost no one today thinks that Marxism, or at least the version put in power in Russia and its satellites under Stalin after World War II, is the solution for any problem.

There have been at least two kinds of responses on the part of radicals to this situation. The first, in Russia and the former Iron Curtain countries and among some intellectuals in the West,[36] is a dogged insistence that Marxism still has the answers – and that the first answer is still to unmask ideology, to show up technocapitalist imperialism for what it is wherever it is, even among supposedly populist leaders in what is left of the old East Bloc.

A second kind of response has been made by Andrew Feenberg,[37] among others. Feenberg takes Marcuse as his starting point. With Marcuse, he takes it to be the case that worker control of technological production has been systematically blocked by technocratic managerialism not only under advanced capitalism but also under "bureaucratic socialism." According to Feenberg, Marcuse failed to carry out the implications of his own theories – as have other insightful thinkers such as Michel Foucault[38] – that capitalist production systems (and their Communist imitators) have built-in anti-worker biases. If we break with Marx on one point – that the unmasking of this bias is an automatic product of historical developments – then workers can freely take control of their own lives, both at work and generally. Feenberg says it is up to the workers of the world to make a choice between an exploitative capitalist system and a world of public ownership and the democratization of the workplace – with a new sort of education and lifestyle spilling over and providing greater freedom in all areas of life.

What, in Feenberg's view, will lead to this remarkable transformation on the part of heretofore compliant workers? What Feenberg offers in place of political revolution is this:

*Deep democratization* implies significant changes in the structure and knowledge base of the various technical and administrative specializations. Furthermore, in advanced societies, where so many relationships outside the sphere of production are technically mediated, self-management is only one dimension of a general attack on technocratic hegemony.[39]

How is this to be brought about? Feenberg says:

Without [a culture of responsibility] . . . those on the bottom of the system are unlikely to demand changes in the distribution of power. To be effective, this demand must meet a sympathetic response from a significant fraction of the technical elites to which it is addressed.[40]

In short, a new order can become a reality if workers are educated to recognize the clear benefits of a new socialized system, and if their consequent demands are met with a sufficiently sympathetic response on the part of at least some technical managers now imbued with a "culture of responsibility."

It could, at this point, clearly be objected that this *postulates* a removal of ideological blinders, on the part of workers and managers, instead of showing how it can realistically be brought about. I believe, however, that Feenberg has made a helpful new beginning in radical theorizing in philosophy of technology, and I think only a modest change in his proposal is needed to make it workable.

What I would propose as the role for radical socialist theorizing today is that it be merged with a Deweyan progressive politicking.[41] According to Dewey, as everyone knows, the solution of urgent social problems – including technosocial problems and even including the problem of technological manipulation of public opinion[42] – is to be sought by way of collaboration among all sorts of activists, from workers and union leaders, to corporate and civic and educational leaders, to intellectuals. Dewey had an ambivalent attitude toward Marxism and toward Communism in Russia; he was extremely leery of violations of civil liberties in the name of democracy, but he recognized the need to unmask the ideological obfuscations of corporate leaders and their cronies in government. Though I am not aware that Dewey ever said this explicitly, the thrust of his thinking on the matter ought to lead us to conclude that the unmasking efforts of Marxist and other radical intellectuals can be a tremendous boon to progressive social activism. It is not necessary that everyone involved be radicalized; it is enough that the radicals among progressive social activists help the rest to see through ideological obfuscations.

Dewey might even say – as he did with respect to so many other supposedly principled approaches – that it would be *wrong* to give any privileged place to radical thought. As with any other set of principles, the demythologizing principles of radicalism work best as *instruments* of progressive social problem solving, not as themselves the solution of the urgent social problems of our technological world.

In all this, there is no guarantee that adequate solutions will be found; all we get is the opportunity to search for solutions, collectively and democratically.[43] Neither is there, in this approach, any guarantee that all ideological obfuscations of all anti-democratic leaders – whether in government or in corporations or within the educational-military-industrial complex – will be unmasked. All I would hope for is that enough of them can be to save the lives of millions of people and the future of the planet.

## CONCLUSION

I believe that Marx Wartofsky is right; we are in a fourth stage in the history of technology, where technological power and political power are so intertwined that political decisions regularly threaten millions of lives and the future of the planet. This has been the situation at least since the development of the atomic bomb during World War II. But today there are many "technological bombs" of a quieter sort – including technological threats to the future of life on earth – that warrant Wartofsky's fear.

I also believe that Wartofsky would agree with me that ideological distortions of public decision making are every bit as important as the willful distortions he mentions. Whether he would agree with me that radical demythologizing of ideological obfuscations can work – even that it can work best – within a movement of progressive social activism, I do not know. But I would hope he could be persuaded on that point.[44]

Finally, I believe that George Allan is right in recognizing a whole range of meanings of praxis. What I would hope is that he might be persuaded that, in our technological world, the threats to humans and to the environment are so great that a special sort of social praxis – a Deweyan progressive social activism enlightened by radical unmaskings of ideology – is the best hope we have for the future.[45]

*University of Delaware*

NOTES

[1] Marx W. Wartofsky, 'Technology, Power, and Truth', in L. Winner (ed.), *Democracy in a Technological Society* (Dordrecht: Kluwer, 1992), pp. 27 and 33.
[2] Like so many other people, I first met Bob Cohen as one of the coordinators of the Boston Colloquium for the Philosophy of Science. This was in the mid-1960s, at a time when my philosophical orientation was undergoing a change. I had been trained in Aristotelian-Thomistic philosophy, but I had taken as my area of specialization the logic of discovery in science. Through roundabout ways, this led me to John Dewey's arguments in favor of an experimental logic and to George Herbert Mead's claim that social praxis is the proper focus for adequate philosophizing. Cohen, as everyone knows, was also interested in social praxis, and we had – and continued to have – many fruitful discussions on the topic. I offer my efforts in this essay as a testimonial to those discussions of the proper nature of praxis.
[3] John Dewey, *Reconstruction in Philosophy*, 2d ed. (Boston: Beacon, 1948), pp. xi–xii.
[4] George Allan, *The Realizations of the Future: An Inquiry into the Authority of Praxis* (Albany: State University of New York Press, 1990).
[5] *Ibid.*, pp. 2–3.
[6] Paul Friedländer, *Plato: An Introduction* (New York: Pantheon and Bollingen Foundation, 1958), p. 6.
[7] *Ibid.*, p. 12.

[8] Plato's Seventh Letter (326b); cited in Friedländer, *Plato*, p. 5.
[9] See I.F. Stone, *The Trial of Socrates* (Boston: Little, Brown, 1988).
[10] See Karl R. Popper, *The Open Society and Its Enemies* **1** (Princeton, N.J.: Princeton University Press, 1971).
[11] See "Descartes, René," by Bernard Williams, in Paul Edwards (ed.), *The Encyclopedia of Philosophy* **2** (New York: Macmillan, 1967), pp. 344–354; also, Bernard Williams, *Descartes: The Project of Pure Enquiry* (New York: Penguin, 1978).
[12] Williams, *Descartes*, p. 25.
[13] Some of the controversy surrounding Descartes's work is summarized by Lucia Palmer in her introduction to Giambattista Vico, *On the Most Ancient Wisdom of the Italians* [*De antiquissima*] (Ithaca, NY: Cornell University Press, 1988), pp. 2–7.
[14] Lewis White Beck, *Kant: Selections* (New York: Macmillan, 1988), p. 413.
[15] *Ibid.*, p. 5.
[16] Controversy over one notorious instance of a philosopher getting involved in a real-life court case can be examined in Larry Laudan, "Commentary: Science at the Bar – Causes for Concern," *Science, Technology, & Human Values* **7** (Fall 1982): 16–19; Michael Ruse, "Response to the Commentary: *Pro Judice*," *ibid.*, pp. 19–23; Philip L. Quinn, "The Philosopher of Science as Expert Witness," in J. Cushing *et al.*, *Science and Reality* (Notre Dame, Ind.: University of Notre Dame Press, 1984), pp. 32–53; Michael Ruse, "Commentary: The Academic as Expert Witness," *STHV* **11** (Spring 1986): 68–73; and Harold P. Green, "Commentary: The Academic as Expert Witness," *ibid.*, pp. 74–75.
[17] Joseph Rouse, "The Politics of Postmodern Philosophy of Science," *Philosophy of Science* **58** (December 1991): 607; the Galison reference is to "History, Philosophy, and the Central Metaphor," *Science in Context* **2** (1988): 197–211.
[18] Allan, *Realizations of the Future*, p. 258.
[19] *Ibid.*, p. 255.
[20] *Ibid.*, p. 292.
[21] *Ibid.*, pp. 294–295.
[22] *Ibid.*, p. 294.
[23] *Ibid.*, p. 16.
[24] John Dewey, *A Common Faith* (New Haven, Conn.: Yale university Press, 1934), p. 27 and *passim*.
[25] But see Robert B. Westbrook, *John Dewey and American Democracy* (Ithaca, NY: Cornell University Press, 1991), pp. 422–428, for the difficulties Dewey encountered because he seemed to open the door to this possibility.
[26] See especially Dewey's *The Quest for Certainty* (New York: Putnam's, 1929) and *Reconstruction* (note 3, above).
[27] See Larry Hickman, *John Dewey's Pragmatic Technology* (Bloomington: Indiana University Press, 1990).
[28] John Dewey, *Liberalism and Social Action* (New York: Putnam's, 1935).
[29] *Ibid.*, p. 92.
[30] Allan, *Realizations of the Future*, p. 4.
[31] See Frederic L. Bender, *Karl Marx: The Essential Writings*, 2d ed. (Boulder, Colo.: Westview, 1986), and L. Easton and K. Guddat, *Writings of the Young Marx on Philosophy and Society* (Garden City, NY: Doubleday, 1967). I take Marx's basic argument to be historical, that once a division of classes emerges, with a class of productive workers exploited by a ruling class, the dynamic of history is set in motion; each stage in history

is characterized by conflict, ending with a revolutionary advance to a new stage; but during each historical epoch, both exploiters and exploited adhere to a ruling ideology that rationalizes the place of each in the social structure of that stage – however much tension remains between workers and exploiters. Only at some future stage, after capitalism and state socialism – when class conflict no longer enforces the demand for ideological obfuscation – can there be even the hope that ideology will no longer corrupt democracy.

[32] See especially Theodor Adorno and Max Horkheimer, *Dialectic of Enlightenment* (New York: Herder and Herder, 1972).

[33] See David F. Noble, *America by Design: Science, Technology, and the Rise of Corporate Capitalism* (New York: Knopf, 1984).

[34] See Herbert I. Schiller, *Culture, Inc.: The Corporate Takeover of Public Expression* (New York: Oxford University Press, 1989), as well as Schiller's earlier *Mass Communications and American Empire* (New York: Kelley, 1969).

[35] Herbert Marcuse, *One-Dimensional Man* (Boston: Beacon, 1964).

[36] See, for example, Peter Osborne (ed.), *Socialism and the Limits of Liberalism* (New York: Routledge, 1991).

[37] Andrew Feenberg, *Critical Theory of Technology* (New York: Oxford University Press, 1991).

[38] Feenberg (*ibid.*, p. 75) cites Foucault's *Power/Knowledge* (New York: Pantheon, 1980).

[39] Feenberg, *Critical Theory*, p. 155.

[40] *Ibid.*

[41] This was already suggested, almost a decade ago, by Morton Schoolman; see his "Liberalism's Ambiguous Legacy: Individualism and Technological Constraints," in P. Durbin (ed.), *Research in Philosophy & Technology* 7 (Greenwich, Conn.: JAI Press, 1984), pp. 229–252.

[42] On the technosocial part, see Hickman, *Dewey's Pragmatic Technology*; for the media manipulation part, see Schiller, *Culture, Inc.*

[43] On such evidence as there is that progressive social activism can ameliorate technosocial problems, at least some of the time, see my *Social Responsibility in Science, Technology, and Medicine* (Bethlehem, Pa.: Lehigh University Press, 1992); see also Michael W. McCann, *Taking Reform Seriously: Perspectives on Public Interest Liberalism* (Ithaca, NY: Cornell University Press, 1986).

[44] I believe Bob Cohen's lifelong willingness to work with philosophers of all persuasions – and with government officials and others throughout the world – is an example for all to follow of how radical demythologizing can go hand in hand with progressive social activism, and also of how such collaboration can achieve good effects even in strongly anti-democratic societies.

[45] I need to end with a bibliographical note: I sympathize much more with Cornel West's reinterpretation of Dewey in *The American Evasion of Philosophy: A Genealogy of Pragmatism* (Madison: University of Wisconsin Press, 1989) than with West's opponent in that book, Richard Rorty, in the latter's claims also to be following in Dewey's footsteps. I am fairly sure that Dewey would have repudiated Rorty as an interpreter of his politically activist philosophy, though I am less sure as to what Dewey would have said about West's "prophetic pragmatism."

SHELDON KRIMSKY

# SCIENCE, SOCIETY, AND THE EXPANDING BOUNDARIES OF MORAL DISCOURSE

The public's acclamation of science was in its ascendancy in the post war period of the 1950s. Scientific achievements were credited with carrying the allied forces to victory in Europe and Asia through the development of radar, the modern technological army, and of course the atomic bomb. Governments throughout the industrialized world were now prepared to invest heavily in science as insurance against future threats to their national security. This change in the government's role was a mixed blessing for scientific institutions. Many disciplines flourished from the new riches of public funds. Some new disciplines were formed out of the war effort and the post-war arms race. But it also meant that scientific research in the American academy became heavily politicized. The image of the lone scientist, broadly educated with the grasp of the large picture, working tirelessly in a makeshift laboratory furnished with hand-crafted equipment, pursuing a path to knowledge according to some ineffable sixth sense, was undergoing a great transformation. The new image was of a strategically planned science consisting of teams of investigators, working on large scale projects, competing for limited funds, positioning themselves in a social structure that would insure the continuity of funding through volatile political times.

While the social structure of post-war science was undergoing a transformation, the linkage of that structure with normative functions made it inevitable that the moral status of science would also undergo change. The relationship between science and ethics has been examined by a number of authors (see for example Merton (1949), Haldane (1957) and Cohen (1974)). For purposes of discussion I cite three relevant aspects of science that bear on moral problems. First, the culture of science possesses norms of behavior that are functional to collective disciplinary goals. A scientific ethics provides rules of practice and principles of self governance within the various professions. Second, through discovery of new knowledge and its applications that expand human possibilities, science is continuously adding to the reservoir of moral problems facing society. And, third, the methods, theories or modes of reasoning in science are cited as models for moral decisionmaking, normative problem solving, or policy choices. It is not accidental that upon introducing new moral problems, science works toward developing the rational means to manage them. Failing to manage these problems makes science vulnerable to external controls. Modern biology is a case in point. Seeking to serve multiple interests during a rapid phase of commercialization, molecular biology became vulnerable to new forms of external governance. The normative transformation that is taking place in molecular biology is illustrative of general shifts occurring the boundaries of moral discourse between science and society.

## AUTONOMY OF SCIENCE

The co-evolution of science and ethics began at least two millennia ago. The nature and character of that relationship has been a subject of continuing scholarly interest pertaining to such questions as: What role has science played in moral reasoning?; how do the ethical norms of science affect the practice of science? Does the normative structure of science provide a model for public life? Can science continue as a self-governing social system?

Perhaps the greatest unity between ethics and science within a system of beliefs can be found in ancient Greek thought. Aristotle posited that virtue was the end to which knowledge was directed; science and morality shared the same end – the contemplation of universal and necessary truths (Ross, 1964), while the greatest virtue was the attainment of moral wisdom. According to MacIntyre (1978, 38) this belief system guarded against conflicts between science and ethics.

During the Middle Ages and through much of the Renaissance the authority of the Catholic Church prevailed on issues of science and ethics. Potential conflicts between science and theology were controlled by a single authoritative view. As the Church began to shift from its Aristotelian orthodoxy accommodating the evidence of the Copernican world view, the protection of science from conflicts with religion was solved by the creation of parallel domains. In areas of moral or theological conflict, science could step aside in deference to ecclesiasticism. The hegemony of the medieval church eventually gave way to scientific authority which blended skepticism, empiricism, and rationality into what John Randall (1926) called "the making of the modern mind." A new culture of science emerged and with it came the methods and ideas to protect that culture from external forms of authority. Thereafter, cultures that have experienced the spectacular growth of science have each faced, in their own way, the dilemmas associated with resolving tensions between scientific interests and public morality.

The new forms of protectionism for scientific practice were expressed in several ways. First, science aligned itself with the most progressive tendencies in society – at the time, industrial entrepreneurship. The moral order was being reshaped by the emergent class of independent capitalists. By associating itself with the new moral order science followed Francis Bacon's dictum that there must be "experiments of light" to discover the causes of things, as well as "experiments of fruit" to apply the knowledge to practical ends (Brown, 1986, 19). Second, science sought to use its own belief structure, the progress it had made in the discovery of physical laws, and the success it had achieved in applying science to technology as a basis of legitimacy for making claims about moral life. The scientists of the Enlightenment believed that a "deductive system of morals comparable to mathematics might be developed" (Randall, 1926, 366). Reminiscent of the unity of science and moral order of the Hellenistic period, the philosopher-scientists of the Enlightenment believed, as noted by Randall (1926, 366), that "The Order of Nature contained an order of natural moral law as well, to be discovered and followed like any other rational principles of the Newtonian world-machine."

Finally, science presented itself as a normative social structure worthy of universal applicability and providing the standard for the moral behavior of lay society. Robert K. Merton's two essays "Science and the Social Order" (1938) and "Science and Democratic Social Structure" (1942) are now legendary in framing the normative conditions of the practice of science. Merton (1957, 595) spoke of an ethos of science as an "emotionally toned complex of rules, prescriptions, mores, beliefs, values and presuppositions which are held to be binding upon the scientist." The widely cited norms in Merton's analysis consist of universalism, communalism, disinterestedness, and organized skepticism. Science was thought to embody its own moral standard and as the "exemplar of applied rationality" (Toulmin, 1980, 59) it was protected from externally imposed standards of moral behavior.

A fourth means through which science sought to protect itself from external control was through a campaign to purge itself of value judgements or non-verifiable claims – commonly known as positivism. The positivist movement in science reached its peak between the First and

Second World Wars. By divorcing itself from ethics and metaphysics, positivists believed that science could be protected from the claims of relativism and subjectivity that plagued secular moral theory (Toulmin, 1980, 45). The French physicist Henri Poincaré (1958, 12) described the separation of domains in his popular book *The Value of Science*.

> Ethics and science have their own domains, which touch but do not interpenetrate. The one shows us what goal we should aspire, the other, given the goal, tells us how to attain it. So they can never conflict since they can never meet. There can no more be immoral science than there can be scientific morals.

Even as science proclaimed its independence of values and metaphysics, some scientists were not shy about applying the intellectual fruits of their fields to influence the development of ethical theory. Physicists and philosophers of science, seeking a rational order in moral reasoning, investigated what Northrop (1947) and Margenau (1964, 138) proclaimed was the "complete parallelism between the formal structure of science and that of ethics." Margenau proposed the formal counterpart in ethics of axioms, protocol facts and rules of correspondence – a carryover of the 17th century ideal that a deductive system of moral reasoning could be developed comparable to mathematical physics. Little progress was made in carrying out the formal analogy between science and moral theory, partly, no doubt, because positivism was so thoroughly discredited.

The principal dilemma for science in the aftermath of the Second World War was to establish an independent and self-governing normative structure while at the same time becoming more fully integrated into the practical and economic life of the nation. The goal was met by giving to scientists the major role in the organization of scientific research programs (Smith, 1990). Price (1964, 20) noted that science became "the only set of institutions for which tax funds are appropriated almost on faith and under concordats which protect the autonomy, if not the cloistered calm, of laboratory."

The visible scientists all too often spoke of their profession as having ascended to a higher moral plane than other sectors of society, particularly politics, law, and business. Jacob Bronowski (1965, 70) was one of several notable scientists who popularized this view:

> ... science has humanized our values. Men have asked for freedom, justice and respect precisely as the scientific spirit has spread among them. The dilemma of today is not that the human values cannot control a mechanical science. It is the other way about: the scientific spirit is more human than the machinery of government.

Two benchmark events of the twentieth century that help shape a new social perception of science were the unprecedented collaboration among scientists in the development of the atomic bomb (Sherwin, 1975) and the perverse role that German science played during the Nazi regime (Proctor, 1988). These events shifted the public image of science from purveyor of knowledge to instrument of mass annihilation and geopolitical power. The nuclear physicists had produced without question the weightiest moral problem since human inquiry into the physical world began: should science pursue an area of investigation that could result in the destruction of human civilization?

While science could no longer claim its moral innocence or its impartiality to political power, it continued to find ways to protect itself from external governance and control. To effect this, a dualism was adopted that distinguished between the pure and the impure aspects of scientific work. Building on Bacon's distinction between "experiments of light" and "experiments of fruit," this view asserted that there are two forms of science. Basic or pure scientists produce knowledge; applied scientists or engineers direct that knowledge to uses society or those in power seek to advance. Within this framework the relevant moral question is: How ought we control the application of knowledge when we are obliged to protect its pursuit? When faced with external criticisms about the nature and direction of research, scientists could always fall back on the dualism of inquiry and application. Inquiry, must remain maximally free in this view. This freedom is part of an inferred social contract between the scientific and the political cultures. In contrast, the application of science to human affairs must be subject to the guidance of informed democratic processes.

## THE MORAL CRITIQUE OF SCIENTIFIC EPISTEMOLOGY

In the last quarter of the twentieth century, science has experienced new restraints on its self-governance. It has also been the target of a new wave of epistemological and ethical critiques from feminists and environmentalists. The new critics of science began turning a moral lens on the scientific method, not for its weakness of rigor but for harboring implicit values. Among these, it is argued that the choice of research programs and the methods used to carry them out were weighted heavily toward certain problems while neglecting others.

Barry Commoner, a refugee from molecular biology, argued in the

1960s that the current trend of mechanistic reductionism in biology is largely responsible for society's neglect of the environment. According to Commoner (1963) too much attention was given to the study of issues that had no relevance to the proper functioning of the planet as the biosphere was facing serious degradation. Moreover, the reward structure in science was designed to give positive reinforcement to the study of molecular biological processes while neglecting the organismic study of natural systems. Similar arguments were advanced during national debates about the war against cancer. Critics of the viral and genetic paradigms of cancer maintained that science was mobilized in an all out front to find a cure that would not disturb the vast sources of industrial pollution (Epstein, 1978). A second wave of moral and epistemological critique was launched by feminist historians and philosophers who characterized the scientific method as gender-laden. In its most extreme form this critical perspective asserted that mechanistic materialism, patriarchy, and the abuse of nature are reinforcing and collateral belief systems. For example, according to Merchant (1980, 291) "Mechanistic assumptions about nature push us increasingly in the direction of artificial environment, mechanized control over more and more aspects of human life, and loss of the quality of life itself." New studies of gender in science (Keller, 1985) gave a new perspective to the 19th century debates about subjectivity/objectivity, mechanism/ vitalism, holism/reductionism. These epistemic ideas were cast in moral categories of misogyny and the degradation of the biosphere. Feminist "standpoint epistemology" was advanced as an alternative to current notions of a value-neutral objective science. The loss of objectivity has moral implications since, as noted by Cohen (1974, 143), its loss negates the "one great quality of the Enlightenment and the humane life" it provides. Not all feminist philosophies are inclined to reject objectivity. While some argue that current scientific conventions produce an androcentric bias in the results of research (Harding, 1986, 25), the possibility was left open for a feminist science with a new set of epistemic and cognitive categories more likely to yield unbiased and objective results. These feminist critiques forged the link between scientific ethics and epistemology.

## LIMITS OF SCIENTIFIC INQUIRY

Exercising greater social control over the direction of research that has served class or gender interests has been a theme in the critical writings of marxists, environmentalists, and feminists. But until recently, the idea of free inquiry had never been critically examined within the mainstream of science. That changed when biologist Robert Sinsheimer (1978) raised doubts about several areas of scientific inquiry including isotope fractionation, the search for extra-terrestrial intelligence, and the pursuit of knowledge that could delay the aging process. Sinsheimer argued that since each of these research programs could result in grave outcomes for human civilization there was time to consider restraints on certain lines of scientific inquiry. Simplification of isotope fractionation techniques, for example, likely will result in the easy manufacture of plutonium. This innovation is a sure path to nuclear proliferation, according to Sinsheimer, and would make atomic weapons accessible to terrorist groups. Other examples cited by Sinsheimer had a more tenuous cause-effect relationship.

The discovery of extra-terrestrial intelligence more advanced than our own will have unforseen effects on the psyche of the human race. And, finally, by slowing down the aging process science will be faced with an issue of trans-generational equity in that the outcome of the research would limit the welfare of future societies. Sinsheimer (1978, 30) concluded: "In a finite world, the end of death means the end of birth."

An entire issue of *Daedalus*, the journal of the American Academy of Arts and Sciences was devoted to the limits of scientific inquiry. Fellow biologist David Baltimore offered a rejoinder to Sinsheimer and declared that any restrictions to the free and open pursuit principle in science is a violation of its essential character. Baltimore (1978, 43) commented that "society can choose to have either more science or less science, but choosing *which* science to have is not a feasible alternative."

In this debate the right of government to selectively fund research was not in question. It is generally recognized that scientists do not have an entitlement to receive public funding to pursue their ideas. Nor was this debate about controls over research experiments that might result in hazards to people or the environment. Federal guidelines on human subjects research and genetic engineering was already established.

However, where the targets of restraint are broad areas of inquiry and not simply experimental systems, the stakes for science are high. The greatest threat is that an influential lobby of private interests could prohibit certain areas of inquiry. Recognizing this, Carmon (1985, 47) argues that pure science is analogous to a type of quasi-speech. "Constitutional protection for new forms of scientific exploration and insight deserves no less deference than that which we ought to accord new forms of political protest."

Moral and political agendas do influence the funding of science despite the best efforts by the scientific community to promote the myth about free inquiry. The federal government under the Reagan and Bush years refused to release funds to study fetal tissue transplants for the treatment of Parkinson's disease. Opponents of this research believed that it would both legitimize and increase the rate of abortions in the United States. In another case, a peer reviewed national survey on human sexual attitudes designed to learn more about AIDS transmission and prevention was derailed because explicit sexual questions offended religious fundamentalists.

## NORMATIVE STRUCTURE OF SCIENCE

Robert K. Merton's formulation of the four norms of science reinforced among its practitioners the view that science had its own universal moral code. "The scientist came to regard himself as independent of society and to consider science as a self-validating enterprise, which was in society, but not of it (Merton, 1957, 605). Bronowski (1965, 70) spoke of values of science as "the inescapable conditions for its practice." The concept of a "scientific ethic" provides the institutions of science with a virtual shield against intrusion by a public morality of scientific practice.

In the 1970s science was faced with new forms of public scrutiny on issues of ethical conduct. Scientists lost authority to set the moral parameters for experiments with human subjects. Disclosure of abuses in medical, psychopharmacological, and psychological experiments were showcased in federal hearings and eventually resulted in legislation. The most visible and politically sensitive cases includes the administration of live cancer cells to elderly patients without informed consent, psychosurgery on incarcerated persons, unreliable data submitted for drug approval by clinical investigators, and the failure to treat syphilitic patients in the infamous Tuskegee study. As a direct result of the publicity of these and other cases, federal guidelines for human subjects research, quality assurance of scientific data, and institutional human subjects review boards were created (Swazey, 1978).

Animal rights became a formidable mass movement in the 1980s after disclosures about abuses in animal care in federally-funded research facilities. These disclosures and a rise of militancy among animal activists brought stricter government accountability for animal care facilities. Neither of these interventions (protection of human subjects and animals) into the laboratory life of science threatened broad areas of inquiry or the fundamental Mertonian norms. But they did extend public oversight into scientific conduct in areas where there was already prior societal involvement. The new animal rights movement was derivative of the antivivesection groups of the 19th century and the protection of human subjects evolved from the ethical imperatives of the Hippocratic oath in medicine.

When genetic engineering was born in the 1970s, scientists faced the prospect of new forms of social accountability. Now the issue was laboratory hazards involving the application of recombinant DNA techniques. Public concern was over the inadvertent creation of an epidemic pathogen even before there was a demonstrable hazard.

In a series of unprecedented actions, more than a dozen local communities were not satisfied with the scope of the federal government's guidelines and enacted laws regulating experiments that involved cutting and splicing genetic material (Krimsky, 1982). These actions were in direct conflict with the self-governing principles of science. This had been the first biochemical technique that had been subject to multi-nation controls.

In his Conway memorial Lecture delivered in 1928, J.B.S. Haldane (1932, 101) noted that "by complicating life, science creates new opportunities of wrongdoing." Since the discovery of recombinant DNA society has been poised for wrongdoing or ethical transgressions in biology. The protective cloak around the biological sciences securing its internal governance, lifted somewhat in response to abuses in human experiments, was bared free when it entered the new frontier of synthetic biology.

Public intrusions into the practice of biology began because of concerns that research scientists were not in a position to police themselves. Popular discourse exploited literary symbols like Frankenstein monsters and Andromeda strains. After more than a decade of federal oversight of the genetics and biochemical laboratories in universities, many scientists cried false alarm. The public's ethical concerns were refocused from laboratory hazards to other issues two of the most prominent are: eugenics and scientific entrepreneurship. Each of these precipitated a new dialogue on the control of scientific research.

## HUMAN GENETIC MODIFICATION

Scientists are rapidly learning how to delete and add functional genes to mammalian cells. This will open the door for molecular eugenics or the modification of the human germ line at the cellular level. If successful, this research could be used on human eggs to improve the human gene pool. The clash in values among social groups over whether scientists should be permitted to engage in human germ line research recapitulates deep moral conflicts in 400 years of modern science. Does progress in scientific research shape what is acceptable moral practice, or does science conform to an independently determined moral code? Will a society now highly suspect and antagonistic to modifying the inheritable genes of humans, eventually accept this process as a new form of control over and liberation from uncertainty in nature.

Another argument against tampering with the human genome is that scientists would be experimenting with the unborn without any possibility of obtaining informed consent. Few would argue today that the human genome is optimally suited for the environment we live in. Our somatic and germ cells are exposed to all sorts of mutagens that are the result of human activities including radioactive isotopes, synthetic organic molecules, and increased UV light due to ozone depletion. But on what basis should we believe that human reason is capable of creating a more optimally suited genome by making pin-point changes in a fertilized human egg? How would we explain to a future germ-line enhanced recipient that his diminished cognitive capacity was an outcome of the effort to improve his life expectancy?

Once science reaches the stage where the techniques for modifying the animal genome have been successfully performed on primates, what will determine whether they will be used on humans? How can society separate the pursuit of objective knowledge about the human genome from the exploitation of that knowledge for political, social, or personal power?

Currently, there is a universally agreed upon informal moratorium on human germ-line manipulation. As of yet, no government is reported to have funded experiments which combine in-vitro fertilization with gene modification. However, like the earth's tectonic plates, the normative boundaries are slowing shifting. There is an uneasy balance between scientific interests and public interests. Initially, scientists found consensus in establishing a moral boundary between somatic and germ cell manipulation, where the latter was prohibited by general agreement, and the former would be carefully monitored under human experiment protocols. However, as techniques for animal germ-line modification improved and as the Human Genome Initiative got underway, a new moral position was beginning to take form in the scientific community.

After a scientific workshop was held on human gene therapy in 1986, the National Academy of Science published a book that outlined four principal categories of human genetic experiments: somatic cell gene therapy; germ line gene therapy; enhancement genetic engineering; and eugenic genetic engineering. The goal of somatic cell gene therapy is to eliminate the clinical manifestation of genetically-caused disease on an individual without affecting the individual's progeny. Germline gene therapy is defined as the insertion of a healthy gene into a human egg. Enhancement genetic engineering refers to the modification of single traits in an individual such as the height of a child. Finally, eugenic genetic engineering applies to the alteration of complex traits involving multiple genes such as intelligence and personality. The shift in the moral boundary for acceptable human gene experiments as illustrated by this classification is from somatic/germ line to therapy/enhancement. While there is a scientific basic for distinguishing between the modification of germ cells and somatic cells, the same cannot be said for the distinction between therapy and enhancement (Krimsky, 1990). Does a person who has high sensitivity to chemical exposure possess "defective" genes in need of therapy or should such treatment, were it available, be considered "enhancing the individual"?

Is extreme shortness an indication of defective genes or part of the normal tail in a population distribution? The boundary between normal and abnormal genes is not rooted in science, but is rather a social construction. Any form of germ line modification will have to face this dilemma and as efforts in this direction are intensified, they will be met with the prospect of greater social control of genetics research applied to modifying the human genome.

## MULTI-VESTED SCIENCE

Among the four institutional norms of science cited by Robert Merton, the shared value of disinterestedness embodies the moral commitment to leave aside one's personal interest when investigating the laws of the natural world. In the idealized situation, scientists have only one interest in the outcome of inquiry and that is to ascertain the truth or falsity of a hypothesis. Fraud, Merton noted, was virtually absent in the annals of science as a consequence of this widely shared norm. If the sole interest of the scientist is the pursuit of truth, conflict of interest among scientists must be an oxymoron. As in cases of fraud, such conflicts were traditionally viewed as deviant cases that do not negate the near universality of the norm.

Fifty years after Merton published his thesis, the social relations and structure of science research has undergone significant change. In particular, it is no longer uncommon for academic scientists in certain fields to have consulting relationships with industry, to hold stock in companies related to their work, to hold multiple patents from their research, or to be involved in the development of a new company. These changes in the biological sciences started almost immediately after the discovery of gene splicing techniques just as industrial genetics had gotten underway. And with these changes came a rapid transformation of the norms of the biological sciences (Etzkowitz, 1989, 14).

For several reasons, entrepreneurship in the biological sciences created a more intense public response and media reaction than similar activities in other disciplines. This was surprising to many biologists who viewed the commercial possibilities in molecular genetics as analogous to what had happened in chemistry, physics, computer sciences, and many other disciplines that had something to offer industry. The public seems to expect a higher standard of ethical conduct in the health sciences than it does in other fields. Since the major portion of health sciences research in universities is publicly funded, scientific investigators are considered more directly accountable to the public interest than other fields.

Scientific fraud and conflict of interest were the subjects of numerous media accounts in the 1980s. After a series of hearings, Congress highlighted these issues in a report titled "Are Scientific Misconduct and Conflicts of Interest Hazardous to our Health" (U.S. Congress, 1990). The report cited scientific studies compromised by financial conflicts of interest and questioned the ability of the scientific community to set its own standards of professional conduct.

Once remote from commercial linkages, the biological sciences witnessed a rapid and aggressive rush among leading scientists and their universities to capitalize on the financial expectations of biotechnology (Krimsky, 1991; Blumenthal, 1986). Many new firms were started by scientists who retained their full academic appointments. The notion of "disinterestedness" in science was subject to a new critical inquiry. Can a scientist whose research is funded by a drug company be disinterested in the outcome? What about a scientist who is evaluating the safety or efficacy of a product that his company is poised to manufacture? Should a scientist who has equity in a company be required to disclose the relationship in a scientific publication related to the firm's commercial interests?

These questions, once the purview of professional ethics, now are at the centerpiece of public policy. The suspicion of the "disinterested" scientist was expressed best by the action of two major scientific journals: the Journal of the American Medical Association and the New England Journal of Medicine that require authors to disclose consulting and equity relationships related to their research.

The changing social structure of science and the growth of reciprocal and symbiotic relationships between academic research and commercial development has resulted in a new public examination of the moral status of science. Despite efforts within the scientific and medical community to retain internal control of the ethical behavior of scientists, the scope of public accountability has been widening. Issues once deemed to be in the domain of professional ethics, have become issues of social ethics. Swazey (1978, 140) notes: "there has been a progressive shift in the locus of control from within clinical investigation to extraprofessionally or bureaucratically mandated laws and regulations."

Scientific fraud, conflict of interest, intellectual property, experiments with animals and fetal tissue, and genetic engineering are among new areas where governments have begun to take a more proactive role in monitoring the behavior of scientists. These trends do not imply that scientists are less moral than their predecessors. Rather, scientists are embedded in a new system of social relations that blur the boundaries between pure and applied research, between private versus public science, and between inquiry for knowledge and inquiry for profit. Scientists are faced with more choices and are situated in more varied contexts than they once were. Changes in the normative structure of science are reflections of changes in the nature and organization of research (Etzkowitz, 1989, 27).

It is a popular misconception that the autonomy of science is narrowed as external events encroach on its self-governance. (Nelkin, 1984, 93). The concept of "scientific autonomy" in this context is treated as an entitlement that comes along with being a member of the professional guild. Not only is science treated ahistorically in this way but "autonomy in science" takes on an essentialist status somewhat like a theory of natural rights. I would argue that the moral autonomy of science expressed in such terms as "freedom of inquiry" does not precede the social context within which science is carried out but is derivative of it. This means that the normative conditions of science do not make sense apart from the political and economic context within which science is embedded.

Once remote from the affairs of public life, there is much more overlap between the moral domains of science and society. As scientists and their institutional cultures have become more deeply woven into the fabric of society through military research, federal grants, and entrepreneurial affairs, the spheres of normative behavior have been pulled closer together resulting in a recalibration of the boundaries of self-governance. The independence of moral spheres is no longer functional in the new system of relations. Thus, it is not as if the moral autonomy of science is threatened, rather a new concept of scientific autonomy and public responsibility is emerging.

*Department of Urban & Environmental Policy Tufts University*

### REFERENCES

David Baltimore, 1978, 'Limiting Science: A Biologist's Perspective', *Daedalus: The Limits of Scientific Inquiry* **107**(2): 37–45 (Spring).
David Blumenthal *et al.*, 1986, 'University Industry Research Relationships in Biotechnology: Implications for the University', *Science* **232**: 1361–1366 (June 13, 1986).

Hanbury Brown, 1986, *The Wisdom of Science*, Cambridge, England: Cambridge University Press.
Jacob Bronowski, 1965, *Science and Human Values*, New York: Harper and Row.
Ira Carmon, 1985, *Cloning and the Constitution*, Madison, WI: University of Wisconsin Press.
Robert S. Cohen, 1974, 'Ethics and Science', in W.H. Truitt and T.W. Graham Solomons (eds.), *Science, Technology & Freedom*, Boston: Houghton Mifflen Co.
Barry Commoner, 1963, *Science and Survival*, New York: Viking.
Henry Etzkowitz, 1989, 'Entrepreneurial Science in the Academy: A Case of the Transformation of Norms', *Social Problems* **36**(1): 14–29.
Samuel Epstein, 1978, *The Politics of Cancer*, San Francisco: Sierra Books.
J.B.S. Haldane, 1932, *Science and Human Life*, New York: Harper & Brothers.
Sandra Harding, 1986. *The Science Question in Feminism*, Ithaca, NY: Cornell University Press.
Evelyn Fox Keller, 1985, *Reflections on Gender and Science*, New Haven: Yale University Press.
Sheldon Krimsky, 1982, *Genetic Alchemy: The Social History of the Recombinant DNA Controversy*, Cambridge, MA: MIT Press.
Sheldon Krimsky, 1990, 'Human Gene Therapy: Must We Know Where to Stop Before We Start', *Human Gene Therapy* **1**: 171–173.
Sheldon Krimsky, 1991, 'Academic Corporate Ties in Biotechnology', *Science, Technology & Human Values* **16**(3): 275–287 (Summer).
Alisdair MacIntyre, 1978, 'Objectivity in Morality and Objectivity in Science', in H. Tristram Engelhardt, Jr. and Daniel Callahan (eds.), *Morals, Science and Sociality*, Hastings-on-hudson: Institute of Society, Ethics, and the Life Sciences.
Henry Margenau, 1964, *Ethics and Science*, Princeton: D. Van Nostrand Co.
Carolyn Merchant, 1980, *The Death of Nature*, New York: Harper & Row.
Robert K. Merton, 1949; 1957, *Social Theory and Social Structure*, New York: The Free Press.
Dorothy Nelkin, 1984, *Science as Intellectual Property*, New York: MacMillan, 1984.
F.S.C. Northrop, 1947, *The Logic of the Sciences and the Humanities*, New York: MacMillan.
Henri Poincaré, 1958, *The Value of Science*, New York: Dover.
Don K. Price, 1964, 'The Scientific Establishment', in Robert Gilpin and Christopher Wright (eds.), *Scientists and National Policy-making*, New York: Columbia University Press.
Robert Proctor, 1988, *Racial Hygiene: Medicine Under the Nazis*, Cambridge, MA: Harvard University Press.
John Herman Randall, 1926, *The Making of the Modern Mind*, Boston: Houghton Mifflin.
Sir David Ross, 1964, *Aristotle*. London: Methuen.
Martin Sherwin, 1975, *A World Destroyed: The Atomic Bomb and the Grand Alliance*, New York: Random House.
Robert L. Sinsheimer, 1978. 'The Presumptions of Science', *Daedalus: The Limits of Scientific Inquiry* **107**(2): 23–35 (Spring).
Bruce L.R. Smith, 1990, *American Science Policy Since World War II*, Washington, DC: Brookings Institution.

Judith Swazey, 1978, 'Protesting the "Animal of Necessity": Limits to Inquiry in Clinical Investigations', *Daedalus: Limits to Scientific Inquiry* **107**(2): 129–145 (Spring) U.S.

Stephen Toulmin, 1980, 'How Can We Reconnect the Sciences with the Foundations of Ethics', in H. Tristram Engelhardt, Jr. and Daniel Callahan (eds.), *Knowing and Valuing: The Search for Common Roots*, Hastings-on-Hudson, NY: Institute of Society, Ethics and the Life Sciences.

U.S. Congress, House Committee on Government Operations, 1990, *Are Scientific Misconduct and Conflicts of Interest Hazardous to Our Health*, Washington, DC: U.S. Government Printing Office.

MIHAILO MARKOVIĆ

# THE CONCEPT OF SCIENTIFIC LAW

If we want to act *both freely and rationally*, we cannot afford either to disregard scientific knowledge of laws, nor to conceive laws in such a fixed, dehumanized, thing-like way that they *a priori* exclude freedom as a dream of an ignorant, naive, romantic person.

A scientific law, in rationalist interpretation, is a true universal statement that expresses necessary connection among variables and has explanatory and predictive power on its own account. Variables here refer indiscriminately to groups of either stones or plants or human persons. Necessity and universality of laws suggest that *all* phenomena covered by the law *without any exception must* occur in the way the law describes and prescribes. That such a statement has explanatory and predictive power on its own account means that any conditions and circumstances of the situation that might perhaps be changed by deliberate human action do not count – human consciousness and will is irrelevant. Truth is interpreted as an absolute unhistorical category: if a lawlike statement is at all true, it is true forever; one single disconfirming instance would prove that the statement was false and did not express any law at all. Thus, either there are no laws whatsoever, or else if there are, they do not make any room for freedom – except in the sense of understanding of necessity and conforming to it.

Many believers in a much more active view of freedom uncritically accept the rationalist's account of the law, agree that the two are incompatible and try to conceptualize freedom in a basically irrationalist way, locating it in a sphere of human existence where no laws hold and no external causes constraint our possibilities. If this is *poetry* it is welcome: there are too many small men around; a greater sense of pride, more boldness, more imagination is badly needed. But if this is *theory* – it is irresponsible, half-truth at best. Only a mad philosopher can forget how strong is the wind, how far the stars; how inevitably the dusk comes after full sunlight and then the night, then the dawn, but again the night. The issue is: how to be free although the wind keeps blowing into our faces, although we cannot reach the stars, although we cannot stop the darkness.

When we think about laws we come across a strange antinomy. Laws are supposed to be universally valid, but the empirical fact is that almost all of them hold only approximately – with many exceptions in individual cases.

The list of those who emphatically deny that a law can have exceptions is quite impressive. Claud Bernard wrote: "There is nothing abnormal in nature, everything occurs according to the laws which are absolute. The word 'exception' is unscientific: once the laws are known there can no longer be any exceptions."[1] Herbert Dingle believed that the only difference between scientific and political laws is in the fact that "human beings sometimes violate one of the social laws whereas the world of external nature is unmistakenly regular."[2] Even such a flexible, historically-minded philosopher as Dewey defined the law as a generalization referring to a characteristic which has been perceived and established without any exception.[3] The next consistent defender of this view in contemporary philosophy is Karl Popper who urges that a lawlike general statement must be rejected when we establish one single falsifying instance.[4]

Nevertheless, as every empirical scholar engaged in collection and tabulation of experimental data knows, observable phenomena do not only sometimes deviate from the laws, but rarely even happen strictly according to the laws. For example, it is a commonplace that hardly any real gas behaves according to the Gay-Lussac law. The law describes a simplified, approximately correct structure. This formula holds only for a nonexistent "ideal" gas. All gases *approximate* this ideal state when their temperature increases and density decreases. Even in the paradigm case of the solar system, planets do not move exactly according to Newton's and Kepler's laws, one of the reasons being that the laws focus on the relation between the sun and each particular planet leaving aside very small and, for theoretical purposes, negligible interaction among planets themselves. Another reason is that real quantitative figures of some physical properties are not as constant as the laws assume but fluctuate around certain mean values.

This kind of consideration led Michael Scriven to assert that "inaccuracy is the key property of all laws."[5] How is it possible to reconcile such a factually well grounded thesis with the view of those who for theoretical reasons define the concept of a law in a way that excludes any exceptions?

This contradiction can be resolved if we make a distinction between the *theoretical* and the *empirical* form of a law. The former is part of an abstract model, i.e., part of a simplified and idealized projection of a segment of reality. The latter is part of an empirically established pattern of data from that segment of reality. The theoretical is the formalization of the empirical. The empirical is the realization of the theoretical.

When a model is applied to phenomena of the real world a theoretician discovers what the empirical researcher has always known: that objects are not grouped into perfectly homogenous, clearly distinct, separated groups but rather into clusters with many borderline cases. Therefore only in the context of an abstract theoretical model in which everything is simplified can we also simplify the concept of a law as to assume its universal exceptionless validity. In a concrete empirical context the laws look much less regular: phenomena fluctuate, and instead of recurring uniformly they undergo unexpected changes in the interactions with phenomena which were neglected in the model; they disperse around a central statistical tendency. Strictly speaking phenomena do not occur according to the theoretical formulations of the laws. And yet the latter are of great cognitive and practical importance. They are structurally similar to real processes. Even when *no* individual event occurs strictly according to the theoretical form of a law *all* events occur in an *approximate* way. This is the sense in which a law "holds" or is "true". But this is no longer the truth of a two-valued logic. When our rational thinking goes beyond immediate empirical reality, when we describe ideal structures which will never be exemplified in their pure forms, we do not assert that something is simply true, or is false, or is true in some cases and untrue in others. We assert something that *could* be true if the conditions assumed in the model would be obtained or is *approximately* true in some important types of real situations.

There is clearly an element of limited human freedom in the very concept of a law. The scientific law is a historical category, a human symbolic creation. To use Stephen Toulmin's[6] example a map must be isomorphic with the structure of corresponding territory. But there are many different actual and possible maps of the same territory. It depends on our practical needs how detailed it will be, and whether it will emphasize communications or morphological properties, or distribution of population, or administrative division, etc. In scientific laws the practical purpose must have universal human character (and this makes them different from ideological *quasi* laws). The natural laws are usually uncontroversial because their practical implications are increased control over nature, which obviously promotes universal human interest. But social laws are controversial when they are formulated in such a way that they ultimately serve the purpose of preserving or revolutionizing existing social forms. A good example is Marx's famous law of the decrease of the average profit rate in capitalism. In Marx's model, technological development implies a faster increase of constant capital ($c$) than of variable capital ($v$). As only variable capital is the source of value and profit, this trend leads to a decrease in the average profit rate. The only alternative for capitalists is to try to squeeze more labor out of the workers – which sooner or later leads to a workers' rebellion. For Marx this law becomes the scientific justification of the idea of revolution. A contemporary scholar who is not committed to revolution but to the preservation of capitalism would build up a different model. Even if he does accept the labor theory of value (which most Western economists don't), he can introduce into the model political and cultural factors which did not exist or play an important economic role in Marx's time: an interventionist state, the political and bargaining power of organized labor, growing role of mass media and overwhelming, easily cultivated general interest in consumption of industrial goods.

Then wages would be kept high and workers would be persuaded to consume instead of rebelling: the system would survive. Again, the model could be made richer by introducing ecological and psychological factors: impossibility of expansion beyond certain limits, revolt against alienated labor, demand for participation and self-management. Consequently the system would have to be transcended.

Surely if laws and models were only subjective conventions invented to rationalize particular interests, they would constantly mislead us, and clearly not all of them do. Some do more, some less. But they could lead us in different directions. They are not mere reflections of reality.

*How to distinguish between a law and any empirical generalization which expresses accidental recurrence of a certain pattern of events?*

Reichenbach tried to answer such a question by developing the theory of nomological statements. He laid down four conditions, each of them being necessary and all of them a sufficient reason to accept a statement as a nomological: it must be exhaustive, proved, general and universal. Reichenbach obviously wished to define nomological statements in such a way as to be able to say: Laws are just nomological statements. If a given statement is analytical, the law in question is logical, otherwise, it is a "law of nature."[7]

In order to have such a simple definition Reichenbach had to make the concept of *nomological* form rather broad. Thus, it includes some conditions which are either insufficiently precise or redundant. Such is, for example, the requirement to prove the truth of the statement in question by deductive or inductive methods. There is no general formula of inductive demonstration. Therefore, inductive probability can hardly be considered a "formal" feature of a certain type of statement. On the other hand, Reichenbach tries to solve in a purely formal and syntactical way problems which require semantical and more informal considerations. For example a *meaningful way of* connecting elementary statements cannot be secured by purely formal requirements (such as Reichenbach's condition of exhaustiveness).

It seems therefore that the concept of *nomological form* should not contain all that Reichenbach has put into it. It should however embrace some *pragmatic* conditions which he overlooked and which were recently emphasized by Von Wright. In his critique of the "covering law" model of explanation, Von Wright has come to the conclusion that the only way to overcome deficiencies of, on the one hand, a too weak positivist interpretation of law (which identifies a law with any empirical generalization) and, on the other hand, a too strong conventionalist view (which makes the law immune to refutation) – is to define a scientific law in terms of "necessary connections between events in nature." And "the idea of natural necessity . . . is rooted in the idea that we bring about things by doing other things."[8]

To this indispensable *practical* root of the idea of necessary connection in a law we must surely add its *theoretical* roots. Thus, it seems to me that the following three requirements have to be satisfied in order to assert that the connection between two types of events ($P$ and $Q$) is necessary and that the statement formulating such a connection is true. We shall see that in each of these three requirements there is an element of human subjectivity, of practical experience, of reference to the already attained level of human culture. The third one is especially interesting for our purpose because it seems to involve an element of choice, of wish, of preference, of cognitive decision making.

The *first* requirement to be satisfied in order to even begin to examine whether the connection between events $P$ and $Q$ is necessary is the following: whenever under certain specifiable conditions we have produced $P$, $Q$ has also been brought about; whenever under some conditions we have removed $P$, we have made $Q$ vanish.

The situation in which we experience a certain regularity in phenomena as the result of our practical action is essentially different from the situation in which we passively observe the succession or joint omission of two groups of phenomena. Hume was right that in the latter case we don't observe anything but succession and, therefore, do not have any right to claim that there was any real link, any interaction among phenomena. However, Hume reduced experience to passive visual observation, he never considered a *practical* situation in which we act and produce changes in our surroundings. In this latter case we have direct experience of objects, of the *force* by which they act on us and resist our efforts. When my fingers are caught in fire they are burnt. That is how I know that fire actually burns paper and does not merely *precede* its burning. In other words, by doing things we come to know that certain links among them are *real*, that they do interact and produce each other instead of merely succeeding one another.

On this ground only we cannot yet be sure that the connection is indeed necessary. But without it we could not be sure that it is not illusory – a mere correlation.

*Second*, in contrast to contingent generalizations, which are theoretically unsupported, nomological statements are firmly entrenched within a theoretical context. It does not make sense to speak of a law outside a given frame of reference – a system. That means that a law must be logically connected with a set of other statements, some of which have been, directly or indirectly, confirmed by an independent initial evidence. We need not specify here the nature of those logical connections. Obviously they can be of various logical types and cannot be reduced only to deductive provability. However, it is important to note two other things: (1) in addition to the more concrete, lower level statements with respect to the given law there are also in the system some more abstract, higher level generalizations. Both groups are needed to *explain* the law, i.e., to establish how it can be derived from the former, on the basis of the latter. (2) Every such explanation implicitly presupposes a very general principle that expresses the idea of order and determination in reality. As we have established earlier, this principle can be formulated in the following negative way: *It is not the case that when a state of the system is given, at the moment $t_o$, any state or no state would follow at the moment $t'$*. The limiting conditions of the system will exclude all possibilities but one definite set which will be considered *determined*. The law itself is a limiting condition within a more concrete system dealing with individual events. The system (object-system) within which a law itself (i.e., a general connection among *types* of events $P$ and $Q$) is determined is a more abstract higher-level system where limiting conditions, on the one hand, exclude all other but states involving $Q$ whenever a state involving $P$ is given, on the other hand, exclude all states involving $Q$ whenever a state not-involving $P$ is given.

The first two conditions allow already to account for the nature of generality which is inherent in every law. This is not contingent, *a posteriori* generality which we establish for the classes of already inspected cases when we do not come across any exception from a certain regularity.

Contingent general statements refer only to *closed* ranges of observed phenomena. Ideological and theological statements appear to be open but whenever "undesirable," "useless," "untypical" or "harmful" data are met, the range of variables would be closed *ad hoc*. In contrast, variables in a nomological statement refer to *open* ranges of cases beyond the limited spatio-temporal continuum in which past observation has taken place. This generality is *projective* in Goodman's sense; the range of application of nomological statements is a projective class which embraces both inspected and *not yet* inspected instances. All laws involve *extrapolations* and *interpolations*, involve claim to successful predictions. But the claim may be invalid in case of free human behavior as we shall see.

The *third* requirement is of a pragmatic character and involves an element of freedom in the very concept of scientific law.

When we deal with contingent, theoretically unsupported generalizations we are ready to abandon them without hesitation when confronted with negative instances. (At least that is how we behave when we are rational; obviously we may cling to some generalizations for emotive and other irrational reasons.)

It is characteristic for nomological statements that, for entirely rational reasons, we hold them so important and fertile that (in case of the discovery of apparently falsifying instances) we will try to preserve them by all empirical or theoretical operations which are compatible with the general criteria of scientific method. The point is that this resistance to change has nothing to do with conservatism, let alone ideological dogmatism or irrational blindness, which erupts when personal or group interests are at stake. This willingness to introduce certain modifications into our total body of knowledge rather than to surrender a law –

belongs to essential features of scientific rationality. From the very fact that a law had been accepted only after very thorough critical scrutiny and testing it follows that any challenge to it must also undergo a comparably exacting critical analysis. And even when it becomes clear that we are confronted with some really new facts and that something in our whole field of beliefs must undergo appropriate change, the most rational solution might be to sacrifice some other elements of the system and to preserve the law. We have a *choice* between the following methods:

(1) *A careful critical examination of empirical statements which are incompatible with the law.* This is an *entirely different* attitude from the one which follows from Popper's epistemology. Hardly any researchers will be immediately tempted to reject a law only because a relevant empirical instance seems to falsify it. They will first challenge data and test their reliability and validity, repeat observations, vary conditions, compare data obtained by different methods. Data may turn out to be accurate but irrelevant if they presuppose conditions different from those under which the law holds.

(2) *The reduction of the range of application of law.* If nevertheless disconfirming instances happen to be both valid and relevant then a change in the theory becomes necessary. But the next harmless change is a limitation of the range of phenomena to which the law is applicable. Those elements of the model which allow us to interpret falsifying data as instances of the law will be eliminated. The model will be somewhat restricted and idealized. This kind of "subjective" adjustment is indispensable not only in social but also in natural sciences. Ohm's law $(1 = (V_1 - V_2)/R)$ would have had to have been rejected long ago because it is clearly falsified whenever the electromotive force (V) fluctuates and where the conductor allows generation of self-induced electrical currents. An alternative to the rejection of the law was the reduction of its range to homogenous conductors with negligible self-induction and to currents with relatively constant electromotive force.

(3) *Elucidation of the differences between the model and empirical reality.* The difference could arise either because of the incompleteness of information contained in the model or because of deliberate simplifications. Thus, when Leverrier discovered some irregularities in the motion of Mercury he did not jump to the conclusion that Newton's law was false. There were at least three more rational alternative accounts for the surprising new fact: (a) our estimate of the distribution of masses

in the solar system could be inaccurate. For example, the observed phenomenon would occur if the mass of Venus was greater by 10% than had been believed; (b) or if there were some unknown small planet in the orbit of Mercury; (c) of if there were some hitherto unknown forces in the solar system which deform the motions of planets.

(4) *Reformulation of the law*. The law could be adjusted by stating explicitly the specific conditions under which it holds. This would be a syntactical rather than semantical change as in (2) and (3). The crucial problem here is what are the limits within which we can still speak about the same law in spite of all modifications of its formulation. The criterion of identity seems to be: Two different formulations express the same law if and only if they differ in the specification of conditions but not in the description of relations among variables.

(5) *Change of meaning of key terms*. The law of the conservation of energy survived all discovery of falsifying cases (such as radioactivity) by changing the meaning and including in its derivation previously unknown hypothetical forms of energy (such as nuclear energy) for the existence of which there was not yet any evidence.

(6) *The transformation of a law into a principle*. Assertions about objects can be reinterpreted and turned into definitions (which are independent of experience). Poincaré has shown that if astronomers were faced with indubitable data incompatible with the law of gravitation, they could nevertheless preserve Newton's formula and give it the status of a principle, i.e., of a statement of conditions under which a statement would be considered a statement about gravitation.

(7) *Revision of a principle*. Principles cannot be refuted by experience and yet they may be revised in the conditions of a profound crisis. Instead of reformulating a series of laws a corresponding correction in very basic conceptual foundations may be preferable. A well known example is Einstein's decision to relativize the concepts of space and time rather than all mechanical laws.

The sequence in which these methods were expounded corresponds to customary priorities of their choice in situations when general validity of a law has been challenged. In the complex structure of our knowledge, embracing a number of strata, from the level of descriptive empirical knowledge to the most general scientific principles to the rules of logic and ultimately to the basic philosophical postulates – most vulnerable and open to constant revision are strata more directly related to experience. The more we approach the theoretical foundations the more reluctant we are: changes there have formidable consequences for the whole building of human knowledge.

The important point is, however, that there is considerable freedom of choice among alternative strategies within certain constraints of a rational behavior. Those theoretical achievements which are, for rational reasons, considered important, fruitful, powerful intellectual weapons need to be preserved and defended. But that could be done in different ways. There is no iron necessity in the dialogue of theory and experience, scientific law and relevant evidence. Even in natural sciences a law is far from being a mere reflection of some fatalistic order in things themselves. It certainly refers to some objective regularity – otherwise it would be completely devoid of explanatory and predictive power. But the law is also relative to our practical experience, to the given historical level of knowledge and to our need for stable systematic organization of all our knowledge. Scientific rationality resists both light-hearted throwing away of its basic achievements (what Popper seemed to suggest), and dogmatic refusal to account for new experience (characteristic of conventionalism).

*Serbain Academy of Sciences and Arts*
*Belgrade*

## NOTES

[1] Claude Bernard, *Introduction à l'étude de la médecine expérimentale*, 1 Partie, ch. 1.
[2] Herbert Dingle, 1951, *The Significance of Science in a Century of Science*, London, p. 307.
[3] Dewey, 1938, *Logic, The Theory of Inquiry*, New York, p. 354.
[4] Karl Popper, *The Logic of Scientific Discovery*.
[5] Michael Scriven, 1961, 'The Key Property of Physical Laws – Inaccuracy', in Feigl and Maxwell (eds.), *Current Issues in the Philosophy of Science*, New York, pp. 91–105.
[6] Stephen Toulmin, *Philosophy of Science*.
[7] Hans Reichenbach, 1947, *Elements of Symbolic Logic*, New York, p. 360.
[8] George Henryk von Wright, 1971, *Explanation and Understanding*, Ithaca, pp. 69–74.

GYORGY MARKUS

# AFTER THE 'SYSTEM': PHILOSOPHY IN THE AGE OF THE SCIENCES

1

Contemporary culture is dominated by the sciences. This does not mean that scientific methods, theories or paradigms would decisively influence everyday thinking today, or would serve as guidelines for the effective orientation in our broader environment, in the "life-world". Just the opposite is true. This is partly because when sciences in the modern sense became autonomous, they distanced and divorced themselves from the habitual, everyday schemes of explanation, and as a consequence both their methods and results became intelligible only for an ever more narrowly delimited circle of expert specialists; and it is also partly because, from the end of the XIXth century on, the idea of a unified "scientific world-view" able to fulfil the task of general cognitive and practical orientation, this great end and promise of the Enlightenment and early positivism increasingly lost its relevance for the development of the sciences themselves. Today such synoptic overviews have been definitely removed from the field of science proper. They are relegated to the sphere of "popularisation" and dissemination of general knowledge, strictly distinguished from genuine scientific practice. The sciences, which in their technical impact unified, homogenized the circumstances of everyday life, the human world, have turned out to be incapable of providing a single, coherent interpretation of reality, comprehensible in its connections – or at least they no longer undertake such a task. "There is no such thing as *the* science, there are only sciences" – stated Scheler already at the beginning of the twenties. If we use the term "culture" in its broad, anthropological sense, there have been few periods in Western intellectual history, in which what was called "science" would have a lesser direct influence upon the forming of a culture than is the case today.

Nevertheless I think it is true, and even evidently true that the sciences – and let me be more precise: primarily the so-called "hard" sciences of nature, and to a lesser degree those social sciences which follow their epistemic model – are the dominant constituents of the *high culture* of contemporaneity. For they are the sole component of this culture in the case of which their becoming autonomous did not mean their simultaneous loss of all social functions, or – to formulate it in a more circumspect way – it did not give rise to deep uncertainties and doubts as to their possible function in human life. The hard sciences, through their broadly understood technical application, became a reality-shaping power whose contribution and progress is necessary for the continuous existence of modern societies. For societies, in which material livelihood is based on the ever more rapid use, and using up, of finite and humanly irreproducible natural resources, in the long run can maintain themselves only under the condition of an equally rapid technical progress. And this can be ensured only through the continuous utilisation of the results of scientific research. In this sense the sciences have become one of the main determinants and steering mechanisms of social change. Of course, one can *evaluate* the human consequences of their role, their impact in widely differing ways. But the *fact* of their social significance cannot be doubted. What is more, it seems also clear that even their negative consequences cannot be remedied *without* their own contributions.

2

"Philosophy is the mother of all sciences". If we accept for a moment this trite (and at best) half-truth of cultural history, we sadly have to conclude that family relations in the field of culture do not seem to shape up any better than is the case in many, more prosaic families. For in the eyes, and from the viewpoint of the sciences philosophy appears like that embarrassing parent who never succeeded in growing up. As opposed to the well-defined concepts of scientific theories which confer upon the latter an unambiguous meaning and make it possible to resolve the dispute between competing theories through their empirical verification and falsification, the truth-value of the speculative conceptual edifices of philosophy remains forever undecidable already due to the fact that their very meaning is the subject of never ending conflicts of interpretation. Therefore, while controversies at the frontline of present scientific research always occur on the basis and background of a consensually accepted corpus of knowledge (the "textbook science") and their resolution contributes to the expansion or modification of this latter, the whole history of philosophy up to the present is characterized by the incessant struggle of opposed doctrines and intellectual sects who do not share even a common paradigm. The cultural form of existence of the specialized sciences is *research* as an ever advancing, collective practice, and even the most significant intellectual achievements are but great *contributions* to it. Philosophy, however, is still primarily embodied in the oeuvre of the "master-thinkers" – in individual and closed theoretical constructs which for this very reason can only be continued in an epigonistic way. Thus in all these aspects philosophy inevitably appears from the standpoint of sciences as a cultural formation that was and remains incapable of overcoming the infantile disorders that have characterized their own formative period, the pre-paradigmatic stage of scientific development. From this viewpoint philosophy is no more than – to use a Husserlian expression – a *Restbegriff*: it designates a mere residuum and remainder; its area of concerns is constituted by problems which could not have been transformed into empirically resolvable scientific questions – either because they are wrongly stated, are pseudo-problems, or because at the present level of our knowledge we lack the appropriate conceptual and technical means that would allow their empirically interpretable and decidable formulation.

3

It would be rather easy – invoking some of the newer ideas and insights of contemporary theories and historiographies of science – to lessen the sharpness of the contrast here outlined which in this form rather pertains to the immanent ideology of scientific research than to the objective characterisation of its practice. Here, however, I would rather refer to the fact that one of the most influential trends of contemporary philosophy essentially accepts the so interpreted opposition, only to radically reverse the values associated with it. For the sciences have paid and do pay a price for their undubitable successes. In part, I have already mentioned it: they became a factor in the effective satisfaction of pressing social requirements only by becoming *monofunctional*, by "freeing" themselves from a whole gamut of tasks which in other cultures were fulfilled by forms of knowledge having the highest social esteem and recognition – first of all from the task of a unified and coherent world-interpretation able to directly orient people in their everyday activities. The modern sciences are constructive, but they are no longer edifying. This is so primarily because they are empirical sciences *of fact*. They are not value-free in the sense that the practice of scientific research presupposes a definite social and cultural environment which makes possible such activities and confers meaning upon them, an environment which is characterized by the dominance of definite values as a matter of fact. But the sciences do not, and cannot investigate the significance and validity of the thus-posited values in general. Thus while they provide an ever growing body of information about that field of possibilities which we can use and exploit in our doings, they say nothing about what it is *right* for us to do.

The philosophical critique of the limits of scientific rationality referred to here does not stop, however, at this point. Its central target is that very concept of "fact" with which the modern sciences of fact operate. From the viewpoint of the sciences a fact is what can be described as a case of some general, conceptually separable and characterizable, causal or functional interdependence, the constituents of which are objects in the broad sense of this word, that is entities which in principle are at our disposal and over which we can gain control, once we have discovered the laws of their potential interaction. The cognitive standpoint of sciences is a radically objectifying one. But not everything which as a matter of fact influences our life is a "fact" in this particular sense, and not all our actions aim at the production or modification of such facts, since we do not have everything even in principle at our disposal and we cannot treat everything that plays a role in life as a mere object. What is more, even science understood as the dynamic process of research today increasingly acquires for us the character of such a "non-disposable". The direction and the rate of its development, the character of its utilisation is less and less determined by the conscious decisions of those – the members of the relevant research community – who seemingly possess the appropriate rational competence for such a task. This all becomes the resultant of disparate, momentary decisions of various state, economic, and other institutions and organizations. In its totality it takes on the form of an anonymous, uncontrollable process. The specialized sciences – so proceeds the critique – are therefore incapable not only of justifying those values the effectuation of which is actually presupposed by their own practice; they also cannot account by their own conceptual means for the real character of their own activity, for all that "happens" with, and through, them.

The philosophical critiques of science, only indicated here, take upon themselves the role of giving voice precisely to what is in this sense "non-disposable" and "uncontrollable". They can designate it, depending on their own conceptual framework, by the names of "Life", the "individual concrete", "Being", "bodily existence" or "generalized textuality". However, in whatever way they characterize it, from their own viewpoint the empirical sciences of fact now acquire the character of a *Restbegriff*, a residual concept: they are what remains from the universal idea of rationality or that of primordial thinking when it is onesidedly and distortively restricted to the description and calculative prediction of the factually-objectively existent, of what exists as a mere object.

4

The adversaries of this dispute are, however, in an unequal position. And not only because of the fact that the empirical sciences, in the consciousness of their indispensable social function, self-confidently can face such a speculative critique, but also because philosophy, being "the mother of all sciences", can hardly regard itself as blameless in respect of the (alleged or real) inadequacies of the modern sciences. From this follows the rather paradoxical situation: generally speaking, the more critical is the attitude of some philosophy towards science, the less inclined it is to deal concretely with science and the more preoccupied it is – with philosophy. This explains the great popularity in contemporary philosophical literature of the rather narcissistic question: "What is philosophy?", "What can philosophy be today?" – the present paper, of course, being just an example of this hackneyed genre.

However, one comes up against considerable difficulties when one tries to answer such a question directly. And not because such answers are unavailable. Individual philosophers and philosophical schools usually formulate what is to be regarded as the genuine subject-matter of philosophical discourse and inquiry quite explicitly. They are forced to do so precisely because the philosopher, in opposition to the scientist, does not have at his/her disposal a consensually accepted body of background knowledge, interpreted in a unified way, which would circumscribe the general outline of the still unresolved and resolvable problems. But the answers given to this question seem to be simply incommensurable. Influential contemporary trends of philosophy assign to it the task of the phenomenological analysis of transcendental subjectivity, the disclosure of the semantic and syntactic structure of the sciences, the revitalisation of gnostic knowledge, the creation of a critical theory of society and even the deconstruction of the tradition of philosophy itself. Philosophy struggles today not only with the fact that its social and cultural function became problematical, it is also in a crisis of identity. "There are philosophies, but there is no such thing as philosophy as such" as Dilthey already formulated the fact of this crisis at the turn of the century.

Dilthey's statement evidently brings to mind the words of Scheler quoted earlier concerning the sciences. The two propositions, however, actually refer to two essentially different states of affairs. Of course, in both cases we are dealing with paradoxical formulations, since multiplicity and diversity can only be ascertained on the basis of some tacitly assumed viewpoint of unity. However, this implicit horizon of unity is, in our two cases, of fundamentally different character. In respect of the sciences it is ensured primarily by the existing interconnections between the different disciplines and specialties, by the partial utilisation in each of the sciences of results and methods developed in other branches of knowledge, and through the widespread practice of interdisciplinary inquiry. As a result no specialized field of research is isolated from the others; they all constitute the plural of *one* science, even though in their totality they do not offer a single, unified and coherent system of cognitions. The universal concept of science as such exemplifies the Wittgensteinian principle of *family resemblance*. The unity of philosophies, on the other hand, is realized primarily through the incessant argumentative dispute and polemics of the opposed schools and trends, in a process in which each of them refers to a largely identical, but again agonistically organized tradition which is, however, articulated, interpreted and evaluated by them in sharply differing ways. If it would make sense at all, one could say that the universal concept of philosophy as such is based on the model of *family animosities*.

5

This present situation of philosophy is the product of a process of decomposition. This latter made untenable its earlier dominant *cultural form* – from the XVIIth to the late XIXth centuries – which still kept philosophy proper in an articulated unity with the only emerging modern theoretical sciences (known at that time by the name of *philosophia naturalis*). The idea of the *system* constituted this cultural form. For the notion of the system is not to be identified with a definite form of literary exposition, indeed it cannot be reduced to some particular type of logical structure as an "internal" form either. In fact, in the history of post-Renaissance philosophy one encounters several system-types differing from one another even with respect to their ultimate principles of construction. In this regard one can clearly distinguish the axiomatic-deductive, the genetic, the transcendental and the dialectical systems. What unifies them is not some abstract identity of their logical cast, but the essential similarity of the way they conceive the role and function of philosophy = science in the totality of human life. And this, in its turn, determines what normative expectations should be, at least ideally, satisfied by such cultural accomplishments, that is it defines the manner in which their meaning and claim to truth is to be comprehended, and therefore also the way they can, or ought to be evaluated and criticized.

6

Here, of course, I can only in the most cursory way mention even the most essential constituents of this idea of the system understood as a cultural form. I would like, however, to make a short reference at least to three important aspects of it.

 a. In its full conceptual realisation the idea of system first of all implies that philosophy = science is, in its manner of being, a chain of cognitions which in some way are *objectified* (and thereby made in principle accessible to everyone). This view seems to be so natural for us that we easily can miss its radical novelty: its break with the classical Greek conception of *philosophia* as *episteme* (the Latin *scientia*) – a conception which even thinkers of the early modern age took largely for granted. For in its original understanding *episteme* meant an acquired, but lasting and firm mental aptitude or habitus undivorceable from the total personality of the individual – a disposition of the soul to insight into truths of the highest type, truths of strictly universal and necessary character regarding the "causes" of the phenomena. Accordingly such knowledge was regarded also as adequately transmissible only in the educative process of the formation of an intellectual and moral character, from person to person. (Incidentally, this explains also the general hostility towards writing characteristic of the classics of Greek philosophy.) The idea of the system destroys this "personalistic" understanding of the socially most esteemed form of knowledge which claimed to be an end in itself. It divorces philosophy – further regarded as a value in itself – from its direct impact upon the life of its practitioners (or recipients), from its personality-forming, illuminative influence, and thereby creates the conceptual preconditions within the framework of which the *modern* conception of the *autonomy* of cultural accomplishments first becomes intelligible.

b. According to those cultural norms and expectations which were implied in the idea of the system, the thus-objectified body of knowledge ought to constitute such a coherent meaning-formation which possesses a purely *immanent sense*, that is, which is understandable and evaluable strictly in and by itself. Thus such knowledge is not only accessible for everyone, but also in principle can be rationally evaluated by everyone with respect to its claim to truth and significance. Philosophy = science understood as system presented its claim to universal truth as the principle of an "epistemic democratism". In this way it not only broke away from the elitism characteristic of the classical conceptions of *episteme*, but it also essentially transformed the basic epistemological requirements and presuppositions associated with its concept. It partly *radicalized* them, partly gave to them (in a curious opposition to the above described tendency towards depersonalisation) a *subjectivist* turn.

In its classical understanding philosophy is not only *episteme theoretike*, but equally *episteme apodeiktike*: it represents the knowledge of necessary and universal truths based upon *proof*, acquired strictly through logical demonstration. This is perhaps the single most important feature of this view deeply influencing the whole subsequent intellectual evolution of the Western inheritance of Greek philosophy: the radical *decontextualisation* of the form of knowledge to which the highest spiritual value is ascribed. For this implied that the value of such knowledge is completely independent of its *source* (both from its relation to the tradition and from the authority or charisma of the person announcing it), from the *way and form* of its formulation and communication (from their poetically evocative or rhetorically persuasive character), further from the direct *utilisability* of its content. Its significance is based solely on the particular character of its truth-value which is guaranteed by the way it can be acquired: through an unambiguously delineated, invariant and interpersonal procedure, through syllogistic inference, the pure form of which had been first clearly recognized and fixed by classical Greek thought. In this way *logos* had been radically demarcated from *mythos* and *epos*; from *metis*, the cunning reason so important, especially in political conduct; from *eikos*, the probable relegated to rhetorics; from the forms of practical know-how, the *techne*; and equally from the description and systematisation of what is observable on the basis of the similarity of its constituents, from *empeiria* and *historia*. It was just this narrowing down, this radical and clear delimitation of the extent of the legitimate and legitimating cognitive grounds which conferred a specific direction upon the admitted discourses concerning the thus-constituted cultural form – they have been unambiguously restricted to discussions of, and judgments upon, its demonstrative grounding, they acquired the form of strictly epistemic critique. In this way classical antiquity first created the *concept of scientificity*.

Classical Greek philosophy, however, did not consistently carry out, to its ultimate end this decontextualisation of knowledge. *Episteme* meant demonstrated knowledge, but the principles universally characterizing the mode of being of all kinds of beings, the principles which as the highest premises make the whole procedure of inference possible at all, were not regarded (at least in the paradigmatic Aristotelian conception) as belonging to the thus-constituted sphere of competence of philosophic *episteme*. They pertain to other faculties of the soul, to *nous* and *sophia*, intellection and wisdom. Their insights are legitimated by the principle of *consensus gentium et philosophorum*, ultimately by the essentially concordant opinion of those whom "everyman" regards as the most competent and the most wise. Thus classical antiquity could comprehend philosophy only in the context of a shared culture, within the framework of, and relative to, a common tradition and form of life. And it is this context which makes intelligible and legitimates the very way it frames its question and the ultimate answers it gives to them. It was this fact which also allowed it to reconcile philosophy's claim to universal validity with its pronounced exclusivity.

The postulates of meaning-immanence and epistemic democratism, however, cannot be reconciled with any admission of a factual context-dependence of knowledge: they demand it to be without any external presuppositions whatsoever. The system ensures this by interconnecting the idea of objective *logical grounds* with that of subjective *certainty*, thereby liquidating the epistemic dualism of the ultimate principles of intellection and the demonstrated truths of *episteme*, "science" proper.

The starting points, the highest premises of the system (in whatever way they be defined) are given in an intuitive self-evidence which excludes even the possibility of doubt. To arrive at such evidences demands genuine intellectual labour, the cleansing of the mind of all its accumulated biases and prejudices; but in principle they are equally available for, discernible by, everyone. These self-evidences constitute the unshakable foundation of all knowledge. Its expansion, the construction and elaboration of the system consists in the total or partial, step-by-step transfer of this evidence through the *construction* of newer and newer truths from the ultimate ones. This is made possible by the *method*: some complex procedure usually not reducible to syllogistic inference alone, but again thought of as a public and interpersonally controllable one. The intellectual building blocks of this method are characterized in terms of such elementary mental operations as every normal person is able to perform. The significance of philosophy = science in this way shifts from the apprehension and contemplation of personality-shaping truths to the *production* of ever new truths built upon a secure and ever-expanding foundation. Therefore the system is both closed and open at once. Closed, because the fixedness of its foundations and of the method from the very beginning predetermines the reach of its discourse and the way it can reach its objects. Open, because the value of the method consists in its productivity, in its capacity to make new, ever more particular phenomena intelligible and explainable from the pregiven standpoint of the system.

c. All this, however, simultaneously means the inevitable collapse of the way in which the role of *philosophia* was conceived and its value legitimated in the classical tradition. In the perception of this latter the ultimate significance of its insight into universal and necessary truths consisted in the fact that such knowledge by its nature relates to that, and to only that, which is unchanging and eternal. Its comprehension therefore elevates the soul above the accidentalities and insecurities that rule the world of everyday experience and opinion, it fosters a spiritual attitude emancipated from the power of *tyche*, from the surrender to what simply happens to us. Precisely for this reason *philosophia* is not some body of knowledge to be learned, but a *praxis*, the highest form of meaningful and happy, salutary life, *bios theoretikos*. In this respect, as to its ultimate end – if one disregards the all important point how it envisages to realizing this end – classical Greek philosophy is closer to the great religions of salvation than to modern science.

The idea of the system disrupts this direct coincidence of theory and praxis, of intellectual apprehension and the good life. It further upholds as a normative requirement the strictly necessitarian character of all scientific truths. Due to its commitment to certainty it even enhances the dogmatism of the classical conception of knowledge. But it no longer seeks what is universal and necessary *above* the sphere of the changing, accidental individual phenomena, *beyond* the realm of transient practical needs and interests. Rather, it now locates the universal precisely in what remains constant and the same *in* the change of the phenomena, in the invariant regularities of their causal interactions, in their *"laws"*. The discovery of such hidden causal mechanisms then makes it possible for us to gain control over their possible effects, to increasingly acquire power and dominion over nature, the ultimate material and object of all our activities. Instead of a "philosophical" way of life based on the intellectual contemplation of the eternal cosmic rationality and accessible only to a few elect, philosophy now aims at, and finds its legitimation in, the active *collective rationalisation* of the conditions of life.

It would be, however, a misleading oversimplification to identify this end solely with an increase in our ability to manipulate natural processes, with the incessant growth of the security and effectivity of sheer survival independent of the meaning of life. The great systems of early modern philosophy regarded power over nature as a fundamentally important, but nevertheless only one aspect of human *freedom*. They understood it as the expansion of the range of possibilities for the realisation of human ends and values that can consciously be chosen on the basis of their recognized validity. For while the idea of the system destroyed the classical project of an *immediate coincidence* between rational knowledge and good life, it still intended and attempted to embrace both physics and ethics by, and within, a single, coherent conceptual construction, even in those cases when their relation was conceived as that of opposition. The understanding of the relation of humans – these free and rational beings able to discover the laws of nature and to modify their activity in accord with the knowledge so acquired – to this very causal order, the place of man in nature, such an understanding casts light not only on the conditions of the realisability of human ends, but also on their intrinsic value and rationality. In the systems of German idealism, which already react to the beginning divorce of the "positive" sciences from philosophy, these two tasks became rather clearly distinguished from each other. The empirical sciences (relegated by Hegel to the sphere of Objective Spirit) are to answer the question of *Können*, of what can be done, while philosophy, in which Absolute Spirit reaches its fulfillment, primarily ought to indicate what should be done, the path to the humanly befitting utilisation of the enlarged scope of activity and choice. Philosophy = science understood as system articulated the grand promise of modernity: to bring, in theory as well as in practice, to an ultimate unity and to fulfill jointly, in harmony, the demands both of self-preservation and self-realisation.

7

It was in the cultural form of the system that early modern philosophy = science achieved its autonomy. First of all, in this form did it become independent from religion and theology, and acquired such an area of the knowable over which it successfully could claim the highest epistemic authority. This cultural form, however, became undone, not least in the result of its own successes: it proved to be in a sense self-destructive.

For those specific characteristics of the system-idea which I summarily outlined above, also made possible a radical revision of the concept of scientific rationality. The emphasis laid upon the nature-transforming function of science allowed it to connect up with technical knowledge, with those "mechanical arts" that earlier were regarded as "servile", and in this way to accept experiment also as a legitimate source of reliable knowledge. Evidence understood as subjective certainty rendered possible the dismantling of that line of demarcation which earlier was drawn between *philosophia naturalis* as genuine science and the observation based *historia naturalis* excluded from science proper, since, at least in some interpretations, the data of the senses also satisfy the requirement of self-evidence. The broad conception of method also made induction acceptable as a scientifically legitimate procedure; and as a result, the probable too could find its way into science. In general, early modern philosophies proved to be capable, while maintaining, and in some respects even radicalizing, the fundamental cognitive features of the classical conception of scientificity, first of all its characterisation as decontextualized knowledge, to overcome to a significant extent its narrowness, rigidity and one-sidedness so often emphasized as its fatal blemish today. It would seem that precisely the clarity and sharpness of the strictly demarcated *logos*-concept conferred upon its tradition such a power and flexibility that permitted it to draw into its orbit much of that through the exclusion of which it had been originally defined.

It is clear that non-theoretical conditions were primarily responsible for the fact that these conceptual possibilities became indeed actualized. However, when this transformation occurred, then science, more exactly the sciences differentiated in respect of their objects and methods, divorced themselves from philosophy. Each great branch of science was now conceived as possessing its own experiential-experimental basis and therefore as having no need for a philosophical legitimation of its knowledge-claims. The concept of scientific rationality embodied in the always-advancing process of research had no more need to be articulated in terms of the concept of some ultimate philosophical *foundation* with indubitable self-evidence. Instead, it could adequately be formulated in terms of graded empirical *confirmation* relative to the achieved level of knowledge. The processes of professionalisation and specialisation, which were essentially completed by the end of the XIXth century, also provided an institutional framework for this dissociation of the various kinds and branches of knowledge. And when the scientific revolutions of the early XXth century – first of all in the basic and exemplary science of physics – made almost unavoidable the recognition of the *fallibilistic* character of all scientific knowledge, the very idea of some "philosophical foundation" for science became – at least *prima facie* – untenable. In this way have gradually come into being the conditions for that antagonism between the sciences and philosophy to which I referred at the beginning of this paper.

8

This antagonism, however, is only one of the symptoms within the sphere of high culture of the more general and fundamental strains and problems of late modernity. First of all it is only a subcase of that value-pluralism and conflict between the socially effective and recognized values which can no more be resolved – at least consensually resolved – by the establishment of a fixed value-hierarchy. On the other hand, this antagonism itself is constituent of that paradoxical situation in which the rapid growth of the quantity of information available about the reality environing us is accompanied by the felt decrease in our ability to directly orient ourselves in the world we live in. And this latter seems to be connected with the fact that under conditions of modernity intentional actions and activities directed at the purposeful transformation of the conditions of life often generate, or are enmeshed in, such anonymous and automated processes, the long-term consequences of which we are unable either to foresee or to bring under our control, perhaps since the degree of their complexity grows even more rapidly than the quantity of available information about them.

The questions: "What is philosophy?", "What can philosophy be today?" have acquired in our days the character of a standard "professional" problem for the philosopher, though evidently these are also personally loaded question concerning his/her intellectual existence. But since the cultural situation, which makes this task of an anxious self-reflexivity unavoidable, is part and parcel of the more general and practical dilemmas created by the contradictions of modernity, the answer to these questions also cannot be independent from one's (explicit or implicit) decisions and choices in regard of these alternatives. One cannot "prove" the correctness of such an answer, though, of course, one is obliged to argue, to offer a rational motivation for it.

In the history and the present practice of contemporary philosophy I think one can observe in a purely ideal-typical sense three fundamental trends and directions in the way these questions are sought to be answered. I would like to add immediately that we are indebted to each of these for insights, or at least for lessons to be learned, which it would be perilous to forget or to neglect. Therefore if now, at the end of this paper, I am to make in some cases critical, and even (for the sake of conciseness) sharply polemical remarks in relation to some of them, I do not intend by this to dismiss or hope to disprove them. I merely want to motivate my own answer and choice.

9

One possible solution to the identity-crisis of philosophy is represented by the program of its "scientifisation", its transformation into a *bona fide* science. Such an objective has been formulated from the moment that philosophy had to face the autonomisation of the sciences. One could trace it back to Kant, and from that time on it has been articulated, with changing interpretations and contents, many times. Any such program elementarily presupposes the possibility of finding such a domain of discourse which is not already "occupied" by one of the positive sciences and which philosophy, in view of its own traditions, seems to be somehow qualified to deal with. Among the plurality of the thus motivated choices of a subject-matter for philosophy, in our days the most influential and significant is that which designates its task as the analysis of science itself. Philosophy understood as the self-reflection of actually existing science reconstructs the structure of scientific theories, methods and procedures of confirmation, the internal logic of the processes of theory-change, and precisely through such a reconstruction also legitimates the claim of science to ultimate cognitive authority as the most effective form of problem-solving.

"Scientifisation" of philosophy certainly has the advantage of being in accord with the institutional status of philosophy as an academic discipline in the company of all the sciences. Its program, however, at least in its history until now, has not met with success (judged, of course, from the viewpoint of its professed end). Philosophy of science has remained philosophy, and has not been transformed into a science. Its whole development has been characterized again by the lack of a background consensus and shared paradigm, the endemic character of disputes between its various schools and trends, in fact by the reproduction of many of the unsolved dilemmas of the philosophical tradition in a new and more particular form. Furthermore even if its program were realized, it would achieve only something which the positive sciences – so it seems – quite well manage without: the scientific legitimation of their cognitive status; but it still would leave unanswered precisely the questions that are today asked with a growing intensity about science and by the scientists themselves: the specific problems of an ethics of scientific research; questions related to the social-political conditions and to the wider consequences of various strategies of development and application of science; in general, those problems which often emerge in the form of practical tensions, and which concern the institutionalized place, organisation and function of sciences in contemporary culture and society. A theoretical approach which takes the present form of science as an immutable facticity and is programmatically restricted to its *internal* analysis alone seems to be in principle unsuited to the critical illumination of such issues.

But perhaps the gravest difficulty faced by such an approach is raised by the fact that the development of philosophies of science has itself cast doubt not only upon the possible significance, but also upon the very meaningfulness of such a program of internal analysis. For in its course serious and weighty considerations have been adduced in support of the view according to which the procedures and criteria making possible the rational validation of scientific theories are themselves dependent upon the historically changing content and general character of the theories in question, and more generally that their demonstrative power presupposes the presence of definite, particular historico-cultural contexts. There is no such scientific method which would contain in itself all the conditions of its cognitive legitimacy. The idea of decontextualisation of knowledge, which deeply permeates the whole of our intellectual tradition, can only be thought of as a never completely realizable *Grenzbegriff*, precisely as an "idea" in the Kantian sense. However, to be able to proceed further in its direction demands a critical effort aimed just at uncovering the relevant "external" contexts and conditions. The program of traditional philosophy of science does not seem to be helpful in regard to such a task.

10

We have already met with the second ideal-typical answer to the identity-crisis of contemporary philosophy. It attempts to give voice to that Other which scientific rationality, reducing everything to the facticity of objects, excludes and in which our finite existence is rooted: to that non-objectifiable ground of our being which is not at our disposal. In a culture, however, which is dominated, and in all its pores permeated, by the sciences, such an end can be realized only negatively, by the "destruction" or "deconstruction" of those deep structures of thought or discourse that as anonymous, largely unreflexive presuppositions have determined the whole direction of Western intellectual development. And since these latter generally received their most pregnant expression and articulation in traditional philosophy, a form of discourse whose potential has now been exhausted, the representatives of this line of thought programmatically describe and characterize their own accomplishment as the "end" of, the *ending with* this tradition of metaphysics, humanism, onto-theology, logocentrism and the like.

This program, however, at least in its history until now, did not meet with success (judged, of course, from the viewpoint of its professed end). Leastwise this seems to be indicated by the fact that every new effort at its realisation has been accompanied by the condemnation of its likely minded predecessors for remaining unconsciously captive of metaphysics.

This was done – and in a rather convincing manner – by Heidegger to Nietzsche, by Derrida to Heidegger, and lately by Rorty to Derrida. The tradition declared to be dead proved to be a rather resilient ghost. At the same time this program obligates. The philosopher subscribing to it ought to realize performatively, in his/her own cultural practice the decomposition and overcoming of traditional philosophical conceptuality and argumentative discursiveness. This first of all seems to efface the boundaries of philosophy as a cultural genre and make uncertain what criteria of criticism are at all legitimate in respect of such accomplishments. This may well be a problem only for the "profession", but not necessarily for the spirit of philosophy. But it can result in an authoritarian dogmatism which represents a danger for this latter, too. For from this viewpoint all argumentative critique must appear, independently of its concrete content, as the mere reproduction of what has been radically ended with, and overcome by, the discourse criticized. And this may create the danger of destroying even that fragile unity of philosophy which is guaranteed by the argumentative dispute, polemic dialogue of its various trends. If the "scientisation" of philosophy seems to transform the philosopher into a specialist with a skill of questionable benefit and able (or intending) to communicate only with experts of the same ilk, the view which demands a principled extraterritoriality for philosophy in respect of the requirements of scientific rationality in general, may turn him or her – since our times are not particularly auspicious for prophets – into a guru applauded by the faithful of his/her particular sect.

The most questionable feature for me of such a style and trend of thought, however, consists in the fact that it seems to leave open only the most extreme alternatives where the relation to the world of contemporaneity is concerned. For the phenomena of modernity appear from this standpoint to be anchored in such a metaphysical deep structure which, as an anonymous occurrence or state, today constitutes the *precondition* of all conscious, intentional choices and actions, be they collective or individual. Therefore, if the present is conceived as the final phase in a history of degeneration and decay pregnant with apocalyptic dangers, it then allows only for the attitude of an empty political attentivism which permanently stands in a resolute readiness for the coming of the "turn", inclined to welcome everything that claims to represent something radically "other"; or, arguing that it is the tradition of our present which made us to be what we are, it suggests the unconditional and wholesale acceptance of the values of modernity, together with all their unresolved contradictions. Between these two attitudes mediation is possible only at the price of willed, deep ambiguities. This characterizes, I think, the view of some of the representatives of French post-structuralism: a merry little apocalypse, from now on and forever after.

11

The third answer-type to the present situation of philosophy – the one with which I would broadly associate myself – does not strive to find an ultimate solution to its crisis, since this latter is only one element and constituent in a much broader socio-cultural complex over which philosophy has no command. Therefore it attempts only to provide a general *orientation in thought* in this situation characterized by the widespread experience of contingency and disorientation, and in this way to keep alive or to cultivate such propensities to which the whole tradition of philosophy ascribes particular significance.

The concept of orientation was introduced into philosophy by Kant. To orient oneself, in its original, straightforward sense, means to find one's temporary place in relation to that from whence one came and to where one intends to go. An explicit need in orientation arises when we are unable to reconcile the recollection of the path covered with the idea of the journey still before us, when we are uncertain whether we are at the right place in respect of our chosen direction or whether our originally elected terminus was – given the experience of our track so far – one where we really want to arrive. All orientative accomplishments demand an answer to the questions: from whence, where and whither. What meaning can be, however, ascribed to the metaphor: to orient oneself in thought in a historical situation? How far is philosophy able to contribute with its cultural means to such a task? And if it can do so, what significance may this have at all?

To orient oneself in the historical present demands first of all its *interpretation* as a particular human situation. Philosophy can contribute to such an endeavour insofar as it attempts, partly through conceptual analysis, partly through the reconstruction of the decisive constituents of our tradition, to shed light on the normative *and* factual preconditions of the now dominant set of practices, on those historically specific, pluralistic and often contradictory, enabling and constraining conditions which make them appear legitimate and rational. Through uncovering their interdependence and collisions, their relation to everyday life-practice, philosophy presents an effort of totalisation within a culture which it cannot presume to constitute as a meaningful totality. In this way it distances itself (or at least, depending on its content, it can distanciate) from some practical and cognitive assumptions and postulates that tend to appear natural today.

The objective of orientation, however, demands something more than a kind of globalizing description or diagnosis of the contemporary state of affairs, the illumination of our present situation from the perspective of a point of departure and arrival. The metaphor of place and path can, however, become deeply misleading here. First of all the philosopher certainly cannot answer the question of "whence" by reconstructing – departing from some antecedent state – the process of an actual historical genesis. This is neither his/her task, nor within the field of his/her competences. In accord with its tradition and the specific character of its cognitive claims, philosophy relates the present not to a factually other historical or cultural particularity, as it is done by the anthropologist, the historian or the sociologist, but primarily to a *universality*; to a general, conceptually characterized paradigm of the relation between humans and their world – of course, as it is conceivable in the light of present experiences and in accord with the contemporary level of knowledge. Philosophy presents the problems of an epoch on the background of existential problems assumed to be universal. Only the articulation of this latter provides it with a general conceptual framework in terms of which it can coherently interpret the phenomena of contemporaneity and legitimate such an interpretation. No doubt, there is a plurality of such paradigms in philosophy today. This cannot be regarded, however, as a sign of its inadequacy, but is rather an essential feature of the cultural function it can actually fulfil. For if philosophy attempts to create some kind of rational connection between our ultimate, though situated self-understanding, on the one hand, and our practical relation to the problems and conflicts of the present, on the other, it can do so now only in the clear awareness of the fact that it is no more possible to choose and designate as meaningful and exemplary a sole conception of the self and life, but it can aspire only to render possible a motivated and reflective choice among their alternatives.

This very formulation, however, may raise quite legitimate doubts as to whether such a conception of philosophy can be rationally vindicated at all. To interconnect in some way a general paradigm of the relation between humans and their world with the articulation of a *practical* attitude to the dilemmas of the present – this does pertain to the task of orientation in thought which ought to answer the question of "where to" as well. Philosophy certainly does not fulfil this requirement by outlining some concrete image of a desired future. And not because utopias are impossible or harmful today, but because the philosopher as such has no specific competences in this respect, for utopian thinking cannot be the privilege or burden of a nowadays thoroughly professionalized cultural practice. Philosophy fulfils this requirement precisely by illuminating the problems of an epoch in the light of existential problems assumed to be universal, in such a way formulating an *evaluative* attitude to the collisions and conflicts interpretatively diagnosed. Undoubtedly every philosophy of this type actually does just that. The question is only: in what way and with what right? For I do not think that it would be possible today to represent convincingly the view which would infer or construct definite values out of some facts – be they of the most general kind.

To the question: how do some philosophies create a motivated connection between a paradigm characterizing a universal human facticity and some concrete, situated value-attitudes, the most elementary answer is that they tend to do it in quite different ways. I think, however, that in this respect they all share at least one common feature. This connection does not have the character of a strict theoretico-logical inference or construction with them. It is essentially of *narrative* nature. After the "system", philosophy can no longer lay claim to the status of positive knowledge; it can only be a theoretized, "conceptual narration" acceptable at the achieved level of scientific inquiry and able (at least in its intention) to be generally orientative in thought. It can create a connection between facts and values by maintaining or suggesting that a "story" which illuminates *our* history can make this latter more meaningful, *if* here and now we take upon ourselves the responsibility to continue it in a definite way and direction marked out by some reflectively chosen values.

From this follows the peculiar, simultaneously maximal and minimal character of the rationality represented by, and demandable from, a philosophy. It is of maximal nature because philosophy ought to fulfil two types of cultural requirements. On the one hand, those the satisfaction of which we can expect from any well told story: that it be meaningful and also engaging by virtue of its relevance to something in our life. On the other hand, since it is a theoretized, *conceptual* narration, philosophy ought also to meet the elementary demands of scientificity: that of conceptual clarity, logical consistency and empirical justifiability of what it assumes to be a matter of fact. A philosophy can always be legitimately criticized for the infringement upon this latter set of norms. But – and in this consists the "minimal" character of the rationality of its discourse – objections of this type, as this is demonstrated by the reception-history of philosophies, usually have a weight only in respect of the concrete form and exposition, not in that of the core conception, the "spirit" of a philosophy. The philosopher, of course, must primarily orient his/her activity to fulfil this latter set of requirements, but the fate of a philosophy ultimately is decided by its ability to satisfy the much weaker and intangible criteria of the first type.

What is the function of philosophy conceived as "orientation in thought", what can philosophy "do" today? I think that – even independently of all considerations concerning the range of its potential audience and the actual effectivity of its impact upon them – it can do rather little. To "orient" oneself is a task for a concrete person in the concrete circumstances of life. Philosophy does not "give" orientation, it can only provide some general guideposts for it, and first of all it can contribute to the cultivation of faculties useful in this respect: the faculty of critical questioning and judgment, of reflexive distancing from one's habitual, social and cultural surroundings, the ability to take responsibility for the choices made. These are useful – for what? I doubt that such propensities would have a particularly high survival- or success-value. They are "useful" to keep alive a tradition which through all historical discontinuities imbues European intellectual history and which usually was borne primarily by philosophy: the tradition of a critical, reflexive self-awareness, which today strives to appraise critically also its own limits; a tradition which, if I would have to designate it by a single word, and in the full consciousness of the historical burden of this name, I would still call *enlightenment.*

*Dept. of General Philosophy*
*The University of Sydney*

MARIA MARKUS

# CIVIL SOCIETY AND THE POLITISATION OF NEEDS

This paper attempts to bring together three social phenomena: civil society, social movements and needs, which have captured my interest in different times and within different context, but whose interconnection became increasingly obvious to me and seems to be also increasingly relevant to the present day social conditions. All of them are closely related to the issue of the possibility of the radicalisation of democracy in contemporary societies.

The problem of needs is embedded above all in the context commonly designated a the 'crisis of the welfare state' in its Western and Eastern varieties. In the West, one of the most interesting concomitants of this process was a *paradigmatic change in the dominant political discourse*.[1] What I mean here is that, in addition to the previously dominant issues of 'rights and obligations', the questions of needs and their interpretation have also appeared on the arena of political discourse. This of course involved some change in the very meaning of the *'political'*.

This change introduced a number of further questions. Firstly, as opposed to legally fixed rights and obligations, which not only lend themselves to universalisation, but in all liberal democratic societies are – at least theoretically – universalised, the universalisation of needs is not only problematic but – as clearly demonstrated by the example of the former state-socialist societies – also antithetic to democratic arrangements. This of course poses a serious problem for the welfare state, especially if its policy is oriented not merely at the creation of minimum safety measures for some of the most disadvantaged groups or individuals but is committed to establishing a more general network of *social rights*. It faces not only a multitude of claims but also the socially stratified distribution of the competences and means necessary for bringing the needs into the arena of public debate in the given political field. In addition, this inequality affects above all those social groups, whose needs are often the most urgent. From the point of view of the realisation of democratic principles, the most negative effect of this situation is the *monopolisation* of the formulation and interpretation of needs, and of the 'appropriate' ways of their satisfaction, by the state apparatuses and various kind of welfare experts. This is especially true of those collective (and individual) needs which cannot be satisfactorily fulfilled through the market mechanisms. And it is exactly this monopoly that is gradually becoming the main object of social contestation. The contesting actors emerge above all in the form of different social movements that critically scrutinise the narrow economic rationalism of the above-mentioned interpretations as well as the very fact of the monopolisation of these processes. The critique alone, however, – important as it is – is not yet a sufficient condition for the successful contestation. In order to produce a real change in this respect, a new type of social self-organisation has to emerge opening a network of horizontal communications, and establishing the new structures of social solidarity. Without such a *horizontal structure* of negotiation and compromise between the competing interests, society can be but an *object* of the policies of different state agencies. It is precisely the creation of such structures that defines the main function of a *democratic civil society*, to a brief discussion of which we will now turn.

1

Perhaps it is worthwhile first to state my conviction (shared by many other authors[2]) that – despite the numerous criticisms directed at the recent 'resuscitation' of this concept – the notion of civil society retained (or perhaps regained) its relevance for contemporary social theory, and it provides an adequate and important conceptual tool not only for the understanding of certain aspects of modernity in general but also of our present social conditions. Moreover, I hope to demonstrate that it retained a significant critical potential in relation to East-Central European and Western liberal democracies alike. This of course does not mean that this concept is easy to define or that it is unproblematic and unambiguous. Those sceptical about its value for social research on the basis of its 'ideological' character and the difficulties with its operationalisation, have their point. The real difficulty with the concept, however, is not – as these critics would have it – that it is 'ideological' or 'speculative'. Concepts are always 'ideological' in the sense that they aim at the ordering and systematisation of reality within some particular theoretical framework. It is also true that the concept of civil society does not lend itself easily to operationalisation and therefore, in order to approach empirically those social phenomena which fall under the cluster of its meanings, one must choose certain partial indicators, that necessarily simplify the social reality. In this respect, however, it is not a new problem in social theory and does not differ from the difficulties connected with, for example, the Durkheimian notion of 'social solidarity' or the Weberian 'spirit of capitalism'.

The real problem with the concept of civil society is its ambiguity and the confusing diversity of meanings associated with this term, a diversity that has its origin in the extremely complex history of this concept. Although I cannot deal here with this history in any detail,[3] it is perhaps important to note that initially the term was to designate societies organised into a state (as opposed to 'barbaric' societies lacking institutionally separated organs of political power unified into the apparatus of a state). It was above all Hegel who created a theoretical rupture in this tradition, separating the concepts of the state and civil society from each other as two distinct (although naturally neither mutually independent nor equal) institutional spheres of social life in modernity. This theoretical rupture has occurred in conjunction with, and as an effect of, certain social-political processes that from the seventeenth century on in social reality itself gradually created, enforced, and legally institutionalised such a division. Hegel himself distinguished three separate dimensions of civil society: the system of needs; classes and corporations; and the civic sphere of public institutions. Zygmunt Pelczynski points out that an especially significant aspect of this understanding, from the point of view of the further development and critical interpretation of the concept, is Hegel's recognition that civil society is not only "an arena in which modern man legitimately gratifies his self-interest and develops his individuality" but also the one in which he (or she) "learns the value of group action, social solidarity and the dependence of his welfare on others".[4] The institutional framework of the so-conceived civil society – though it historically varies – includes certain stable elements, such as a constitutionally guaranteed set of civil rights, the independent public sphere of various voluntary associations and an open communicative network.

As it was indicated by Habermas in his early work on *The Structural Transformation of the Public Sphere*,[5] one of the main components of the historical development of civil society was the creation and expansion of an *independent public sphere*.[6] This public sphere which emerged between the private realm and the state, has elevated private individuals (that is, individuals holding no office) into a *public body*, debating and exchanging opinions concerning matters of general interest. Although at its inception this sphere (as it emerged in Western Europe) did not function as exclusively political, but rather grew out of, and was interconnected with, the literary public, from its very beginning it critically questioned the political norms of the absolutist state and so, directly and indirectly, contributed to the formation of new political principles and mechanisms. Its *exclusionary* character, that is, the fact that at this stage it encompassed mainly the ascending bourgeoisie acting as a 'universal class', ensured its relative *homogeneity*. It was above all this specificity of the early structure of civil society that not only allowed, but in a sense made necessary, the acceptance of principles of *equality and general accessibility* of the public sphere. However, the fact that this 'disregard' for privileges and social differentiation in principle was made possible by the *practical exclusion* (or minimalisation) of effective social differences, made these principles contradictory and fragile from the very beginning. When the extension of the franchise, education, the press, and other means of public communication resulted in the expansion of this public body, it lost its homogeneity and became an arena of competitive interests and social conflicts. Using Arendt's expression, the place of 'political issues' was taken over by 'social' ones, a situation which, according to Arendt, already in itself meant a decline and atrophy of civil society. Although I don't share this assumption, it is true that the existing structures and principles of civil society proved themselves unable to cope with the dramatically increased differentiation and pluralisation of its composition. Since the market mechanisms on their own are only able to *coordinate* economic activities but cannot fulfill a task of *social integration*, the task of the *pacification of society* had to be solved by the institutional intervention of the state. It has been, however, convincingly argued by Habermas,[7] that specific institutional structures of the state provide adequate means only for *system integration*. The basic principles of legitimation in political democracies demand, however, that the system integration be build upon the consensually established norms of social integration. The increasing interference of the state in these latter processes, including the production of new patterns of meaning, can be accomplished only through such an extension of its competences which cannot be but paternalistic and more or less explicitly oppressive, treating its citizens as the *objects* of its administrative policies and not as *political actors*.

As an effect of these processes, civil society in contemporary mass democracies represents above all an arena of institutionalised competition among various interest groups, all acting separately and in potential opposition to each other as pressure groups upon the state. It is probably unnecessary to add here that the organisational level, the relative strength, and thus also the bargaining position of different groups in this situation is grossly unequal. Naturally in this competition too the stronger is the winner. But the role of the state is not that of a passive registrar or mere executor of the pre-given outcome. It is in fact the state which establishes the *terms of compromise* and realises it through its policies. Obviously it is in the legitimating interest of the state to achieve in this process a minimum of consensus. Such consensus is facilitated by the fact that its subjects are connected as a rule to several interest groups, since one and the same individual, in his/her different social roles and function, can have an array of quite divergent interests. Furthermore, the state has also at its disposal powerful means of influencing 'opinions' and thus of the 'fabrication' of consensus. The emergence of the welfare state – now independently of the scope of provided by it services – created a number of new problems, not least from the point of view of such a production of consensus.

Even when the welfare provisions did not go much beyond the establishment of a *minimal security*, as was the case in some countries, they still moderated some of the most negative effects of market mechanisms. In several Western societies, however, the functioning of the welfare state included also the recognition of some more general '*social rights*' of all citizens, like a universal right to education or to adequate health services. These policies meant a more or less conscious effort, if not to reverse, then at least to restrict one of the most drastic effects of the 'great transformation' – described so graphically by Karl Polanyi: the subsumption of the whole of social life under the economic imperatives.

The conservative counteroffensive of the last two decades against the welfare state attempted to dismantle just these achievements. Although this attempt, initiated in the name of an unhindered economic rationality equated with general progress of well-being, has not really been successful, its costs in human suffering and increase of inequality was enormous. The initial popularity of such policies, however, indicates some genuine problems with the practices and institutions of the welfare state. In welfare state societies some aspects of the relation between macroeconomic processes and social desiderata or imperatives became a matter of conscious political decision and intervention. Obviously such decisions sensitively affect the whole society, each of its groups and strata. In a democratic society, therefore, they must somehow correspond to social standards of justice, equity, dignity, and to life aspirations as represented by different groups. Since under contemporary conditions these standards are increasingly pluralised, such correspondence would require a prior confrontation and debate between these pluralised conceptions on the basis of which a kind of agreement can be reached, which does not attempt to homogenise the different expectations but is able to establish consensually some common *frame of reference*. Within such a common frame of reference an assertion of differentiated (and even conflicting) interests could be combined with some forms of cooperation and solidarity. If civil society has no appropriate forms and mechanisms for such horizontal negotiations through which it could learn to combine the routine of social cooperation and solidarity with self-assertion and autonomy, the state alone is in a position to choose the terms on the basis of which these decisions are made and compromises reached. This not only means a further restriction of the social autonomy, but it also distorts the very fabric of social life, through what Habermas described as the 'colonisation' of the life-world. One of the important reasons for the impotency of civil society in relation to these trends is connected to the largely outdated, but still very much alive, belief that the task of social integration requires a homogeneisation and an *all-encompassing unity* of society. The acceptance of plurality and solidarity, rather than uniformity, only very recently started to be considered seriously as the possible bases for social integration.[8]

Beyond these most general social implications, the extension of the state's competences also had some more practical consequences. Given the organisation of apparatuses of the modern state as complex administrative bureaucracies, the extension of welfare provisions and services was accompanied by the (usually disproportionate) expansion of the state bureaucracies administering them. This, of course, meant also a corresponding growth in the costs of maintaining these apparatuses. As a result the expansion of services often went together with a decline in their social effectiveness in relation to their costs. This tended to leave everybody dissatisfied: the 'average' tax-payer, because a growing burden did not result in the promised and expected social effects, but often the 'beneficiaries' as well, who found themselves mere objects of a bureaucratic control. This leads us directly to the *problem of needs* and the social mechanisms of their construction.

Before, however, turning to this second component of the argument, let me for the sake of clarity to summarise briefly some of the main points concerning the understanding of contemporary civil society represented here.

1) I understand by civil society primarily the whole network of those voluntary and particular (that is not all-encompassing) associations and organisations, together with the social instruments of opinion formation, articulation, and expression, which ought to be distinguished from the specific institutions of the state and economy.

2) Civil society cannot be identified with the private sphere, in opposition to the state-organised public one. It is a public sphere, or multiplicity of public spaces, a structure of self-organisation of society, outside the institutional framework of the state but not independent of it. The political organisation of the state and the legal guarantees and frameworks provided by it constitute a precondition of a more-or-less autonomous civil society, although – historically speaking – the liberal democratic state does not precede the emergence of civil society. Furthermore, although the exact boundaries between the spheres of state and civil society cannot be easily and unambiguously established either in a descriptive or a normative sense (these are historically changing and depend upon the concrete social situations), the separation of the two spheres is a pre-condition of any democratic order.

3) Similarly, civil society cannot be equated simply with the non-political areas of social action (as opposed to the state-dominated political sphere). Different segments or dimensions of civil society can be – and usually are – oriented towards different areas of social life: politics in its broader sense, economy, culture, etc. Although it is voluntary-associative and pluralistic, this does not mean that it can express and promote only particularistic interests. It often reflects critical concerns over general issues kept off the official agenda, and promotes their public discussion.

4) The most important features of civil society, whatever other historically concrete characteristics it may display, are: formal equality of civil rights, conscious acceptance of (or at least a tolerance for) the plurality of interests, and the public-transparent character of its functioning.

5) Finally, one may identify the major function of civil society as the continuous confrontation of the state's activity with the aims and aspirations of the structured population not only mediating between the

two but attempting to influence the former in accordance with the latter. This function involves the elaboration of those normative structures through which group identities and the encompassing collective identity of a given society are defined, including the definition of its traditions, the norms of social behaviour and an elaboration of the common frame of reference, stimulating at the same time a continuous process of a collective learning.

2

The contemporary phenomenon of the *politicisation of needs* cannot be reduced to the – already not so new – separation of the satisfaction of certain needs from the household and from immediate market relations, and to the fact that this function is taken over by the administrative apparatuses of the welfare state. This of course made them *public* and subject to public scrutiny, but not yet political, in the proper sense of the word. The genuine *politicisation of needs* has occurred only when – in connection to the abovementioned process – the question of needs has been brought into a *public discourse* (or perhaps rather discourses) and became a subject of *contestation*. Such contestation has to be understood not purely in terms of negotiations concerning the satisfaction of needs (its modes and levels), but extends to questioning of their interpretation as well as of the competence of different public institutions and bodies to take over such interpretation. Politicisation of needs, in this sense, means not only – and perhaps even not above all – that they compete for the distribution of social resources, but not less importantly, it also means the *recognition of the contingent nature of needs* and the awareness of their interconnection with the process of individual and collective *identity formation* and with *different ways of life*.[9]

When, for example, today different groups of women take up the issue of social provision of child care, this is not simply asking society to help in the fulfilment of their parental functions. It also thematises such issues as: What is the equitable distribution of responsibilities in this respect? Are there any particular social limitations of life-aspirations connected with motherhood? What modes of life should an appropriate child care accomodate? And so on. This of course opens a whole new discourse (or even a whole series of discourses) between women representing divergent positions in this respect, between women (or parents) and the representatives of child care agencies, and in the more general public forum. After all, the answers to these questions have more or less direct implications for all other members of society as well. In such debates the participants not merely confront their opinions and positions with one another, but the opinions themselves became queried and positions relativised. The autonomy of the individual- and collective-identity formation can be ensured only by such a continuously renewed self-reflection, accomplished in a discursive confrontation with others. This naturally also stands for the process of need-interpretation. If it is not to result in new forms of external control, this too can be accomplished only in a *participatory social discourse* and not through definitions imposed from the 'outside' (or, as a matter of fact, arrived at monologically).

But if needs are basically contingent, and if, at the same time, they pertain to socially formed self-identities, can they be universalised? What type of needs (and on what basis) can be transformed into *claims*? How do such claims relate – if at all – to the rights and duties (or obligations) with which they are often connected? Is there any *objective* mode for establishing what are the *'real'* *needs* and for the prioritisation of their satisfaction? If not, what then is, or could be, the basis of some social obligation to satisfy needs or their claimant portion which cannot be appropriately satisfied through market preferences?

Such and similar questions, as well as the theoretical attempts to provide some sort of answer to them, have of course, a long history within political theory, moral philosophy and sociology, and there is no possibility in this paper to provide an even cursory systematisation of the relevant views and debates. With the establishment, expansion, and the later ensuing 'crisis' of the welfare state, however, at least some of these problems have gained *practical* import. Empirically provided 'solutions' to these problems are implicitly or explicitly reflected in social policies, and as such are to a degree shaping the main directions of present contestations. These contestations attempt to truly politicise needs which have been subjected by the welfare state to an *administrative-bureaucratic management*.[10] They challenge above all the anti-participatory character of this latter, which is firmly embedded in the general institutionalisation of a technocratic rationalism, sensitive only to narrowly defined economic imperatives. Within this system, the definition and the interpretation of needs are subjected to a specific logic, which demands, among others, some sort of (in practice mostly pragmatically accomplished) *evaluation* of needs and their *subdivision* into those, satisfaction of which is incorporated into the state's social policies, and those which are considered to be the citizens' private business and referred solely to the market mechanisms for their satisfaction.[11]

The scope and the content of these two basic categories of needs varies not only from country to country but also fluctuates within one particular society in connection with social pressures, the economic and political situation etc. In all cases, however, the task of some sort of subdivision remains. The point here is not to deny the justification for a certain distinction among needs, or to negate the relative scarcity of social resources which necessarily limits the scope of those needs which society is able to meet. This latter in itself justifies certain hierarchisation. What is to be questioned is the *mode* of, and the *involvement* in, such a subdivision and hierarchisation.

The *bureaucratic-administrative model* of need-construction and -interpretation, with which we are dealing here, and which only partially displaced the previously dominant model of their *traditional-social construction*, displays a number of specific characteristics from this point of view.

One of those characteristics most relevant for our discussion is the already mentioned subdivision of needs into two distinct types. This, as I already indicated, involves an intrinsically evaluative process. Yet there is a continuous, although not quite successful, attempt by the administrative apparatuses to present this subdivision as based predominantly on objective-rational considerations, independent of the normative-moral ones. This is facilitated by the tacit assumption about the nature of needs as essentially quantitative. At the same time, an attempt is also made to disconnect as far as possible the administrative definition and interpretation of needs from that of the 'clients' themselves. This *double 'uncoupling' process* is practically accomplished through a recourse to the 'scientific', or even purely technical, definitions, based on alleged expertise, which then allow the characterisation of certain needs as *real* or *true*, that is *needs* which can be formulated in an *objective* (expert) manner, and as such, contrasted to the *subjective wants* or desires of an individual.[12]

The basic intention of such uncoupling is to reduce the vulnerability of the state-produced definitions and interpretations of needs, by reducing their contestability, under the conditions of increasingly pluralistic value-systems.

Independently of the question of the ultimate success of such a procedure, it is intrinsically deeply problematical. First of all, because it puts under question the *competence of the subjects* themselves to recognise and to define their own needs, and engages in a more or less extensive process of what Agnes Heller calls a 'paternalistic imputation of needs'.

As convincingly argued by Plant and others,[13] needs *cannot* be discovered in some kind of *objective* manner by sociologists, social workers, psychologists, educators or whoever else. Even if a distinction between needs and wants is justified, it cannot be accomplished on the basis of the latter being considered as a sort of a conscious psychological state and the former as corresponding to some 'objective state of things', to which the subject does not have any privileged access. Moreover, whenever such a distinction is made, it not only reflects but promotes the idea and the norms of one specific and privileged ('normal') mode of life, which the 'objectively' defined needs 'ought' to follow.

I do not wish, of course, to argue that all needs, and the means required for their satisfaction, are always equally recognised by their subjects. Agnes Heller, for example, in her earlier work distinguishes between 'needs as project' and 'needs as lack'.[14] This distinction, however, which includes not only a descriptive but also a normative dimension, gains its meaning and significance only in conjunction with the assumption that needs do not appear in isolation, but form specific 'clusters', or *systems of needs*, attached to *different forms of life*. The normative underpinning of this distinction allows it to serve as basis for the critique of a social system that does not allow the differentiation and pluralisation of life alternatives within which the needs-as-lack could be transformed into needs-as-project. It implies that such a process of need-definition and need-interpretation which is insulated from the subject, and/or attempts to homogenize needs, necessarily encroaches upon the autonomy of individuals and groups by imposing upon them a definite interpretation of their identity and a specific way of life. Furthermore, if the constitution of needs is understood as an *intersubjective/communicative process*, the constructed systems of needs are also *symbolic constructions*, directly related to value-systems.[15] This means that the new forms of control and domination created by the anti-participatory, bureaucratic type of need-definition include also a more-or-less subtle intervention into the symbolic structures of life-world and into social processes of meaning-construction.[16]

Elaborating on Habermas' thesis concerning the 'colonisation of the life-world', Offe points out that:

Economic and political regulation is no longer limited to the manipulation of external constraints of individual behaviour but intervenes, in the service of technocratic standards of rationality and coordination, into the symbolic infrastructure of informal social interaction and the production of meaning through the use of legal, educational, medical, psychiatric, and media technologies.[17]

It is above all the conditions created by such type of intervention that are challenged and contested by the process of politicisation of needs. However, the chances of the different groups to gain the autonomy of self-interpretation and identity-formation, and to successfully challenge the definitions imposed upon them, are far from being equal. It is obvious that in stratified societies, characterised by domination and subordination along multiple social divisions (between classes, genders, ethnic distinctions, age groups etc.), these chances are also differentiated. But the sources of this particular inequality are even more complex. As it has already been asserted, the systems of needs are connected to values, and are, in this sense, symbolic constructions. This, however, means also that "language is constitutive for needs",[18] that is, needs are only accessible to us in a communicatively interpreted form, and that a challenge to the imposed definitions requires an access to the appropriate means of public communication. Consequently, the autonomy of identity formation and self-interpretation depends also upon the distribution of skills, competences, and appropriate means necessary for participation in the public discourses. At the same time, the ability of different groups to express themselves in public, both in terms of their mastery of the appropriate (i.e. socially accepted and dominant) vocabularies and in terms of their access to the socially available means of communication and opinion formation, is strongly *stratified*. The inequality of competencies is perhaps one of the main mechanisms which renders 'mute', or at least obstructs, the self-expression of particular social groups and strata. Although the possibility of introduction of new vocabularies and idioms of discourse is somewhat more open across social divisions, it is also restricted by the already mentioned inequality of access to the public means of communication.

Nancy Fraser analyses these forms of inequality introducing the concept of 'the socio-cultural means of interpretation and communication' – *MIC* (of which the needs-talk is only a subcategory), by which she understands "the ensemble of culturally and historically specific discursive resources available to members of a given social collectivity".[19]

She presumes that in modern societies, MIC are not only plural and multidimensional, but they include hegemonic and counter-hegemonic elements. As was already mentioned, it is above all the hegemonic portion of the means of interpretation and communication, in relation to which different social groups are in unequal position. Therefore in the process of politicisation of needs contestation is not limited to the substantive claims formulated within these vocabularies but, by providing new interpretations and introducing new vocabularies, it attacks also this symbolic dimension of social inequality.

These various forms of contestation of the state apparatuses' and the experts' monopoly of needs-interpretation are the catalytic points around which new political identities emerge and are being crystallised. Challenging the bureaucratic-technocratic logic of the need-interpretation, politicisation of needs not only democratises the discourses concerning needs, but also *transforms* the '*clients*' of the welfare policies into *political identities*, be it individuals, informal groups, or organised social movements.

It could thus be said that, while the *bureaucratic-technocratic construction of needs* at least partially took over the role which was exclusively fulfilled by their *traditional social construction*, taking place quasi naturally in the process of socialisation, as a part of sharing a specific mode of life, now this former is itself challenged by the politicisation of needs, meaning the transformation of their formulation and interpretation into matters of *public debate*.

Thus we have here *three differentiated, and historically specific*, modes of need-construction,[20] the consecutive development of which does not, however, displace totally the earlier ones, but rather increases the complexity of the whole process of the definition and interpretation of needs and of the means for their satisfaction.

The traditional construction of needs shares one common characteristic with the bureaucratic one in so far as both attempt to create identity – so to say – *externally*. This externality is, however, quite different in these two cases. The spontaneous process of socialisation – insensitive as it may be to the autonomy of individuals – is much more organically related to the actual conditions of life of the concerned persons. This, of course, may well lead to the full interiorisation and passive acceptance of these conditions, and in this sense has a rather conservative character. At the same time, the bureaucratic construction of needs, with its division between 'objective needs' and 'subjective wants', the satisfaction of which is realised through different social mechanisms, on the one hand, increases in principle the freedom of choice between different activities and enjoyments (even if these are always limited by the boundaries of the concrete life-situations). It also acts to some (limited) extent in the direction of equalisation of the life-chances, but, on the other hand, in so doing it abstracts from the actual differences in life-conditions, and attempts to reinforce one particular mode of life as the norm of 'normality'. This means that it is unable to create such conditions which would allow the fulfillment of those demands for autonomy and freedom of choice that it itself has aroused. Although in this respect it may be perceived as more oppressive, it also creates a particular dynamics, that Agnes Heller characterised through the concept of 'dissatisfaction',[21] and which sets the conditions for self-reflection and thus for possible challenges to this model.

It is in response to this oppressive aspect of the bureaucratic interpretation of needs that the process of politicisation of needs has not only emerged but is gaining momentum today. However, since, the reality challenged by this process is itself a contradictory one, its contestations also point in many different directions, and only the participants themselves can decide which alternative (or alternatives) to choose. I certainly do not intend to suggest the desirability of substituting the plurality of challenges by the choice of some common single form of 'good life'. Rather these choices have to be envisaged as contributions to the consensual establishment of the already discussed 'common frame of reference', the most significant 'vehicle' of which can be identified with what in recent literature is often denoted as the *New Social Movements*. It is they that provide a framework within which not only the construction of new identities is negotiated, but – in connection with it – a *new type of political subject* is also emerging.

3

The contemporary social movements[22] are extremely heterogeneous social phenomena, even if we abstract here from the movements of the 'traditional/classical' type, some of which still maintain their significance today. This heterogeneity ranges from the organisational forms, through political orientations, relation to modernity, up to differences in the willingness and ability of the separate movements to listen and to understand their potential partners and their opponents. Despite this heterogeneity, however, they do share some formal and substantive characteristics, which allow us to formulate certain generalisations concerning their potential for revitalisation and restructuration of civil society and thus for democratisation of public life. Whatever other characteristics they display, their membership is usually based not upon common economic situation, social status, class, or occupation. In general it is organised not along the line of existing social stratification, but rather around various specific social 'issues',[23] life-form preferences, marginalised interests, and so on. They generally attempt not so much to promote some pre-given and well-defined interests, but rather to question certain accepted values and meanings and to articulate new ones around which they create and organise the informal structures of communication, opinion exchange, and direct mobilisation. By so doing they expand the socially relevant practical discourse, increase social self-reflexivity, and broaden the scope and autonomy of social choices. Although the specificity and significance of these movements cannot be reduced to the process of *politicisation of needs*, this latter, as well as the *politicisation of identity formation* connected to it (on the individual and collective level) and the increasingly reflexive manner of these processes, constitute one of their most innovative characteristics. They practically pose the question which is theoretically formulated today by a number of writers within the democratic tradition: the question of the possibility for *normative grounding of a plurality of forms of life* and the systems of values and needs representing them.[24] It is thus not accidental that many of the authors[25] point to the centrality of 'identity-formation' as to a crucial specificity of the contemporary social movements. Of course, the framework of the movement in itself does not ensure the democratic, autonomous character of this process. Even less does it ensure an ability and/or willingness of the particular movements to extend the dialogue *beyond* their own boundaries. Fraser,[26] for example, shows how the process of interpretation and reinterpretation of needs and identities generates an *internal* group solidarity. But the new identities seem in her account to be established in an exclusively 'oppositional' manner (in an opposition to all others 'outside' the given group). This process therefore seems to create ultimately a structure similar to the one represented by 'traditional' movements (formed around pre-established social identities): once the new group identity is formed, the movement acts as an organ of political pressure for promoting the claims and interests connected with it *against* all other groups. It is not recognised that the re-constituted group-identities have to be negotiated also within the framework of a broader social identity in an ever more open dialogue which is able to recognise mutual claims, without subordinating one partner of the dialogue to the other, that is in a *process constitutive of civil society*.

To be sure, far from all contemporary social movements (perhaps not even a majority) consciously accept plurality as a desirable phenomenon. Moreover, not all of them are seeking to establish new identities or new solidarities. Some are not even prepared to enter any discourse in which their own position would have to be open to a possible scrutiny and contestation. Many of these movements (especially some numerically less significant ones) represent mainly an attempt 'to salvage what remnants of tradition and traditional identities can be salvaged,' considering this to be the only option for re-establishing a public order and authority.

However these 'defensive' tendencies of social movements (which, by the way, are to a lesser or larger degree present virtually in all of them) cannot be condemned simply as unambiguously conservative. They not only reflect the complexity of the situation but also represent a way of coping with this complexity, with the task of protecting society against its 'systemic colonisation'. At the same time, *what is to be defended* and *what needs change*, actually constitutes one of those main issues that requires confrontation and negotiation between the different actors in the field of civil society, a negotiation that would be able to establish some common ground on which decisions about the defensive and offensive targets could be made. It is through such confrontations, which have no backing in any formal authority, that a *collective learning process* (within and between different groups) can continuously take place, expanding the horizons of all participants and serving as a possible base in alliances and solidarities among different groups. These alliances, however, would have to be able to accept differentiation of interests and ways of life not on the grounds of mere tolerance but on the basis of mutual understanding that could lead to the mobilisation of the only social resource which is not, and cannot be, controlled from above – social solidarity. Solidarity, according to Habermas, is the only force that can make the mechanisms of the state and economy "sufficiently sensitive to the goal-oriented results of radically democratic formation of public will." But, he adds that "(t)he socially integrating force of solidarity would have . . . to be in position to assert itself against the forces present in the two other steering resources, money and administrative power".[27] Asserting itself 'against' this power does not mean that civil society can make itself independent of these systemic structures. For although the activity of the social movements is oriented above all towards 'pushing back' the state's interference and penetration into the social-cultural world, they themselves not only build upon the formal-legal structure secured by the state, but – as convincingly argued by Arato and Cohen[28] – they must also establish at least some channels through which the new concerns can find their place in the state's policies and the new forms of activity can acquire appropriate legal guarantees, without incorporating the movement into the establishment. It means that the task of the democratisation of civil society must be interconnected with the democratisation of the formal political structures. In other words, using Castoriadis' vocabulary, they have to keep open the possibilities of the constant revision of the already instituted framework and to remain engaged in instituting new forms of social imagination not allowing the 'instituting society' to be enslaved by the 'instituted one'.[29]

Since this is a continuous task, this story is open-ended. It does not yet have a moral (and perhaps it never will) but it offers us some hope.

*School of Sociology*
*University of New South Wales*

NOTES

[1] Perhaps the clearest account of this change has been provided by Nancy Fraser in: *Unruly Practices*, University of Minnesota Press, 1989, esp. chpts 7 and 8. The concept of 'politicisation of needs' is borrowed from her work.

[2] It is impossible to footnote here even the most significant contributions to the contemporary debate on civil society, but one of the most recent and most systematic defences and elaborations of this concept and its transformations can be found in Jean L. Cohen and Andrew Arato: *Civil Society and Political Theory*, MIT 1992.

[3] I briefly addressed some aspects of this diversified history in two of my earlier papers: 'Constitution and Functioning of the Civil Society in Poland', B. Misztal (ed.): *Poland After Solidarity*, Transaction Books, 1985, and 'Formation and Restructuration of Civil Society: Is there a General Meaning in the Polish Paradigm?', *International Review of Sociology* XXI/1-3, 1985. Important insights into this history of the concepts are offered by the already mentioned work of Cohen and Arato.

[4] Zygmunt Pelczynski: 'Solidarity and the "Rebirth of Civil Society" in Poland, 1976–1981', in John Keane (ed.): *Civil Society and the State*, Verso, 1988, p. 364. A more elaborate argument is provided by Pelczynski in his Introduction to the volume edited by him: *The State and Civil Society*, Cambridge U.P., 1984.

[5] Originally published in 1962, it has been translated into English and published by MIT Press only quite recently, in 1989.
[6] There has been a growing number of criticisms concerning the assumed singularity of this sphere (see e.g. Craig Calhoun (ed.): *Habermas and the Public Sphere*, MIT Press, 1992, esp. papers by Benhabib, Fraser, Eley and Ryan) and certain aspects of these criticisms were acknowledged by Habermas himself (*ibid.*). Although I do not wish to enter this debate here, historically speaking, the sphere analysed by Habermas was qualitatively different from its 'competitors' (e.g. the plebeian one) in so far as it was constituted around a new type of media which made it possible to transcend the narrowly local character of the public concerns, thus creating conditions for the qualitatively different public discourse. Nevertheless Habermas' exclusive focus on this single type of early modern 'publicity' is not unproblematic, since his analysis also has a 'normative' dimension. (see e.g. McCarthy's argument in Calhoun, op. cit.)
[7] Jürgen Habermas: *Legitimation Crisis*, Beacon Press, 1975, or his later argument in *The Theory of Communicative Action*, Polity Press, 1987, vol. II, VI/1-2.
[8] Development of this idea can be traced to the the 'pluralistic solidarity' slogan of the Polish Solidarity. Unfortunately, however, this slogan, important as it had been, was just a slogan, which well served the task of mobilisation but did not survive the test of time and did not prepare the Polish people for the real acceptance of the social plurality and diversity. At the same time, it did stimulate valuable theoretical reflections which could well find their way back into the social practices.
[9] A similar problem, although in a much broader philosophical context, is discussed by Agnes Heller in her *Beyond Justice*, Blackwell, 1987 (see esp. chpt. 5). Also relevant for this aspect of my discussion are works of Nancy Fraser, especially her: "Social Movements versus Disciplinary Bureaucracies" (CHS Occasional Paper, no. 8, University of Minnesota, 1987), or its later version in *Unruly Practices*; valamint K. Soper: *On Human Needs*, The Harvester Press, 1981.
[10] See Fraser, 1987, op. cit.
[11] I am abstracting here (for a while) from other existing distinctions between needs, e.g. a distinction that separates the 'consumptive' (or distributive) type of needs from non-consumptive needs, which again include at least two separate categories: those satisfiable only in the private/intimate sphere and those which are (or can be) the subject-matter of political action.
[12] A distinction between 'needs' and 'wants' does not have, to follow this line of course. Heller, for example, uses a similar terminology when making an important distinction between 'needs as wants' and 'needs for self-determination' (Agnes Heller: "On Being Satisfied in a Dissatisfied Society" in Feher and Heller: *The Postmodern Political Condition*, Columbia U.P. 1988).
[13] R. Plant, H. Lesser, and P. Taylor-Gooby, 1980, *Political Philosophy and Social Welfare*, Routledge.
[14] Agnes Heller, 1985, 'Can "True" and "False" Needs be Posited?', *The Power of Shame. A Rational Perspective*, Routledge and Kegan Paul.
[15] Agnes Heller: 'The Dissatisfied Society', in *Power of Shame*, op. cit.
[16] Analysis of the various forms of such intervention occupies an increasingly central place in recent social theory and for some authors (e.g. Habermas or Melucci) it provides an important focus in their understanding of modernity. See e.g. J. Habermas: *Theory of*

*Communicative Action*, op. cit.; A. Melucci: *Nomads of the Present, Social Movements and Individual Needs in Contemporary Society*, Hutchinson, 1989.

[17] K. Offe, 1985, 'New Social Movements: Challenging the Boundaries of Institutional Politics', *Social Research* 52(4): 846.

[18] J. Habermas, 1985, 'A Philosophico-Political Profile' (Interview with NLR), *New Left Review* 151: 92.

[19] Fraser, 1987, op. cit., p. 7.

[20] I deliberately omit here a yet another, historically emerged, type of need-construction, which became known as the 'dictatorship over needs' (see: F. Fehér, A. Heller, and G. Márkus: *Dictatorship over Needs*, Blackwell, 1983).

[21] A. Heller: 'On Being Satisfied in a Dissatisfied Society', op. cit., pp. 21–22.

[22] There is an extensive on-going debate in the sociological literature concerning the nature and functions of the contemporary social movements. The positions represented in this debate range from considering them as drastically novel and denoting a totally new social/historical configuration (see e.g. Touraine: 'Social Movements: Special Area or Central Problem in Sociological Analysis?', *Thesis Eleven*, 1984/9); through a more moderate and balanced acknowledgement of their social significance and/or novelty (Perhaps the majority of the authors quoted here, whose positions are still quite diversified in many other aspects of their analyses); up to the total scepticism, doubting both (e.g. G. Olofsson: "After the Working-Class Movement? An Essay on What's 'New' and What's 'Social' in the New Social Movements", *Acta Sociologica* 31/1, 1988).

It is not my task here to reflect upon the above propositions or to elaborate my own position systematically. The observation introduced here deal only with those particular aspects of the contemporary social movements that have some direct relevant to the general topic discussed here.

[23] Offe makes an interesting observation that the demands of these movements are not class specific but rather either strongly universalistic or, to the contrary, highly particularistic. (C. Offe: op. cit.) These, of course, are not always clearly separated, as it is often the case that e.g. a residential group organised initially in the defence of its quite particular interests, is gradually transformed, in the course of action, and merges its activities with a broader environmentalist movement.

[24] For simplicity's sake, let me refer here only to two authors, whose position in this respect is quite compatible, to Jürgen Habermas in his *Theory of Communicative Action*, and to Agnes Heller in her *Beyond Justice*.

[25] See e.g. Jean Cohen: 'Strategy or Identity: New Theoretical Paradigms and Contemporary Social Movements', *Social Research* 52/4, 1985; Alberto Melucci: "The New Social Movements. A Theoretical Approach", *Social Science Information* 19/2, 1980.

[26] N. Fraser, op. cit.

[27] J. Habermas, 1986, 'The New Obscurity: The Crisis of the Welfare State and the Exhaustion of Utopian Energies', *Philosophy and Social Criticism* 11(2): 15.

[28] A. Arato and J. Cohen, 1984, 'Social Movements, Civil Society and Problems of Sovereignty', *Praxis International* 4(3).

[29] C. Castoriadis, 1980, 'Socialism and Autonomous Society', *Telos* 43. It is quite possible to draw a parallel here with Touraine's conception of social movement as an ongoing 'struggle for the control over historicity'.

ABNER SHIMONY

# CYBERNETICS AND SOCIAL ENTITIES*

## 1. PERSPECTIVE

The ideas of cybernetics have a long pre-history (Mayr, 1970), but they were crystallized in the 1940's by A. Rosenblueth *et al.* (1943), N. Wiener (1948), and others. The conception that was formulated was broad in scope, and to many enthusiasts it promised a unitary view of a number of natural, social, and engineering sciences (Heims, 1991). The present paper is concerned with the applicability of cybernetics to social theory. I have the impression, unsupported by a quantitative census, that the popularity of this idea among social scientists has diminished since the 1950's, perhaps because of the *a posteriori* realization that cybernetics is no more likely than any other compact set of ideas to simplify the immense array of social phenomena. Whether or not this statement is a correct summary of intellectual history, it is salutary as a maxim of caution about social theorizing. Nevertheless, I remain optimistic about the thesis that the general ideas of cybernetics, if appropriately specified, are illuminating in social theory. My purpose is to propose appropriate specifications, and thereby to outline a defense of the thesis. A full defense would require the detailed discussion of many problems of social theory in the light of the cybernetic thesis, and that is beyond the scope of this paper.

## 2. GENERALITIES CONCERNING CYBERNETIC SYSTEMS

I shall mean by 'system' a concrete entity, not a kind of thing, and I shall assume that it has fairly definite spatial and temporal bounds, but not necessarily spatial continuity. With any system S there is associated a set of *states*, each of which is a full specification of the variable properties of the system within the constraints imposed by its defining and invariant properties. A *cybernetic system* has a distinguished subset G of states, called 'goal states,' together with mechanisms for maintaining or approximately achieving them. G does not consist of the disordered or high-entropy states that would be likely to occur in the absence of the special design of S just because of the general laws governing the components, but rather of ordered or low-entropy states. Furthermore, there is in S some *internal representation* of the set G of goal states, and a *monitoring device* for determining whether the actual state lies in G, and if not by how much and in what direction does it deviate. Finally, there are *control devices* by which S operates upon itself in order to correct for any deviations which the monitoring device discerns between the actual state and G.

Some comments are needed in order to clarify this rather complicated and condensed definition.

(i) A cybernetic system is *teleological*, for if the system S is disturbed so as to place its state outside G, then the monitoring and control devices will act so as to bring S approximately back to G, at least with a high probability. Consequently, the set G serves as an end or "telos" for S. After the strenuous campaign of the great seventeenth century physicists to rid physics of the Aristotelian doctrine of final causation, it appears that cybernetics has reinstated teleology. The reinstatement, however, is in no way a reversal of the history of physics. Cybernetics has *demystified* the final cause and has given it a *derivative status*, by subsuming it under the causality of general natural laws. The monitoring and control devices of a cybernetic system provide the special boundary and initial conditions which eventuate – by the operation of general natural laws – in the restoration or maintenance of goal states.

(ii) Similarly, cybernetics has demystified *holism*, at least in a large and important class of cases. There is no doubt that a cybernetic system behaves holistically. First of all, the goal states are ordered states, in which correlations are sustained among parts that frequently are spatially far separated. Furthermore, the subsystems which make the achievement of goal states highly probable – the devices for internal representation, monitoring, and control – are designed either by an engineer or by natural selection to mesh with each other in intricate ways. Nevertheless, when one examines specific instances of cybernetic systems, one often finds non-holistic explanations of holistic behavior. In the case of mechanical, thermal, and electrical servo-mechanisms, each part of the cybernetic system operates in accordance with physical laws, constrained by local boundary and initial conditions, but the design of the system ensures the proper meshing of these conditions, so as to eventuate in the cooperative behavior of the system as a whole. Much more remarkable is the fact that the holistic behavior of the organic cell is well understood in terms of the physical interactions among its constituent molecules (Monod, 1971, especially ch. 4; Ptashne, 1986). Molecular biology has provided very strong evidence against the need for new laws on a biological level that are irreducible in principle to the laws of physics. When we come to the holistic behavior of multicellular organisms, the evidence for the reducibility to physics is less compelling, just because of the complexity of the systems, but most biologists see no obstacle in principle to the progressive realization of the reductionist program until we arrive at the level of mentality. Many scientists and philosophers (myself included) are anti-reductionist concerning mentality, since feelings and thoughts differ in character from the neurophysiological entities with which they are correlated. The complex discussion of holism can be summarized by saying that the general conception of a cybernetic system does not entail that the holistic behavior of the system is inexplicable in terms of the laws governing its parts and their interactions; whether the mystery of holism can be exorcized in a specific kind of cybernetic system depends upon its detailed character. In physical servo-mechanisms and in simple biological systems the mystery can be exorcized, whereas in systems endowed with mentality it seems to me to persist. Whether the holism of social cybernetic systems is explicable in terms of the laws governing individual human beings, once their mentality is acknowledged to be ontologically fundamental, is one of the central concerns of this paper. A positive answer will be given in Section 5.

(iii) Two of the terms employed in the characterization of a cybernetic system have mentalistic overtones, namely "internal representation" and "monitoring device." My intention, however, is to make no commitment to a mentalistic interpretation of these expressions. It is clear, for example, that these terms should be interpreted physically when one considers the operation of a direct acting thermostat (*The Way Things Work* 1967, pp. 20–21). The goal states, which are those in which a room has a temperature close to some designated value, are represented by the setting of a graduated dial which controls the compression of a spring. That this setting actually serves as an internal representation depends upon its meshing with the monitoring and control devices. The compression of a spring prevents the closing of a valve unless the pressure in a tube of liquid overcomes the spring's constraint, and this pressure is a function of the temperature. Clearly all these steps are physical. On the other hand, I do not exclude the appropriateness of a mentalistic interpretation of "internal representation" and "monitoring device" in certain types of cybernetic systems. An urgently hungry food-gatherer will consciously have in mind a class of goal states, those in which hunger is appeased, and the monitoring of the deviation of the actual state from the goal states is performed by conscious hunger pangs aroused by proprioceptive nervous signals. Nor do I wish to preclude the possibility that the internal representation and the monitoring are mentalistic but subconscious, as postulated by psychoanalytic theory.

(iv) I do not wish to convey the erroneous impression that cybernetic systems are the only systems that exhibit collective behavior. For example, in crystals in thermodynamic equilibrium, the atoms are spontaneously arranged in regular arrays, maintaining order over distances millions of times the distance between nearest neighboring atoms. This fact does not contradict the proposition that thermodynamic equilibrium is a statement of maximum entropy and hence of maximum disorder, given the constraints on the system. The binding of atoms in the crystal releases energy to the surrounding vapor and thermal bath, thereby increasing entropy globally. Nevertheless, it is surprising that short range forces, together with statistical considerations, can account for the emergence of the long range order of a crystal, with a qualitative character entirely different from that of its microscopic constituents. Another example of non-cybernetic orderly behavior is provided by simple machines, not equipped with monitoring and control devices. These do indeed have goal states from the standpoint of the designer and user, and when the machines are kept in good repair, the goal states are achieved as a direct result of the principles of construction and the conditions of normal use. These cases of non-cybernetic collective behavior must be kept in mind as a caution against too hastily attributing cybernetic character to orderly social systems, which is the main subject of investigation of this paper.

## 3. SOME BIOLOGICALLY IMPORTANT SPECIFICATIONS OF CYBERNETICS

In this Section four specifications of the general idea of a cybernetic system will be considered, and their realization by multi-cellular organisms will be discussed briefly. The biological examples are intended as a propaedeutic for the application of cybernetics to social systems in the following Sections.

(a) *A cybernetic system S may be hierarchical in the sense of containing and essentially relying upon components $S_1$, $S_2$, . . . which are themselves cybernetic systems.* A component $S_i$ will have goal states of its own, internal representations of these, and devices for monitoring and controlling deviations, but the supervening system S will make use of $S_i$ for the purpose of carrying out its own cybernetic program. The outstanding biological example of a hierarchical cybernetic system is a multi-cellular organism, its individual cells being the cybernetic components $S_i$. A cell $S_i$ in S has much of the autonomy of an individually existing protozoan – metabolizing nutrients, synthesizing proteins, repairing organelles, etc., and doing so in accordance with the program of goal states encoded in its DNA. Of course a certain amount of the autonomy of $S_i$ with regard to gathering nutrients, getting rid of wastes, and exercising motility is sacrificed in the *quid pro quo* that has somehow been contracted with the supervening organism S as a result of evolutionary development. Since the life strategy of all but the simplest multi-cellular organisms depends upon specialization of functions, the supervening organism depends on cells like $S_i$ for specialized contributions to S's program, but a necessary condition for these contributions is the maintenance of the vitality of $S_i$ in its own nearly autonomous fashion.

An entirely different kind of hierarchical cybernetic system is a group of social animals, like colonies of ants and termites or herds of musk oxen. The component cybernetic systems $S_i$ in these groups are the individual animals, each with its own goal states, etc. The autonomy of social animals differs greatly from case to case. In some the individual animal can live for a long time separated from a group, even though social living expedites the achievement of its goal states. In others, like the agricultural ants, cooperation of the group is essential to ensure nutrients to the individual members. The hierarchical character of even the most closely cooperating animal society differs essentially from the hierarchical character of a multi-cellular organism, in which each cell has identical DNA (except for occasional errors of replication, but the difference between one cell and another in a multi-cellular organisms is due less to these errors than to the differences of gene expression). In spite of the fact that each individual animal in an animal society has its own DNA, the program for cooperation encoded in each can be very rigid. The society as a whole has goal states that may require the sacrifice of the lives of individual members. Moreover, not only do the warrior ants of a tightly organized colony sacrifice their lives for the colony, but even the members of a more loosely organized group, like a band of hyenas or musk oxen, take individual risks for the benefit of the group. That the mechanism for achieving the goal states of the society is embedded in the cooperative instincts of individual animals is consistent with characterizing the society as a whole as a cybernetic system, since the general definition of a cybernetic system did not restrict the mode of monitoring the deviation of the actual state from the goal states or the mode of control whereby this deviation is diminished.

It is worth remarking that the plea just made for broadly construing the concept of cybernetic system should not be used to obliterate a conceptually valuable distinction between cybernetic organization and passive symbiosis. A good example of the latter is the symbiosis of termites that ingest cellulose and the bacteria in their abdomens that digest it. Neither could survive without the other. But the symbiosis is passive, in that no mechanism exists individually or cooperatively to monitor and control deviations from states favorable to the symbiosis. Rather, there is an arrangement resulting from evolution whereby the termite's "doing its own thing" and the bacteria's "doing their own thing" ensures their mutual survival, without cybernetic mechanisms. Likewise – and indeed *a fortiori* – the maintenance of a rough equilibrium between predators and prey in a part of the ecosystem is epiphenomenal upon the teleologies of the separate species, and does not depend upon devices of monitoring and control that characterize genuine cybernetic systems.

(b) *A corollary of the hierarchical structure of a cybernetic system is that – within limits – there is replaceability and interchangeability of the subordinate cybernetic components*, whether these are cells of a given type in a multi-cellular organism or individual animals in a society. Excessive losses, can, of course, be disastrous to the supervening cybernetic system, and in the case of social animals the loss of certain individuals (such as a queen bee) is more traumatic globally than the loss of others. Often there are mechanisms in the supervening cybernetic system for compensating for losses of components, for instance by the enhancement of growth rates.

(c) *Cybernetic systems vary in their tightness of integration.* This thesis is too little emphasized in most discussions of cybernetics for two reasons. First, the instances of cybernetic systems in engineering are almost by conception tightly organized, because efficiency and precision are obvious desiderata of good design in devices that serve specifically envisaged human ends, and these desiderata are best achieved by strictly specifying the goal states and the procedures for monitoring and control. Second, in biology the most dramatic instances of cybernetic systems, the ones which catch one's imagination, are tightly integrated systems like the "chemical machinery" of the cell (Monod, 1971, p. 62) and colonies of social insects. The fact remains, however, that biology provides a wide diversity of degrees of integration of cybernetic systems, as one would expect from the diversity and opportunistic character of evolutionary development. Lions, for example, are only sporadically social, and even though they cooperate with their bands in social phases, they are capable of prolonged solitary existence. It should be emphasized that the distinction between tightly and loosely integrated cybernetic systems in biology should not be conflated with the distinction between systems in which all cells have the same DNA and those in which the component systems have their individual DNA. There are protozoan colonies in which all members are descendants of the same single protozoan and hence have almost exactly the same DNA, but their integration consists of little more than a common retractile motion when one cell of the colony is touched – a very low degree of integration. On the other, social insect colonies are rigidly integrated, even though each separate animal has its own DNA.

(d) *The components of a cybernetic system may have machinery whereby other components as well as the system as a whole are represented.* This feature is absent in the classical engineering devices of cybernetics, such as thermostats and mechanical governors, though it can probably be attributed to some sophisticated computers. By contrast, it is common almost to the point of universality in biological cybernetic systems. In a single cell the DNA represents the entire range of functioning of the cell, the gene for a protein being a part of the DNA that represents all replicas of that protein. In the case of a multi-cellular organism, the DNA of each cell represents the entire organism, the differentiation of expression of the genes resulting from the different environmental conditions of the individual cells. Because of the potentiality for all of these diverse expressions, the DNA contained in a single cell represents in a clear sense both the cooperative behavior of the organism and the specialized behavior of the components. In the case of social animals, the representation of other members of the group and of the group as a whole is embedded in various ways in the psychology of the individual animals – in an array of instincts for imprinting, in the capacity to learn the communication system of the species, in the ability to learn by imitation, and in the power to master the sensory clues that distinguish members of the in-group from members of outgroups. The capacity for culture is one of the essential biological traits of human beings, and even though its operation is still little understood, one can confidently say that it is inseparable from the power to represent the actual and potential behavior of other human beings.

## 4. SOCIAL ENTITIES AS CYBERNETIC SYSTEMS

The preparatory work of the preceding Sections has provided the concepts needed for stating the central thesis of this paper: that *human social groups and institutions are typically hierarchical cybernetic systems, loosely to moderately integrated, whose functioning depends essentially upon the cybernetic character of individual human beings and upon their capacity to represent other human beings.* A thesis of such broad scope needs extensive exploration in order to demonstrate that it is neither stronger then the empirical evidence warrants nor so weakened by qualifications that it lacks explanatory power. Nothing so ambitious will be attempted here. I shall undertake only to sketch the applicability of the thesis to two types of cases, but I hope that these will suffice to indicate that it is plausible and illuminating. No claim to novelty is made in presenting the general proposition that social entities have a cybernetic character, for this was said by the cyberneticists themselves and, in somewhat different language, by functionalists and related social theorists. Nevertheless, I believe that my formulation of the above thesis has the virtue of holding certain ideas in balance in spite of tension among them. As a result, I believe that my formulation achieves a sharpness of focus that is lacking in other formulations, even when they are as judicious and well informed as those of Malinowski (1944) and Nagel (1961, pp. 520–535).

The two types of social entities that will be analyzed are the business corporation and the family. The first is an entity based upon an *associative relation* (*Vergesellschaftung* in the terminology of M. Weber, which in turn was an adaptation of F. Toennies's *Gesellschaft*):

The orientation of social action within it rests on a rationally motivated adjustment of interests or a similarly motivated agreement, whether the basis of rational judgement be absolute values or reasons of expediency. (Weber, 1961, p. 219)

CYBERNETICS                                                                 189

The latter is a *communal organization* (*Vergemeinschaftung* in Weber's terminology, adapted from Toennies's *Gemeinschaft*):

The orientation of social action . . . is based on a subjective feeling of the parties, whether affectual or traditional, that they belong together. (op. cit., pp. 218–219).

There are such great differences between these two types of social organization that the details of cybernetic analysis can be expected to diverge greatly, but the applicability of my thesis to both of them supports my claim of the generality of the thesis.

a. *The corporation*: The business corporation evolved as a tool serving the ends of individuals – to pool capital in ventures which are too large for the individuals, to manage complex economic ventures cooperatively, and to limit the liability to individuals in case of failure. But the association which evolved to serve these ends has a structure and a collective behavior which have become detached from the aims of individual shareholders. The corporation aims at profit, and its aim is less mixed with other motivations than is commonly the case with individuals pursuing economic ends. A corporation has proximate goal states, consisting of the successful achievement of various plans of production, marketing, and acquisition, but all of these are subsumed under the ultimate goal state of economic profit. The impersonality and concentration of ultimate goal of the corporation is proverbial ("it has no soul to be damned, and no body to be kicked" – Edward Thurlow). The corporation possesses other general attributes of a cybernetic system. There is an internal representation of the proximate goals, in the form of corporate plans, which are presented at shareholders' meetings and elaborated in the directives that filter down through the chain of officers. There are monitoring devices in the form of accountancy and inventory reports, which report the actual state of the corporation and permit a comparison with the goal states. And there is a mechanism for achieving the goal states in the *modus operandi* of the corporation: the execution of the basic productive work, the dismissal of personnel who are insufficiently productive and aggressive and their replacement by others who are expected to perform better, and the planning of new ventures.

The specifications of cybernetic systems listed in Section 3 are clearly exemplified by the business corporation. It is obviously a hierarchical system, consisting ultimately of individual human beings, but having also substructures, including the board of directors, the higher management, the departments and subdepartments, etc. That the individuals are cybernetic systems with their own goal states, distinguishable from those of the corporation but tacitly relied upon by the corporation as a whole, is obvious. But the substructures may also themselves be cybernetic systems – for example, when the board of directors votes itself very high salaries, pensions, and stock options, which are not consonant with the goal states of the corporation as a whole (unless one accepts the rationalization that these benefits greatly enhance performance).

The question of the degree of tightness of integration of the corporation as a cybernetic system is complex. The voluntary aspect of the association (*Gesellschaft*),of which the corporation is an exemplary case, attenutates the basic emotional and biological bonds among the members. A shareholder in the corporation is legally free to sell his or her stock, and an officer is legally free to resign. On the other hand, there are certain factors working towards tightness of integration. There are the external legal constraints of corporate law and the internal constraint of the character of the corporation. And there are the desires for economic gain and for the power and prestige of office which (proverbially and often actually) have the effect of bending the goal states of individuals in the corporation to the corporate goals. There is a selection process governing advancement on the corporate ladder that favors those individuals whose goals are effectively subordinated to those of the corporation. In sum, the degree of integration of a corporation is loose in comparison with the cybernetic organization of an individual organic cell or an ant colony, and nevertheless relatively tight when compared with other human institutions.

Finally, there are various mechanisms whereby the corporation as a whole and its members are represented in each individual. Even the least sophisticated of employees is aware of his or her place in the chain of command, knowing from whom orders are taken and to whom they can be given. The formal aspects of the corporate organization are typically made available to all employees in brochures, particularly when there are profit-sharing plans for the purpose of enhancing morale and identification with the corporation. And the informal power structure is known to a greater or less degree by officers and employees, depending upon psychological sophistication and access to information.

It is reasonable to claim, I believe, that any *Gesellschaft* with a definite mission and a definite charter will exhibit, *mutatis mutandis*, the lineaments of a cybernetic system that have been summarized for the business corporation.

b. *The family*: The family is the archetypal communal organization or *Gemeinschaft*. The "subjective feel of the parties . . . that they belong together" is deeply engrained in human psychology when the parties are members of the nuclear family (mother, father, and children). Although this sentiment is diluted in the extended family (in a way that is relative to the culture, as we know from the variety of systems of kinship and consanguinity), it nevertheless does usually extend a considerable distance beyond the nuclear family. The prolonged period of helplessness of the human child, as compared to almost all other animals, entails that emotional attachment of the parents and the child to each other is biologically very advantageous. There is a large body of literature that interprets family sentiment as unconscious rationality, designed or hit upon by evolution as a strategy for achieving the personal goals of protection and nurture. Such considerations may have the effect of playing down the sharpness of the distinction between *Gemeinschaft* and *Gesellschaft*, but for the present purpose this question is unimportant. The central point is that the extended family is a system with goal states of its own, distinguishable from those of its individual members, even though these goals may ultimately be intertwined. The extent to which the family has its own goal states depends upon the culture and upon the status of the family within the culture, but whenever it does happen, the family is a cybernetic system. Perpetuation of the family is typically the primary goal, so that over and above the desires of individual members of the family to have children, there is a pervasive feeling that it is important that the family extend to future generations. When biological propagation is frustrated, there is typically recourse to adoption. Other characteristics of the goal states of the family are respectability, honor, power, maintenance of status, and wealth. Often it happens that the maintenance of the status of the family requires some member to fill a military or civil or ecclesiastical position, contrary to the individual's proclivities or ambitions. There are mechanisms for monitoring the deviation of the actual state of the family from its goal states, such as the oversight exercised by individuals to whom that role is traditionally assigned (e.g., the "head of the family") or by self-appointed individuals who have strongly internalized the family's goals. And there are mechanisms for repairing deviations of the actual state from the goal states – e.g., shaming or arousal of guilt in backsliding individuals who fail to maintain respectability, or the threat of disinheritance.

The representation of the members of the family in the psyche of one of its individual members has a deep biological basis. The infant is programmed to recognize at an early age its mother or her surrogate, a psychological trait similar to imprinting in lower animals but more subtle and flexible. In spite of their great psychological adaptability, young children exhibit different attitudes and behavior towards familiar and unfamiliar persons; and because of the residence pattern sanctioned by the culture, this trait can and often does give special places in the mental map of a child to the father, the siblings, and other kin. The extent to which the representation of individual family members is transformed into the representation of the family as a unity, transcending its current membership, is largely a matter of the general culture and of the specialization of that culture inculcated by the household authorities. Variability in this respect is a major factor in the variability of the tightness of integration of the family.

The small primitive societies that have been the focus of theorizing of social anthropologists are much more complex entities than the family, with a variety of legal, economic, political, medicinal, and religious institutions. Nevertheless, they are cybernetic systems in very much the same way as the family, as theorists like E. Durkheim recognized in speaking of their "organic solidarity" (1961, pp. 208–213).

## 5. THE ONTOLOGY OF SOCIAL ENTITIES

The ontological status of a whole composed of parts is a philosophical problem that recurs at many junctures: e.g., the relation between physical space and the points that compose it, the relation between a composite quantum system and the sub-systems that compose it, the relation between a macroscopic body and its microscopic constituents, the relation between an organic cell and its constituent molecules, the relation between an organism endowed with mentality and its constituent organs, and the relation between social groups and the individual human beings composing them. I have elsewhere (1993) defended several propositions about the multi-faceted problem of wholes and parts: (i) that the ontological relation between wholes and parts is not necessarily the same in each recurrence of the problem, and indeed one can anticipate that in some cases the whole is completely explicable in terms of the properties of its parts and their interrelations, whereas in other cases the opposite is true; (ii) that parodies of both the holistic and the reductionistic positions should be avoided, especially the fatuousness that any kind of interaction among component parts automatically entails holism; (iii) that ontological and epistemological problems should not be conflated, and especially that the difficulties of solving the many simultaneous equations governing the interaction among $n$ bodies shows nothing about the holistic status of the system composed of those $n$ bodies.

Of the various junctures at which the whole-parts problem appears, two are particularly relevant to the principal concern of this paper. The first is the relation between an organic cell and its constituent molecules. Two generations of intensive work on the molecular biology of the cell have made highly probable the thesis that the entire biological functioning of the cell – metabolism, growth, repair, and reproduction – can be accounted for fully in terms of the physical properties of its constituent molecules; indeed, the cell has become the archetypal example of a chemical cybernetic system (Monod, 1971; Ptashne, 1986). There is, of course, no pretense of calculating from first principles the details of the construction of a protein under the direction of a gene in the cell's DNA, but that lacuna is epistemological rather than ontological. Nor is there any pretense that any conceptual problems of gene expression are solved, but the unsolved problems do not look like insuperable obstacles to the "progressive research program" of physically explaining the biological processes of the cell; other conceptual difficulties that *prima facie* were as intractable as those remaining unsolved have already yielded to analysis. In sum, the teleological behavior of the cell and its miraculous integration can be understood quite well in terms of interactions among molecules. The cell is a cybernetic system with goal states that are represented in its DNA, which acts as a social molecule, and the mechanisms of monitoring deviations of the actual state from the goal state and of diminishing such deviations are molecular. (A clear and vivid schematic presentation of a segment of cellular cybernetics is given on p. 74 in Monod, 1971.)

The spectacular success of the molecular biology of the cell has encouraged physicists to maintain that it is only a matter of time until the psychological processes of higher animals, including human beings, will likewise be understood entirely in physical terms; and progress along two fronts, computer science and neurophysiology, is cited as a further reason for optimism about the physicalist program. I strongly dissent. There is an obvious obstacle to physicalism in psychology that was completely absent in cellular biology: namely, that the crucial *explananda* in psychology are not complex motions and behavior or material transformations, but rather the sensations, emotions, and thoughts that somehow accompany the physical processes of the nervous system and yet are qualitatively entirely different from them. It seems to me that a minimum condition for an adequate ontology is what may be called "the Phenomenological Principle": that the ontology must be rich enough to account for appearances. Physicalism in psychology violates this principle, because neurological processes cannot have the subjective appearance of sensations, emotions, and thoughts unless their *esse* is different from what current physics characterizes them as being.

It is beyond the scope of this paper to elaborate either of the two foregoing elliptical arguments, one favoring a program of reduction and one against. Rather, I inquire what these two signposts, pointing in opposite directions, suggest concerning the ontological status of social entities.

The answer seems to me unequivocal. There is no obstacle in principle to understanding social groups and institutions in terms of their individual human constituents. The cybernetic analysis of Section 4 is in full agreement with holists and superorganicists (like Kroeber, 1948, ch. 7, and White, 1949) that social entities are systems with much autonomy, that they are typically independent of their specific composition and can survive essentially unchanged when there are drastic replacements of personnel, and that they have goal states distinguishable from and often in conflict with those of individuals. But contrary to the holists and superorganicists I maintain that these collective properties of social entities are comprehensible in terms of individual human psychology – once it is recognized, of course, that individuals are genetically and culturally programmed to interact with the represent other human beings. As cybernetic systems, social groups and institutions have mechanisms to monitor deviations of their actual states from goal states and for effecting changes so as to reduce such deviations. But, as seen in the examples of the corporation and the family in Section 4, these mechanisms are clearly embodied in individual human beings, who envisage group goals and exercise persuasion and discipline in order to realize them.

When one contemplates social phenomena, one finds nothing contrary to the Phenomenological Principle in the thesis that social entities are ontologically reducible to their individual human constituents. The General Will, the Manifest Destiny, the Great Fear, the Solidarity, etc., etc. are all real collective phenomena, but there is no evidence that their loci of concrete feeling are anything but the minds of individual human beings. One is not obliged to postulate a new category of existence, like the superorganic, in order to account for the appearance of these things, in the way that the category of mentality must be postulated in order to account for psychological appearances.

Finally, the mode of integration of social entities is clearly supplied by individual psychology, namely, by the representation in the individual psyche of the existence and the roles of other members of the group, and sometimes of the group envisaged as a unity. It is not a paradox, but rather a wonderful fact about human beings, that social entities exist as autonomous systems because their individual human constituents are social animals. Benjamin Franklin aptly summed up both the cybernetic character of the infant American republic and its ontological reducibility to its citizens when he said, "We must indeed all hang together, or, most assuredly, we shall all hang separately."

*Departments of Philosophy and Physics*
*Boston University*

### NOTE

\* Dedicated to Robert S. Cohen, who has studied the relation between science and society for many decades.

### REFERENCES

E. Durkheim, 1961, 'On Mechanical and Organic Solidarity', in T. Parsons, E. Shils, K. Naegele, and J. Pitts (eds.), *Theories of Society* I, New York: The Free Press of Glenco, pp. 208–213.
S. Heims, 1991, *The Cybernetics Groups*, Cambridge, MA: The MIT Press.
A. Kroeber, 1948, *Anthropology* (new edition), New York: Harcourt, Brace and Company.
B. Malinowski, 1960, *A Scientific Theory of Culture and Other Essays*, New York: Oxford University Press.
O. Mayr, 1960, *The Origins of Feedback Control*, Cambridge, MA: The MIT Press.
J. Monod, 1971, *Chance and Necessity*, New York: Vintage Books.
E. Nagel, 1961, *The Structure of Science*, New York: Harcourt, Brace and World.
M. Ptashne, 1986, *A Genetic Switch*, Cambridge, MA: Cell Press.

A. Rosenblueth, N. Wiener. and J. Bigelow, 1943, 'Behavior, Purpose, and Teleology', *Philosophy of Science* **10**: 18–24.
A. Shimony, 1993, 'Some Proposals concerning Parts and Wholes', in *Search for a Naturalistic World View* **II**, Cambridge, England, and New York: Cambridge University Press.
*The Way Things Work* (1967), New York: Simon and Schuster.
M. Weber, 1961, 'Types of Social Organization', in T. Parsons, E. Shils, K. Naegele, and J. Pitts (eds.), *Theories of Society* **I**, New York: The Free Press of Glencoe, pp. 218–229.
L. White, 1949, *The Science of Culture*, New York: Grove Press.
N. Wiener, 1948, *Cybernetics*, New York: Wiley.

KRISTIN SHRADER-FRECHETTE

# RISK MODELS AND GEOLOGICAL JUDGMENTS: THE CASE OF YUCCA MOUNTAIN*

After a detailed study of the conflicts surrounding government regulation of chemical carcinogens, Harvard University risk assessors recently concluded that scientists and regulators do not "face up to the value judgments that must be made in . . . [risk] regulation."[1] Overselling the role of science in risk decisions, they are rarely explicit about the nature of the ethical and policy judgments underlying such decisions. As a consequence, "informed public discussion" of risk rarely takes place.[2]

Continuing the tendency to "oversell science" in risk assessment, in early 1991 Environmental Protection Agency (EPA) Administrator William Reilly announced a major initiative to anchor EPA decisions more to "the scientific understanding of risk" than to the public's risk "perceptions."[3] Reilly's rehearsing the theme of "the public is irrational, but scientists are rational risk evaluators" is nothing new. Many technological risk assessors have scoffed openly at the public's alleged irrationality toward risks such as nuclear power. Alvin Weinberg has accused members of the public of engaging in a contemporary "witch hunt" caused by "environmental hypochondria"; he argues that risk assessors need to bring the public to its "senses" so that it can realize that there is no scientific basis for much public fear and risk aversion.[4]

Weinberg's and Reilly's overselling the scientific component in quantitative risk assessment (QRA) is problematic for the very reasons that the Harvard researchers emphasized. Ignoring the subjective judgments present in any QRA not only fails to make this subjectivity explicit but also renders public, democratic discussion of alternative judgments impossible. It allows science to coopt democracy.

1. OVERVIEW

Although QRA ought to play an important scientific role in environmental policymaking, this essay argues that the emphasis of Reilly and many risk assessors – on scientific QRA rather than on public views of risk – is misguided. It is misguided in part because all science and all QRA involves a particular type of normative judgment, methodological value judgments. Because the public typically has as much right as risk assessors to determine the acceptability of such value judgments, it is important for the public to reclaim the normative questions that recur in QRA.[5]

After giving an example of how sensitive risk assessments can be to methodological value judgments, I explain both the nature of these judgments and argue that they are unavoidable in all science and QRA. To illustrate some of the reasons why QRA ought not become the sole prerogative of risk assessors, rather than also the public, and some of the reasons that QRA cannot be wholly scientific, next I examine several important methodological value judgments made in assessments of the proposed Yucca Mountain high-level radioactive waste repository. After providing an overview of US policy regarding high-level nuclear waste and the selection of Yucca Mountain as the only proposed site for a waste repository, I discuss several typical methodological value judgments used in the Yucca Mountain assessments. The essay reveals their subjective character and argues that the public likely would not accept these expert judgments. I conclude by arguing that, once we recognize the pivotal role methodological value judgments play in QRA, then we can no longer argue for diminishing the role of public perceptions in evaluating environmental risks. Moreover, in the specific case of Yucca Mountain, the presence of such questionable judgments in the QRA's may explain, in part, the reason why 80 percent of Nevadans are opposed to the facility, even though Department of Energy (DOE) risk assessors favor the site.

## 2. SCIENCE AND SUBJECTIVE JUDGMENTS

Recent risk assessments done by the Ford Foundation-Mitre Corporation (FFMC) and the Union of Concerned Scientists (UCS) illustrate how sensitive QRA conclusions are to methodological value judgments. Both the FFMC and the UCS assessments used identical data – probabilities and consequence estimates – associated with the risk of commercial, nuclear-fission reactors. Yet, their risk conclusions were contradictory. The Ford Foundation assessment recommended use of nuclear energy, and the UCS study advised against it. How could two studies that agreed on the probabilistic and scientific data reach opposed conclusions? The answer is that the two studies used different methodological rules and different methodological value judgments to interpret the data. The Ford group used the Bayesian expected-utility rule, whereas the UCS analysts used a maximin rule. The moral of the story is that, even when they are based on the same data, risk-assessment conclusions can be highly sensitive to subjective methodological judgments about how to evaluate given risks.[6]

Many of the flaws that threaten the objectivity of risk assessments at radwaste sites occur in part because all science and all quantitative risk assessment (QRA) are laden with unavoidable methodological value judgments, at least some of which are questionable. Many of these value judgments concern how to estimate and evaluate particular risks. They deal with factors such as how to simplify the data and how to assess model reliability, sampling, extrapolations, human error, credible worst cases, and so on. Methodological value judgments, for example, include the judgments that a particular model adequately represents the situation, or that the number of samples is sufficient, or that certain types of worst cases could occur. While not all methodological value judgments are avoidable in QRA, and although one can never be absolutely certain whether or not a particular methodological value judgment is correct, nevertheless the reliability of any particular such judgment often can be assessed on the basis of the evidence for and against it. For example, it is possible to assess the methodological value judgment – that a particular hydrogeological model is reliable – by evaluating the evidence for and against it, based on analogous cases, past events, short-term predictions, and so on. A substantial part of this evidence will include estimates of the range of uncertainty associated with the model.[7] The methodological value judgments – that inject some degree of subjectivity into all science – occur at all three stages of risk assessment: risk identification, risk estimation, and risk evaluation.

Such value judgments likewise appear in the risk management occurring after assessment. Indeed, as a prominent group of risk assessors recently concluded: "while it is common to separate the technical steps of exposure and dose-response assessment from risk management, an important conclusion of this study is that risk-management decisions are part of nearly every risk assessment step." The group of risk assessors then illustrated, in a lengthy table, "for each risk assessment step, the types of risk management decisions typically made in response to selected uncertainties." In other words, they showed that methodological value judgments arise even at the allegedly purely technical stages of risk assessment, because even scientists must make management decisions about how to deal with technical uncertainties.[8]

Methodological value judgments are particularly troublesome in science and QRA because they are often not recognized as value judgments. In fact, proponents of the "standard account" of risk assessment believe that it is more neutral and value free than it is. They typically maintain that risk assessment is a largely non-qualitative, *scientific discipline* to be perfected along hypothetical-deductive lines,[9] and that, if they merely discover the correct algorithms, then they will have the power to predict and assess risks in a wholly neutral way, much as Einstein and Bohm hoped for deterministic, predictive power for quantum mechanics. Proponents of the standard account of risk assessment also believe that it is possible to separate the allegedly purely technical steps of risk assessment from risk evaluation and risk management.

Critics of the standard account argue, however, that QRA – and especially its third stage, risk evaluation – is not merely a *scientific investigation* but also a *political procedure* to be negotiated among scientists, engineers, policymakers, and the public.

They argue that because of the many unavoidable methodological value judgments involved in the process, risk assessment is neither always nor wholly objective.[10] Just as physicists do, assessors must make value judgments about which data to collect, about how to simplify myriad facts into a workable model, about how to extrapolate because of unknowns, about how to choose statistical tests to be used, and about how to select sample size. They must make value judgments in determining criteria for NOEL – no-observed-effect level,[11] in deciding where the burden of proof goes, which power function to use, what size test to run, and which exposure-response model to employ.[12] Although such judgments are also methodological assumptions, it is important to recognize that they are judgments about methodological *values*, values such as reliability, simplicity, predictive power, completeness, explanatory adequacy, and so on. Because such assumptions are *value judgments*, they can be wrong. They represent partially subjective aspects of allegedly objective risk estimates and evaluations.

Because risk assessors must continually make methodological value judgments at every stage of their analysis, proponents of the standard account of risk assessment err when they claim to be able to separate some allegedly purely technical aspect of risk assessment. They also err when they claim that risk assessment can be isolated either from risk management or from public perceptions of risk. Because every part of assessment involves methodological value judgments, and because even allegedly scientific judgments have implications for risk-management recommendations, risk assessment is not completely separable from risk management and public perceptions of risk.[13]

One of the main goals of this article is to reveal the way that risk assessments of proposed high-level radioactive-waste repositories are laden with methodological value judgments. Although no scientific enterprises are completely free of subjectivity, at least with respect to methodological value judgments, and although not all value judgments are problematic, obviously the best science and risk assessment are the least subjective. Once we are aware of the unavoidable methodological value judgments inherent in all risk assessment, we shall be forced to assess which judgments are subjective in a damaging sense. We shall also be forced to recognize more clearly the role of public and democratic decision making in shaping our evaluations of acceptable risk, especially uncertain risks. The more questionable are the subjective methodological value judgments of scientists and risk assessors, the more ought members of the public to decide how much uncertainty about risk they are willing to bear. The public must decide how safe is safe enough, how safe is fair enough, how safe is voluntary enough, and how safe is equitable enough. In other words, to the degree that risk assessors' value judgments involve both ethics and uncertainty, risk analyses ought to be open to challenge from the public likely to be affected by such assessments. Our account of the methodological value judgments inherent in risk assessment of repositories is thus one part of a much larger argument that risk evaluation and radwaste policies are not the sole prerogative of scientists, engineers, or policymakers. Such policies are not their sole prerogative because risk decisions are not wholly objective accounts of science and because risk decisions often typically affect everyone. They are laced with values, and on questions of value facing citizens in a democracy – particularly value questions affecting human welfare – the public has the right to play a role in policy determination.

## 3. NUCLEAR WASTE POLICY AND YUCCA MOUNTAIN

To see the role played by value judgments in QRA, we shall examine state-of-the-art risk assessments, those associated with evaluating the world's first proposed permanent nuclear waste facility, in Yucca Mountain, Nevada. As early as 1955, researchers representing the US

National Academy of Sciences (NAS) recommended permanent isolation of high-level radioactive wastes in mined geological repositories, a position the NAS spokespersons hold today.[14] This basic approach to disposal of high-level radioactive wastes is still being pursued in virtually every nation in the world.[15]

The most fundamental reason that virtually all governments and nuclear-risk experts have pursued a policy of developing repositories for permanent geological disposal of high-level radioactive wastes is that they wish to maximize waste isolation. Other arguments in favor of permanent geological disposal are that it minimizes both costs and hazards, especially transport risks to and from a storage facility. Still other reasons for permanent disposal are that we, members of the present generation, should solve the high-level radioactive waste problem, not merely store the waste and thus leave the burden to members of future generations.[16] The underlying assumption of this rationale for disposal is that only a *permanent* geological repository addresses important ethical obligations to future persons. The technical disadvantages of permanent geological disposal of high-level radioactive wastes are the lack of experience with long-term isolation and the difficulty of knowing geological features and processes at the great depths and over the long time periods required. Some persons also oppose permanent geological disposal because they claim that it is impossible to assure isolation of the wastes underground. Other arguments against permanent disposal focus on technical uncertainties, on political difficulties associated with siting the facilities, on ethical problems related to imposing such a risk on members of future generations, and on the importance of the retrievability of the waste, so as to leave open the options for future storage or disposal.[17]

In 1982, the US Congress passed the Nuclear Waste Policy Act (NWPA), perhaps the single most important piece of legislation affecting high-level radioactive-waste disposal. The act mandated permanent disposal of radwaste, a policy that had for years been the conventional wisdom. Containing timetables for the DOE to accomplish permanent, underground disposal of high-level waste, the NWPA provides guidelines for site selection of possible high-level radioactive-waste repositories.[18]

Under the guidelines of the 1982 NWPA, the DOE selected a number of sites as potentially acceptable for the first permanent high-level radwaste repository in the US. In 1987, the choice of sites was narrowed to Hanford (Washington), Yucca Mountain (Nevada), and Deaf Smith (Texas). After much political compromise, the US Congress passed the Nuclear Waste Policy Amendments Act of 1987; one of its main provisions was to mandate study of only one site, Yucca Mountain, Nevada. Other special features in the act are the requirements to create a Nuclear Waste Review Board in the National Academy of Sciences; to ship spent fuel in NRC-approved packages, with state and local authorities notified of shipments; and to provide an analysis, between the years 2007 and 2010, of the need for a second repository.[19] Only if the Nevada site is found unacceptable will other possible locations be considered. Currently scientists and engineers are studying the hydrogeology, seismicity, volcanism, and climate of the Nevada location. Because 80% of Nevadans oppose the facility, the DOE plans for Yucca Mountain remain in question.[20]

### 4. SUBJECTIVE JUDGMENTS ABOUT THE YUCCA MOUNTAIN RISK

One obvious question raised by the contradictory Ford-Mitre and UCS conclusions is whether similar contradictions would arise if one employed different methodological value judgments to interpret other risks examined in QRA. One way to answer this question would be to examine state-of-the-art risk assessments, some of the best of which are being done for the world's first proposed permanent, high-level nuclear waste repository, in Yucca Mountain, Nevada. We shall argue that Yucca-Mountain assessors use a number of questionable methodological value judgments to evaluate repository risks. Because many of these value judgments are both necessary to, and typical of, justifications for permanent geological disposal, problems with them raise questions both about the whole enterprise of permanent burial of high-level radioactive wastes and about the allegedly scientific character of QRA.

Because the QRA's of the proposed repository are among the most expensive, thorough, state-of-the-art studies, they provide excellent cases in terms of which to examine the scientific character of QRA. In this section – the main body of the article – we illustrate that Yucca Mountain QRA's contain questionable methodological value judgments that threaten both the strictly scientific status of QRA and that argue for greater public participation in risk evaluation.

### 4.1. *Risk Magnitude and Risk Acceptability*

One of the most basic judgments typical of repository risk evaluation is that the magnitude of the radiological hazard alone determines risk acceptability and that a particular magnitude of risk is acceptable regardless of other considerations. Many risk evaluations, at Yucca Mountain and elsewhere, rely on the methodological value judgment that predicting hazardous events having a certain low probability of occurrence is sufficient to provide a guarantee of facility safety. Obviously, however, no particular level of probable safety is ever known to be enough. No premises about some level of safety are alone sufficient to justify a conclusion about site acceptability, because site acceptability is a function of normative and evaluative premises. It is not a logical consequence of particular empirical claims. Hence value judgments about the acceptability of some level of risk cannot be justified on purely empirical grounds. One assessor, studying ground motion at Yucca Mountain, based his "conservative estimates" of risk on the value judgment that predicting ground motion that had a one-in-ten chance of occurring was a sound basis for seismic design of the repository.[21] However, we could just as easily argue that we ought to design for a one-in-a-hundred or a one-in-a-thousand chance of particular ground motion. Other groups of risk assessors concluded that there had been no "significant" fault-related movement on the Yucca Mountain Site in the last 500,000 years,[22] and that moderate earthquakes would likely occur at Yucca Mountain at recurrence intervals of tens of thousands of years.[23] The obvious question this raises, however, is whether 500,000 years of no "significant" fault-related movement or 10,000 years of no moderate earthquakes is enough. Should there have been 5,000,000 years of no such significant movement? Or more or less? Choosing a level of designed safety obviously involves value judgments.

Some risk assessors calculated the probability of a volcanic disruption hazard at Yucca Mountain as $10^{-6}$ per year,[24] while others have estimated this risk as between $10^{-8}$ and $10^{-9}$ per year,[25] with the probability of volcanism exceeding $10^{-4}$ over 10,000 years.[26] Not only do these numbers cover a wide range, but there is no certain way to be sure that the possibility of volcanism is ever gone; volcanism at an individual center may last 500,000 years, contrary to what many Yucca Mountain assessors have assumed. Moreover, Yucca Mountain is relatively free of faults, but the paucity of faults does not guarantee safety from future volcanic disruption.²⁷ Hence, the obvious question is: why is the methodological value judgment – that the risk of volcanism is acceptably low – reliable? Even the DOE peer reviewers, such as K. V. Hodges, have themselves argued that it is "patently absurd" to think that there are reliable predictions for geologic events 10,000 years in the future.²⁸

The annual probability of various accidents involving radionuclide releases at Yucca Mountain, according to some risk assessors, is between $10^{-6}$ and $10^{-9}$.²⁹ Obviously these volcanic and accident risks are small, per year, but the inference that such values are "low enough" represents a highly questionable methodological value judgment (in part) because, over centuries, the probability is much higher. For example, if the probability of volcanic disruption is $10^{-6}$ per year, then for 10,000 years, that probability would be $10^{-2}$, an extraordinarily high risk. The value judgment about the acceptability of such a risk is questionable in part because, although the per-year risks may appear low, the risks to future generations, whose members have no say in contemporary decisions about Yucca Mountain, are higher than those that we would likely accept for ourselves. Indeed, a value judgment that such future risks are acceptable arguably fails to take account of citizens' concerns about issues such as due process, equal protection, compensation, and the rights of future persons. The apparent failure to take adequate consideration of the rights of future generations, in making the ethical judgment that a given level of risk is acceptable, is particularly apparent in the radiation standards to be used at Yucca Mountain. European scientists are quite critical of the fact that US radiation dose limits do not extend beyond 1,000 years for the individual and 10,000 years for the population. European regulatory protection against radiation, however, extends up to 1,000,000 years – 990,000 years longer than US protection.³⁰ Moreover, European waste containers have longer projected lives – 100,000 years, and not the 300 to 1,000 years of the US canisters. This is one reason that the Swedish nuclear program has been so successful, for example, with its emphasis on long-lived engineered barriers.³¹ The fact that other countries are making much more conservative plans for managing and containing high-level radwaste suggests that typical US levels of acceptable risk, whether from volcanism or radwaste canisters may not be adequate. Hence, the value judgment that given levels of risk are scientifically ethically, socially, and politically acceptable is highly controversial. One reason for the controversy is that such judgments appear to rest on a highly questionable, short-term, utilitarian justification.

In assuming that a given number of years of canister reliability – or a given number of future cancers caused annually by Yucca Mountain – is acceptable, assessors presuppose that the magnitude of a risk, alone, is grounds for acceptance. Obviously, however, equity of distribution, compensation, free and informed consent, as well as other social considerations, also play a major role in whether a given number of cancers, for example, is acceptable. After the first several hundred years, cancers caused by the Yucca Mountain repository are predicted to be roughly 100 to 1,000 per year, approximately the same number that would occur as a result of naturally occurring uranium-ore bodies.[32] It is not clear, however, why this number of cancers is acceptable, particularly if the magnitude alone is judged as grounds for acceptance. Future persons affected by the waste will have little say in their contracting cancer, and they will have no possibility of exercising their due-process rights. Indeed, there appears to be no adequate vehicle for compensating them. In the absence of such ethical and social considerations, the value judgment – that a given magnitude of risk or number of cancers caused by Yucca Mountain is acceptable – appears questionable.

In assuming that a given level of risk, or a given number of years of travel time of radioactive waste, is acceptable, risk assessors also appear to presuppose that low-probability events are very unlikely to occur. The likelihood of such events, over the long term, is not known with real accuracy, however, because there is no existing frequency record, and because we have no guarantee that the future will be like the past. If the future is not like the past, then allegedly rare events (like seismic activity) could shorten the hundreds of years required for radwaste migration to take place at repositories such as Yucca Mountain. As several USGS geologists put it, speaking of the Maxey Flats radioactive waste facility: extreme events of low frequency may perform the work of thousands of years of creep and slope wash. For example, an eight-hour deluge of 28 inches of rain may cause several thousand years' worth of "normal" erosion. Therefore, infrequent events threaten Maxey Flats' integrity more than do continuous processes.[33] The same could be true at Yucca Mountain and at any other disposal site. Indeed, the DOE peer reviewers for the Yucca Mountain *Early Site Suitability Evaluation* (ESSE) warned that geology is an explanatory, not a predictive, science. They also argued that no reliable predictions regarding seismic or volcanic activity at the site were possible. If no reliable predictions are possible, then it is also not possible to know that a given level of risk is the case and that it is "low enough."

Another problem with the value judgment that given repository risks are low enough – for example, $10^{-6}$ or $10^{-9}$ – is that many risk estimates are subject to considerable uncertainty. Uncertainties of six orders of magnitude are "not unusual" in risk assessments characterized by incomplete data on long-term accident frequency.[34] If this is true, and if similar uncertainties are applicable to repository-risk figures, then an annual risk of $10^{-6}$ of an accident involving a radionuclide release might really lie between approximately $10^{-3}$ and $10^{-9}$. An annual risk of $10^{-3}$ is arguably not a low risk; indeed, it is approximately three orders of magnitude higher than the risks that are normally regulated by government. Hence, one good reason for questioning the value judgment that a given level of risk – estimated for repositories like Yucca Mountain – is acceptable is that, because of uncertainty in the estimates, the real risk could be quite high. Indeed, the real risks associated with radioactive waste storage at past US facilities have not been insignificant. According to government estimates for the Hanford facility, for example, just the normal radiation releases from this site could result in an annual exposure of 580 person-rem.[35] According to some of the lowest government estimates, this level of exposure from only one site could cause approximately 12 cases of cancer and 116 genetic deaths over a 100-year period.[36] Given that exposures could go on for thousands of years, that other estimates for cancer and death are much higher, and that the risk is likely to accelerate, the magnitude of possible deaths and cancers becomes quite high. Admittedly, the Hanford facility is geologically and technologically quite different from the proposed Yucca Mountain repository and from other facilities likely to be used in Europe. Nevertheless, government projections of future Hanford leaks suggest either that there is no way to control the releases, or that they are too expensive to contain. Either of these suggestions could be quite damning for permanent geological disposal. If human error helped amplify the risk at Hanford, then human error could likewise help amplify the risk at other sites. For all these reasons it is not obvious that the magnitude of hazards predicted even for the newest proposed facilities, such as Yucca Mountain, is acceptable. Nor is it clear that the risk associated with previous radwaste facilities is acceptable.

## 4.2. Risk Reductions and Risk Acceptability

Yet another faulty methodological value judgment often present in evaluations of repository risks is that by using very conservative design techniques and by reducing uncertainties as much as possible, these reductions will be sufficient to insure safety.[37] One problem with such a judgment is that, in a situation where uncertainty is very great, as at Yucca Mountain, it is often not possible to know whether a given strategy is "conservative enough." Obviously, in a situation that is quite dangerous, even the most conservative strategy may not be conservative enough. The most conservative strategy for hang gliding, motorcycle riding, or bungee jumping, for example, may not be conservative enough to insure adequate safety. Hence, there may not be good scientific reasons to assume that, because assessment strategies for a repository are conservative, they insure adequate safety. One reason they may not is that, given the heterogeneity of values for basic hydrogeological parameters at many sites, it appears that experts often believe that their assessments are conservative when they are not.[38] For example, one official DOE document admits that "generally, categories of processes or events that have a small probability of occurrence at the site will be eliminated from consideration."[39] After this statement, the DOE policy document enjoins scientists to choose assessment values such that "the realistically conservative analyses might use a conservative value that is within one standard deviation of the mean value."[40]

However, choosing a value that is within one standard deviation on either side of a mean value (for an accident probability) would include only approximately 68 percent of all the cases in the frequency distribution. It is hardly conservative to have an analysis that takes account of only 68 percent of the cases, particularly if the problems likely to arise come from outliers, events associated with catastrophic failures of the repository. For example, we know that there are eight orders of magnitude of variation in the saturated hydraulic conductivities of the various tuffs at Yucca Mountain.[41] If one takes a number – one standard deviation above the mean values – as a conservative estimate of the saturated hydraulic conductivity of the tuff, this might not take adequate account of repository safety. This allegedly conservative number might represent only three or four orders of magnitude of variation in the tuff. The remaining 32 percent of values that were not considered might include the tuffs for which saturated hydraulic conductivity was even four or five orders of magnitude higher. Hence, choosing an allegedly conservative value, up to one standard deviation above the mean, does not represent the extremes that could actually occur at Yucca Mountain and at other sites. And if not, then the methodological value judgment – that risk reductions and conservative techniques are sufficiently conservative – could lead to errors in the final assessment of several orders of magnitude.

It is especially difficult to know that a given repository design is conservative enough, or that a given risk reduction is large enough, because the DOE has contradictory "fundamental goals" in meeting regulatory requirements for safety at Yucca Mountain. Its allegedly conservative stance toward safety is undercut both by schedule constraints and by economics. The DOE explicitly seeks to "minimize financial and other resource commitments" but also to "protect public health and safety."[42] Likewise, its two other (of four) goals are to "comply with applicable laws and regulations" and yet to "maintain an aggressive schedule." Obviously, short-term economy is at odds with health expenditures, and following health regulations is at odds with maintaining a tight schedule. The fact that these four goals conflict makes it less likely that assessors can make a reasonable methodological value judgment that designs are conservative enough or that risk reductions are great enough. Given these four goals, government officials could justify even contradictory policies. Moreover, much the same situation is likely to occur at permanent waste facilities in other nations. There is always a tension between safety and economic efficiency. Given this tension, it is questionable whether repository siting, design, and management will always reflect adequate risk reductions.

Of course, it is important to point out that, in some cases actual risk assessments at Yucca Mountain have *not* been conservative. For example, in an assessment of seismic and faulting hazards,[43] scientists have based their designs on the assumption that the repository can withstand quakes that occur once every 10,000 years. Because moderate quakes are likely to occur during this period, and stronger quakes approximately once every 100,000 years, if the calculations were wrong by only one order of magnitude, then the repository might not be able to withstand a likely quake. Hence, such designs appear to need even more conservatism, more than allowing for an order of magnitude of error. Because risk assessments characterized by incomplete data on long-term accident frequency are typically wrong by four to six orders of magnitude,[44] it might be more reasonable to make the value judgment that allowing for errors of higher orders of magnitude represents a more conservative stance.

To some extent, the DOE itself is responsible for failure to design and perform repository assessments that are adequately conservative. After admitting that its Yucca Mountain approaches to geohydrology, geochemistry, and waste containers were not conservative, the DOE officially defended its "realism" by arguing that "evidence available" has been "sufficient to generate considerable, if not complete, confidence in the minds of responsible investigators."[45] In other words, the DOE simply appealed to its own authority to defend its lack of conservatism. Obviously such appeals provide no empirical evidence to justify non-conservative quantitative risk assessment (QRA).

## 5. CONCLUSION

Judgments about how to evaluate the risk posed by permanent geological disposal of high-level radioactive wastes are sometimes highly questionable, in part because they are judgments about methodological, political, social, and ethical values. Moreover, to the degree that these value judgments are typical of QRA in general, to that extent they argue against leaving risk evaluation to scientists alone. If QRA is irrevocably bound up with value judgments, then the public has an interest in helping to decide such questions of value. Value judgments are not the sole prerogative of scientists.

*Distinguished Research Professor of Philosophy*
*University of South Florida*

### NOTES

* Thanks to the US National Science Foundation, Ethics and Values Studies, for Grant ISP-82-09517 and the US Department of Energy/Nevada Nuclear Waste Project Office for Grant 685873, 9.3.7, 4-C, both of which supported work on this essay. Any opinions expressed in this article, however, are mine and do not necessarily reflect the views of either the National Science Foundation, the US Department of Energy, or the State of Nevada. For constructive criticism of earlier drafts, I am grateful to Bill Freudenburg, Steve Frishman, Reed Hansen, Carl Johnson, Charlie Malone, Naomi Oreskes, Gene Rosa, Joe Strolin, and Richard Watson.

[1] John D. Graham, Laura C. Green, and Marc J. Roberts, In Search of Safety: Chemicals and Cancer Risk, Cambridge, Harvard University Press, 1988, p. 198.
[2] *Ibid.*, p. 198.

³ See W.K. Stevens, What Really Threatens the Environment?, New York Times, January 29, 1991, at C4:
"William K. Reilly . . . told a Senate committee in Washington on Friday that in many cases, the public and Congress are at odds with scientists over which environmental threats are the most serious. He has begun a campaign to reassess the priorities, and he leaves no doubt that his money is on the scientists."
⁴ A. Weinberg, Risk Assessment, Regulation, and the Limits, in Phenotypic Variation in Populations, A. Woodhead, M. Bender, and R. Leonard (eds.), pp. 121–128 (1988). For discussion of the views of other risk assessors who criticize public or lay views of risk, see K.S. Shrader-Frechette, Risk and Rationality, chs. 1–4 (1991). See also D.T. Hornstein, Reclaiming Environmental Law: A Normative Critique of Comparative Risk Analysis, 92 Colum. L. Rev., pp. 604–605 (1992). Both Shrader-Frechette and Hornstein argue that risk assessors need to take account of public views of risk because the public often focuses on important ethical, social, political, and legal aspects of risk that are typically ignored by traditional risk assessors.
⁵ See Hornstein (note 4), p. 565.
⁶ For discussion of the Ford-Mitre and UCS studies see Shrader-Frechette (note 4), pp. 100–101.
⁷ See, for example, E. Reichard et al., Groundwater Contamination Risk Assessment, pp. 101ff. (1990); hereafter cited as: Reichard, GCRA.
⁸ Reichard (note 7), pp. 177–179. Much of this discussion of value judgments in quantitative risk assessment is based on Shrader-Frechette (note 4), pp. 53–74. For a discussion of risk assessment and its components, and an analysis of needed improvements in risk assessment, see Shrader-Frechette (note 4), pp. 5ff., 169–219.
⁹ R. Rudner, The Scientist Qua Scientist Makes Value Judgments, in Introductory Readings in the Philosophy of science, E.D. Klemke, R. Hollinger and A. Kline (eds.), p. 236 (1980); B. Allen and K. Crump, Aspects of Quantitative Risk Assessment as Applied to Cancer, in Quantitative Risk Assessment, J. Humber and R. Almeder (eds.), pp. 129–146 (1987).
¹⁰ P. Ricci and A. Henderson, Fear, Fiat, and Fiasco in Woodhead et al. (note 4), pp. 285–293; R. Setlow, Relevance of Phenotypic Variation in Woodhead et al. (note 4), pp. 1–5.
¹¹ M. Schneiderman, Risk Assessment: Where Do We Want It To Go? What Do We Do To Make It Go There? in Humber and Almeder (note 9), pp. 107–128; E. Foulkes, Factors Determining Target Doses in Hazard Assessment of Chemicals, J. Saxena (ed.), pp. 31–47 (1989).
¹² K. Busch, Statistical Approach to Quantitative Risk Assessment in Humber and Almeder (note 9), pp. 9–55; see also Setlow (note 10).
¹³ Reichard et al. (note 7), p. 177.
¹⁴ C. Fairhurst, National Research Council and National Academy of Sciences, Statement in US Congress, The Federal Program for the Disposal of Spent Nuclear Fuel and High-Level Radioactive Waste, Hearing Before the Subcommittee on Nuclear Regulation of the Committee on Environment and Public Works, US Senate, 101st Congress, Second Session, October 2, 1990, p. 18 (1990).
¹⁵ D. Deere, US Nuclear Waste Technical Review Board, Statement in US Congress (note 14), p. 18. The position in favor of permanent geological disposal is also confirmed by A. Blowers, D. Lowry, and B. Solomon, The International Politics of Nuclear

Waste, p. 318 (1991) and by the US National Academy of Sciences (Commission of Geosciences, Environment, and Resources, National Research Council), Rethinking High-Level Radioactive Waste Disposal, pp. v, 6 (1990). See Waste Isolation Systems Panel, Board on Radioactive Waste Management, A Study of the Isolation System for Geologic Disposal of Radioactive Wastes (1983). US ERDA, Final Environmental Statement: Waste Management Operations, Hanford Reservation, Richland Washington, ERDA-1538, vol. 1, pp. X–74, ii. 1–57 (1975); US AEC, Comparative Risk-Cost-Benefit Study of Alternative Sources of Electrical Energy, WASH-1224, pp. 3–83 (1974); See also I. Amato, Dangerous Dirt: An Eye on DOE, 130, Science News, pp. 221 (Oct. 4, 1986).

[16] Blowers et al. (note 15), pp. 318–319.

[17] R. Murray, Understanding Radioactive Waste, pp. 127, 142 (1989); Blowers et al. (note 15), p. 318.

[18] For discussion of the 1982 Nuclear Waste Policy Act, see L. Carter, Nuclear Imperatives and Public Trust, pp. 47ff., 73ff. (1987) and US Congress, High-Level Nuclear Waste Issues, Hearings Before the Subcommittee on Nuclear Regulation of the Committee on Environment and Public Works, US Senate, 100th Congress, First Session, April 23, June 2, 3, 18, 1987 (1987). See also Radioactive Waste Legislation, Hearings Before the subcommittee on Energy and Environment of the Committee on Interior and Insular Affairs, House of Representatives, 97th Congress, First Session, June 23, 25; July 9, 1981 (1981).

[19] For discussion of the 1987 Act, see J.D. Raeber, Federal Nuclear Waste Policy as Defined by the Nuclear Waste Policy Amendments Act of 1987, 34 Saint Louis University L.J., pp. 111–131 (1989).

[20] R.H. Bryan, Governor of Nevada, Statement in US Congress, Nuclear Waste Program, Hearing Before the Committee on Energy and Natural Resources, US Senate, 100th Congress, April 29 and May 7, 1987, Part 3, p. 41 (1987), provides the 80-percent figure. See also M. Yates, DOE Reassess Civilian Radioactive Waste Management Program in Public Utilities Fortnightly, pp. 36–38, esp. 36 (1990); R. Loux, Will the Nation's Nuclear Waste Policy Succeed at Yucca Mountain? in Public Utilities Fortnightly 27, p. 52 (Nov. 22, 1990).

[21] J. King, Approach to Developing a Ground-Motion Design Basis for Facilities Important to Safety at Yucca Mountain in International Conference for High-Level Radioactive Waste Management (1990) (Item 4 in US DOE, DE91000566).

[22] US DOE, NNWSI History in Bibliography of the Published Reports, Papers, and Articles on the Nevada Nuclear Waste Storage Investigations, January 1985, NVO-96-24 (Rev. 5), pp. 1–30 (1985).

[23] J. King et al., Assessment of Seismic Hazards at Yucca Mountain in American Nuclear Society Annual Meeting (1988) (Item 1 in US DOE, DE90006793).

[24] B. Crowe, Volcanic Hazard Assessment for Disposal of High-Level Radioactive Waste (1986) (Item 61 in US DOE, DE88004834).

[25] L. Metcalf, Preliminary Review and Summary of the Potential for Tectonic, Seismic, and Volcanic Activity at the Nevada Test Site Defense Waste Disposal Site (1983) (Item 142 in US DOE, DE89005394).

[26] J. Emel et al., Postclosure Risks at the Proposed Yucca Mountain Repository: A Review of Methodological and Technical Issues, NWPO-SE-011-88, p. 10 (1988).

[27] E. Smith et al., Regional Important of Post-6 M. Y. Old Volcanism in the Southern Great Basin: Implications for Risk Assessment of Volcanism at the Proposed Nuclear

Waste Repository at Yucca Mountain, Nevada, Report no. 10, Annual Report for the period 7/1/87 to 6/30/88 Submitted to the Nuclear Waste Project Office, pp. 1–37 (1988).

[28] K.V. Hodges, Statement in J.L. Younker, S.L. Albrecht, W.J. Arabasz, J.H. Bell, F.W. Cambray, S.W. Carothers, J.I. Drever, J.T. Einaudi, D.E. French, K.V. Hodges, R.H. Jones, D.K. Kreamer, W.G. Pariseau, T.A. Vogel, T. Webb, W.B. Andrews, G.A. Fasano, S.R. Mattson, R.C. Murray, L.B. Ballou, M.A. Revelli, A.R. Ducharme, L.E. Shepard, W.W. Dudley, D.T. Hoxie, R.J. Herbst, E.A. Patera, B.R. Judd, J.A. Docka, L.R. Rickersten, J.M. Boak and J.R. Stockey, Report of the Peer Review Panel on the Early Site Suitability Evaluation of the Potential Repository Site at Yucca Mountain, Nevada, SAIC-91/8001, p. 384 (1992).

[29] L. Jardine et al., Preliminary Preclosure Safety Analysis for a Prospective Yucca Mountain Repository in Waste Management '87: Waste Isolation in the US, Technical Programs, and Public Education, R. Post (ed.) (1987) (Item 249 in US DOE, DE90006793).

[30] Emel et al. (note 26), pp. 40–41; J. Emel et al., Nuclear Waste Management: A Comparative Analysis of Six Countries, NWPO-SE-034-90, pp. 4–5 (1990).

[31] Emel et al. (note 30), pp. 5–10.

[32] W. Williams, Population Risks from Uranium Ore Bodies, EPA 520/3-80-009, pp. 1–23 (1980).

[33] W. Carey et al., Hillslope Erosion at the Maxey Flats, Radioactive Waste Disposal Site, Report 89-4199, p. 34 (1990).

[34] L. Cox and P. Ricci, Risk, Uncertainty, and Causation in The Risk Assessment of Environmental and Human Health Hazards, ed. D. Paustenbach, p. 1026 (1989).

[35] US ERDA (Energy Research and Development Administration), Final Environmental Statement: Waste Management Operations, Hanford Reservation, Richland, Washington, 2 vols. (ERDA-1538) vol. 1, p. x–74 (1975); see K. Shrader-Frechette, Nuclear Power and Public Policy, ch. 2 (1983).

[36] US AEC (Atomic Energy Commission), Comparative Risk-Cost-Benefit Study of Alternative Sources of Electrical Energy (WASH-1224), pp. 3–83 (1974).

[37] See J. Tillerson et al., Uncertainties in Sealing a Nuclear Waste Repository in Partially Saturated Tuff in Proceedings of an NEA/CEC Workshop (1989) (Item 88 in US DOE, DE91000566).

[38] See J. Lemons and D. Brown, The Role of Science in the Decision to Site a High-Level Nuclear Waste Repository at Yucca Mountain, Nevada, USA, 10 The Environmentalist, p. 7 (1990).

[39] US DOE, Office of Civilian Radioactive Waste Management, Yucca Mountain Project Bibliography, 1988–1989, DOE/OSTI-3406 (Suppl. 2) (DE90006793), pp. 3–13 (1990).

[40] US DOE (note 39), pp. 3–8 and 3–9.

[41] R. Peters et al., Fracture and Matrix Hydrologic Characteristics of Tuffaceous Materials from Yucca Mountain, Nye County, Nevada, SAND84-1471, p. i (1984) [Item D170 in US DOE, NVO-96-24 (REV. 5)].

[42] M. Cloninger et al., Waste Package for Yucca Mountain Repository: Strategy for Regulatory Compliance in Waste Management '89 (1989) (Item 149 in US DOE, DE90006793).

[43] J. King et al., Assessment of Faulting and Seismic Hazards at Yucca Mountain (1989) (Item 15 in US DOE, DE90006793).

[44] See note 34.

DOROTHY WEDDERBURN

# SOME REFLECTIONS ON INEQUALITY AND CLASS STRUCTURE*

This lecture, delivered in London in 1979, is offered for reprinting in this volume in honour of Bob Cohen, on the occasion of his retirement. Concerned as it is with an analysis of contemporary social and economic processes it may appear to be rather remote from Bob's own field. But through his fearless search for truth and his ability to synthesise insights from different academic disciplines, particularly across the science/social science/arts divide, he has provided inspiration for me, as well as many others in our own specialisms. It may also be asked whether these observations on British Society from fourteen years ago still have relevance. I have, therefore, appended a brief postscript to indicate some of the ways in which they may.

I have selected three themes for this lecture.* First, I wish to explore the objective structure of economic inequality in Britain today as revealed by the work of the Royal Commission on the Distribution of Income and Wealth (1974–1979) and to evaluate the way in which the Commission has related this to the extensive social and economic changes which have occurred in the post-war period. Second, I shall consider how these findings might illuminate questions being addressed by contemporary sociologists about the nature and extent of social mobility in Britain and its significance for the formation of social classes. Third, – and here I shall move into an area of some speculation – I shall consider the probabilities attaching to certain industrial and political responses to the sum of these changes – how they have been and are perceived – and whether those perceptions and responses have contributed to weakening any stable normative under-pinning of the industrial order. Such an exploration is vital if we are to assess realistically Britain's capacity to absorb the strains imposed by our endemic industrial problems in a period of world recession, as well as of rapid technological change.

It is interesting to consider why a standing Royal Commission was set up in 1974:

to inquire into existing and past trends in the distribution of income and wealth, and containing, as a preamble to its terms of reference, the declaration that its activities were:

> to help to secure a fairer distribution of income and wealth in the community.

It was the first time in the post-war period that governmental concern and interest in inequality *as such* had been voiced in this way. Generally the period had been dominated by the doctrines that economic growth combined with full employment and appropriate social policies would suffice to improve the lot of those at the bottom of the income distribution, and would produce an inevitable trend towards greater equality. Even the 'rediscovery' of poverty in the sixties by people like Titmuss, Townsend and Abel-Smith, led first, to debates about improved social policies, and only second to the expression of doubts that:

> any equalizing forces at work in Britain since 1938 can be promoted to the status of a natural law . . .

Richard Titmuss's conclusion in 1961 that:

> Ancient inequalities have assumed new and subtle forms: conventional categories are no longer adequate for the task of measuring them,

had little political impact.

By the early seventies the parameters of discussion were changing in the face of two important developments, one economic and one political. First the assumption of *continuous* growth was shown to be untenable as rates of productivity increase slowed down and as unemployment rose, not only in the UK but also in other industrialised countries. Second, and not necessarily independent of the first, the Labour Party was seeking a rapprochement with a critical and dissatisfied TUC. The end of the 64–70 period in office had been marked by increasing rank and file trade union pressure as well as opposition from the General Council of the TUC, both to proposals to legislate in search of greater, so-called, 'peace' in the industrial relations field (*In Place of Strife*) and to incomes policies. After their 1970 defeat, the Labour Party in opposition sought to hammer out the lineaments of a more radical policy with the TUC – what eventually became the 'social contract' – in which the issue of incomes policy (or wage restraint) was placed firmly in the wider context of income and wealth distribution. In 1974 the Labour Party's electoral promises included a statement that they would:

> use taxation to achieve a major re-distribution of both wealth and income,

# INEQUALITY AND CLASS STRUCTURE 217

and specifically declared their intention to introduce an annual wealth tax, as well as speaking of the need for a Commission to advise on income distribution.

So it was perhaps not surprising that when the Royal Commission (with Lord Diamond as its Chairman) was eventually established in 1974, the popular press described it as:

a time bomb ticking away beneath the feet of the rich,

and the CBI, when presenting evidence in 1975, could assert that the terms of reference were such as to prejudice the outcome of any enquiries from the very beginning. Yet when, following the return of a Conservative government this summer, the Commission was wound up, after the publication of eight weighty official reports (as well as background papers and evidence), it is doubtful how many people even knew that the Commission still existed, let alone what those eight volumes contained. Its influence upon policy has been negligible. This lack of impact stems from two factors. The first was the conduct of the Commission itself which defined its role as that of a fact finding, not a policy recommending body. The second was the change in the political climate as the labour government, facing increasing economic difficulties, turned away from its radical election promises. The cynic might be forgiven for going along with A. P. Herbert's view that:

Royal Commissions are not so much for digging up the truth as for digging it in.

But, in this case, whilst the density of the language and statistical material in the reports means that those who use them must be prepared to dig, they provide a more comprehensive, consistent and accurate picture of the distribution of economic resources in this country over a 30-year period, than has ever been available before and from which a number of important conclusions may be drawn.

The first is that despite a 30-year period of immense demographic, social, economic and industrial change, the basic *structure* of the distribution of personal resources has remained remarkably stable. We can think of families (including unmarried individuals) as being arranged in order of the size of their total income from all sources before tax and then examine the share of total income which accrues to various layers. Thus we find that in 1949 the wealthiest families in the top half of the distribution, received three quarters of all personal income, and the bottom half received only one quarter. Thirty years later the picture is the same. Moreover, the *after* tax shares are also basically the same. There is one noticeable change. The share of the richest 5 per cent of tax unit/families has fallen quite substantially (from 24 per cent in 1949 to 16 per cent in 1976) and most of this is concentrated at the top 1 per cent (about 300,000 or so families). But this fall has been balanced by very small increases in the share of all other groups in the top half of the distribution leaving the bottom half where they were. Clearly, whether you call this a move to a more equal distribution or not depends upon whether you attach more weight to the decline in the shares at the top, or to the stability of shares at the bottom. The traditional measure, the Gini coefficient which gives equal weight to all those movements, registers a fall from 41.1 per cent in 1949 to 36.5 per cent in 1976–77. As for wealth, there again a decline in the percentage of total wealth owned by the richest individuals can be discerned but I find more remarkable the continuing high degree of concentration. One half of all individuals own over 90 per cent of all personal wealth, and 20 per cent own more than three quarters. Can we evaluate both this overall stability and the significance of what changes there have been?

The stability is surprising for two reasons. First, many people believed that even if economic growth was not itself inexorably leading to greater equality, then some governments, at least, during this period thought that they had been pursuing egalitarian objectives, which might have been presumed to result in re-distribution. Second, irrespective of whether the movement was towards or away from greater equality, some change in the distribution was likely, given the quite remarkable scope and depth of social, demographic and economic changes in the post second world was period.

It is, of course, hazardous ever to ascribe more change to one historical period than another. But for our concerns, today, the second world war period up to the mid-seventies, witnessed a rise in productivity in manufacturing industry greater than we know ever to have been sustained over such a period before. There was an unprecedented increase in the average standard of living – a doubling in just under thirty years, that is in less than a generation, when the previous doubling had taken more than fifty years. It is important to note here, particularly for the purpose of our speculations later on, that there has been a change of direction from the mid-seventies on. A fall in real personal income such as occurred in 1975 and 1976 and a slowing of the rate of increase since, will administer a severe shock to expectations built up over such a long period of time. There was also, of course, inflation, at an accelerating rate, reaching a peak of 25 per cent in 1975–76, but scarcely falling below double figures in the last ten years.

Other important changes were occurring in the population structure. Although the ratio of the working population to total population remained remarkably stable, the age composition of the total population changed as a result of an almost doubling of the proportion of people over the age of 65 compared with pre-war. Activity rates for men have fallen in the post war period, particularly among the young, as they have made increasing use of educational opportunities, and also among the elderly who have shown a trend towards earlier retirement. As for women, their employment and particularly that of married women has risen dramatically. In the quarter century between 1951 and 1976 nearly all the increase in the labour force consisted of women – some 3 million, and mostly of married women.

The composition and distribution of employment has changed far more rapidly in this post war period than in the interwar years. Between 1951 and 1975 there was a massive increase in service employment – 3.5 million – but a sharp fall in employment both absolutely and relatively in production industries, that is manufacturing, public utilities and construction. Occupational shifts were also considerable. The distribution of male employees by occupational group had remained fairly stable between 1921 and 1951. But between 1951 and 1978 the proportion employed in semi- and unskilled manual jobs halved (from 45 per cent of the total to 22.8 per cent) while foremen, higher and lower professional and managerial occupations, doubled. The inter-war period had seen a growth of clerical employment among women. But this was slowed down and then halted post war. Between 1951 and 1975 women's semi-and unskilled manual employment was almost halved, whilst growth occurred in those same white collar occupations in which male employment had increased.

The Commission attempted to measure the effect of some of these changes upon the income distribution. The consequences of the occupational and industrial shifts were found to be slight because of the wide dispersion of earnings within occupations and industries. But the influence of four other factors was important. These were the pattern and extent of marriage, the proportion of the population which is elderly, the proportion of the 15-to-24 year-old age group in full-time education and the extent of female employment. For example, other things being equal, an increase in the proportion of old people in the population and of adults in full-time education will increase the proportion of low income recipients, while an increase in married women working will tend to increase the proportion of families in the middle and upper range of the distribution as two incomes are combined in one family unit. The calculations suggest that over the twenty year period studied, if no other forces were at work, the four factors together would have made the income distribution more unequal, with a smaller share of total income going to the bottom 50 per cent of families. But, as we have seen their share was in fact stable and there was overall a small movement towards greater equality because of the fall in the share of income received by the top income receivers. This, it appears, owed little to the social and demographic trends discussed here.

Thus for the great majority of families, where earnings are the principle source of cash income, their relative position in the distribution bas been influenced by a complex interplay of factors, largely offsetting in character. This is not, of course, inconsistent with considerable movement for particular individuals and families within the income distribution in the course of a life-time or from generation to generation, a point to which we shall return. But the most important thing to remember in any case is that, for families of all types and composition, probably the most important factor making for change in this period was the increase in living standards consequent upon general economic growth.

But if we wish to understand the social implications of inequality we must concentrate in some detail on what has been happening to those we might call 'the rich' and 'the poor'. Both groups were the subject of special studies by the Commission. Before we do this, however, we should be clear about one thing. So far we have been discussing changes in the amount of inequality (i.e. redistribution vertically) in direct personal cash income. There is another and important meaning of redistribution. That is the redistribution which results from the complex operation of the total system of personal taxes (direct and indirect) and of benefits in cash (social security and supplementary benefits) and kind – (the health service, education etc). The work of J.L. Nicholson which he has now been pursuing for many years, shows that in its totality this system results in a large measure of redistribution, both vertical and lateral. As he says, there are few households not affected by the system (i.e. who are not net gainers or losers) and that for many the gain or loss is considerable. The net effect upon the total post tax and benefit distribution is to make it more equal but the degree of redistribution has not increased very much over the post war period, despite the fact that, what might be called 'the interference', i.e. the percentage of income taken in direct tax and returned in benefits in cash and kind has increased greatly. The impact of this complex system on the actual standard of living enjoyed by families is considerable and again we shall have occasion to return to this point.

Let us now, however, return to examine the structure of the money income which forms the basis of our distribution, and the structure of asset ownership and let us first examine the rich. Here it would be desirable to be able to combine the data on income and wealth for the same families but unfortunately the Royal Commission was unable to complete its work in this area. However, commonsense and internal evidence enable us to deduce that there is a very considerable overlap between those with the highest incomes and those with the greatest wealth. And as soon as we look at the top stratum – (say the top 1 or 2 per cent) we find that their sources of income and the composition of their assets is quite distinctive. One characteristic is that the rich often combine two or three sources of income, e.g. employment, self-employment and investment income. Another is that the self-employed (including in this category close company directors) are particularly well represented. Yet another is that a quarter of all pre-tax income is derived from investments, compared with 2 to 3 per cent in the bottom half of the distribution. In fact the top 1 per cent of income recipients receive 30 per cent of all personal investment income and the top 10 per cent well over a half.

The qualitative differences in the character of assets owned by those at the top of the wealth distribution parallels that in the character of income. Those who are the richest 1 per cent in terms of total assets own 54 per cent of all listed UK company shares owned by persons, and these shares account for a fifth of their total wealth. (The bottom 80 per cent hold less than 1 per cent of their wealth in shares). The richest own over a half of all land (excluding dwellings) and this accounts for another 15 per cent of their total wealth. In 1975 the *average* value of the assets owned by the wealthiest 1 per cent of individuals was over £130,000, whilst the average for the 80 per cent at the bottom (which will include people with no wealth at all or even in debt) was a mere £1,400. There is little doubt that there is a quarter to half a million families whose control of economic resources is different in kind to the rest of the population. But it is, of course, also their *share* of the total (not, it must be noted, the absolute amount) which has been declining. Does this represent an erosion of their position, and if it does, to what factors can this be attributed?

As to the decline of the share of the top wealth holders, undoubtedly estate duty (and its successors) have played some role in reducing the extent to which large wealth holdings can be maintained intact from generation to generation. But there are good reasons for believing that the statistics may overstate the extent of decline because of the opportunities for rearranging asset holdings. There are considerable difficulties in dealing satisfactorily with the valuation and inclusion of overseas wealth holdings, trusts or other re-arrangements of property within the family, and with certain forms of wealth such as paintings, antiques and jewelry.

In 1926 Josiah Stamp began his British Association address by saying:

> It will probably not be disputed that one of the fundamental institutions of our modern life which is likely to come under criticism and challenge in the next twenty or thirty years is inheritance.

It is not our purpose here to criticise or defend. The evidence from the Royal Commission and of recent studies is, however, that inheritance is the social institution which accounts for the enormously high concentration of wealth at the top end of the scale in Britain, much greater, for example, than in the USA.

As for income, there are some obvious reasons why the relative share at the top should have fallen. There has indisputedly been a long-term trend for a narrowing of the dispersion of earned income at the very highest levels. Income from dividends in real terms has not kept pace with the increase in income from other sources, whilst personal ownership of shares has also declined in favour of institutional holdings – pension funds, insurance companies and the like. But there are also some good reasons for believing that the statistics overstate the decline. We know that both tax avoidance and evasion has increased over the post war period, as marginal rates of income tax have risen. The Royal Commission demonstrated in its very last report, that the self-employed are particularly well placed to benefit. But those with employment incomes have also benefitted by the extension of the cluster of benefits, known as fringe benefits, which are reflected to only a limited extent in the figures of income distribution. It is true that these are not confined to the very wealthiest as I am discussing them now – indeed, the total cost to employers of discretionary benefits has increased from 10 per cent of total labour costs in 1960 to 20 per cent in 1977. But there is no doubt that their value is greatest at the highest salary levels. The latest Hay/MSL data show that the total cost to employers of superannuation and other fringe benefits at salary levels of £20,000 and above in industry in 1978 represented more than an additional third of average salary. They will be worth more than that to the recipient. Here we have a clear example of the way in which, to quote Richard Titmuss again:

Ancient inequalities have assumed new and subtle forms.

Then we turn to the poor. The Commission devoted a whole, and probably its largest volume, to low incomes, which they defined as the bottom 25–30 per cent of the income distribution. This group has received approximately 10 per cent of pre-tax income over the whole period. The Lower Incomes Report, however, is the only one to try to make allowance for differences in the income 'needs' of families of different size and composition and thus the data are not strictly comparable with those quoted above for the rich. On this equivalent income basis, however, the Commission find, perhaps not surprisingly, that 60 per cent of the families in the bottom quarter of the income distribution relied almost exclusively on state benefits, while those that did have some earnings often had them supplemented by state benefits. They comprise those who have always been the disadvantaged groups in our society – the old, the disabled, the long-term sick, one parent families and (an increasingly important category since the mid-sixties) the long-term unemployed and some wage earners with large families. More than one elderly person in three and one third of *all* children are to be found in the lowest quarter of the equivalent income distribution. Thirty years of a so-called welfare state has not shifted them from the bottom of the pile nor has it improved their relative position. What it *has* done, however, is to ensure that they have benefitted from economic growth. For the real value of incomes going to the bottom 25 per cent has moved roughly in line with GNP, rising at least up to 1975 and 76, when they fell, again in line with GNP. The linking of some social security and supplementary benefits to the price index and to movements in average earnings has been vital for ensuring a share in growth. This has of course been a political decision and so it can be changed. But growth has not ensured any automatic redistribution to improve the *relative* position of the poorest.

To summarise this evaluation of the evidence from the Royal Commission on the Distribution of Income and Wealth we may say first, that over the post war period major demographic, social and economic changes *have* affected the distribution of resources between individuals and families but in such a way that changes making for increased inequality have been largely offset by others making for greater equality. Hence the need to disaggregate the statistics. Second, a closer examination of the top and bottom of the distribution reveals a stability in the *sources* of wealth and the *origins* of poverty which assumes a class character. The underlying factors making for privilege or exclusion are still to be found in the basic workings of the economic system. Ownership of wealth still confers cumulative advantage while inability to participate in the labour market still condemns to the bottom of the pile.

At this point some of the audience may object, with disappointment, that this is an unremarkable conclusion to reach and that I am arguing that *'plus ça change'*. That is not the case. It must be clear that I am using 'class' here in a special sense. For whilst it is undoubtedly the case that the working class in the traditional sense of that word – manual wage workers in industry – have a high probability of finding themselves in the bottom part of the income distribution when old, sick or unemployed, the great majority of the working class are to be found in the middle levels of the income distribution. They are to be found there in part, at least because of the contribution of wives who are in outside employment and thus contribute to family income. The Commission cited evidence that the proportion of working families in poverty would increase three-fold if married women did not go out to work. Such phenomena have social significance and require closer investigation when we move to consider the relationship between the distribution of economic resources and society's perception of and response to that distribution.

Moreover, when I speak of ownership of wealth conferring cumulative advantage, I do not wish to imply any simplistic relationship between the personal ownership of particular assets such as land and company shares with the ability to control economic decision making. The heavy concentration of the ownership of these assets which are in the personal sector contributes greatly to the inequality in the distribution of personal wealth. But today only about one third of ordinary company shares are owned by the personal sector and the share of personal wealth in total national wealth has been falling and probably now constitutes no more than a half. It is, of course, possible that the same elite group controls British industry as did so thirty years ago, but that remains to be demonstrated – it cannot be derived from these data and this in itself adds a new dimension to traditional concepts of class.

As for the interpretation of the findings about the poor, it could be argued that so long as the individuals at the bottom end of the distribution have shared equally in economic growth then we have as a society done well. But this prompts two questions. Are the levels of income received by those I have called the 'poor' adequate, and what will happen now that economic growth has slowed down or even ceased? I cannot embark on a debate about definitions and measures of poverty here. In my own work over the years I have been acutely aware of the dangers of perpetuating poverty by using a relative poverty standard, such as supplementary benefits, which increases in real terms over time when total real income rises. For then, without some basic change in the shape of the income distribution, the poor will always be with us. The Royal Commission showed that if the 1961 supplementary benefit standard was increased to allow only for price increases, and was then applied to the 1975 income data the number of the poor would be halved compared with using the contemporary measure. But fortunately for us more work has been undertaken in the last four years to examine the kind of standard of living which supplementary benefit levels imply – for example, the nutritional adequacy of the diets for children that could be purchased by the supplementary benefit allowance for children, the stocks of clothing owned by supplementary benefit recipients and the extent to which they have to borrow to make ends meet. Much of this work has been carried out by the Supplementary Benefits Commission itself. All of it makes a depressing tale. After examining it no-one could I think doubt that the 15 per cent or so of people living below or just a little above supplementary benefit levels, were to use the Supplementary Benefit Commission's phrase, 'deprived of an income which enables them to participate in the life of the community'.

This evidence must also be related, I believe, to other evidence of social inequalities. Post neonatal infant mortality still shows a threefold social class gradient; mortality for men rises from 77 in social class 1 to 137 in social class V; dental decay is more than twice as high among children in social class V as in children in social class 1. Other examples may be cited. I would be the first to admit that economic resources alone are not responsible for such outcomes. Education, access to information, general environmental conditions all contribute. But it is hard not to conclude that the Britain of 1980 is still a society which leaves much to be desired in terms of ending deprivation.

As for the impact of the slowing down of economic growth, not only does it mean that with present policies unchanged (or even cutting back on levels of spending) there will be no improvement in the position. Worse than that, there is likely to be a deterioration. The marked increase in long-term unemployment which is now occurring will mean that the absolute numbers of people (and of children) forced down to these levels of deprivation will increase. This, together with the frustration of the accumulated expectations of those in employment, is I believe the 'time-bomb' ticking away beneath all our feet as we enter the '80s.

At this point I turn to my second theme, the relationship of these findings about the distribution of economic resources to the findings of contemporary sociologists about the nature of social mobility and class formation in Britain today. For there are, I believe some interesting parallels. If the picture I have presented is accurate – a general overall increase in living standards, with considerable movement in the middle ranges of the distribution, but with structural stability in the position of a small stratum at the very top, and a rather larger stratum at the bottom, this is not unlike recent findings on social mobility of John Goldthorpe and his colleagues at Nuffield. They are in the process of publishing the results of their authoritative study which is the first in this country for 23 years.

They use their own scheme of seven class groupings derived from occupations (so that class 1 for example covers high grade professionals, top administrators in public and private companies and government etc. While class VII includes semi-and unskilled manual workers in industry). Mobility is measured both in terms of sons' position relative to fathers' and in terms of individual work life mobility. There are four main conclusions of significance for my argument. First, that the mobility experience of men between class positions has been more extensive and diverse than has hitherto been assumed, and that, in particular, there has been a considerable amount of upward mobility over the last 50 years. Second, that the top social class has recruited quite substantially from the sons of fathers who were themselves in lower social classes. But because this class consists of occupations where employment has expanded, particularly rapidly since the second world war, it is possible for such recruitment to co-exist with a considerable measure of inter-generation stability. Third, that three, so-called, 'intermediate classes', which contain rank and file white collar workers, small proprietors, lower grade technicians – an extremely heterogeneous group of occupations – are characterised by much mobility, both upwards and downwards. The expansion of employment in these occupations has been satisfied by the recruitment of men from lower class origins, thus contributing to overall upward mobility. But whilst intermediate occupations serve as stepping stones for some people to even higher social groups, by the nature, of what Goldthorpe calls, the marginality of these occupations for some they also serve to lead back to lower class occupations over the course of a working life-time.

Fourthly and finally, although inter-generational social mobility has hitherto been under-estimated, there is evidence of homogeneity of origins in two strata – one at the very top and one at the bottom of the class structure. Within the top class, there is evidence of the existence of a small elite group with a much higher degree of self recruitment than in the class as a whole. At the same time there is evidence that in the remainder of the 'top class', once positions have been attained – from whatever origins and by whatever route, there are powerful guarantees of continuity at that level. At the bottom of the structure, on the other hand, there is also stability. There may be between a fifth and a quarter of the employed male population who are themselves the sons of manual wage workers in industry, and will never have had any experience of being anything but manual wage workers. In other words we see the emergence of an industrial working class stratum, homogeneous not only in terms of work life experience but also between generations.

It is tempting to identify Goldthorpe's elite stratum with the group which I have called 'the rich'. My 'poor' however are not identical with any stratum of homogeneous manual wage workers, because with few exceptions the poor, as we saw, possessed specific characteristics of disadvantage. Nonetheless, it is also clear that the probabilities of such disadvantage afflicting industrial wage earners of the kind described above are higher than for other occupational groups, and in this sense, at the very least, there is a clear link.

So we may ask, how has the sum total of these economic and social changes been perceived by the individuals experiencing them, and how, if at all, have they shaped attitudes and behaviour? Here I will concentrate upon the manual wage workers and the intermediate classes, that is about three quarters of adult males. I would argue that the total effect of these changes has been to blur the delineations of class as an economic phenomenon – what I have called the structuring of class positions – but without in any way reducing (on the contrary even perhaps heightening or extending) the sense of powerlessness and uncertainty which the great majority of employees experience in employment today.

One important reason for this, I submit, is that there has been a weakening of the link, once to be seen as very direct, between the individual's monetary returns for participation in the production system and his or her experience as a consumer. Today the sum total of an economic class position has become increasingly opaque, mediated by a proliferation of institutions and processes which transform the contracted money wage or salary into something unrecognisable in terms either of its immediate or long-term purchasing power. Moreover, comparisons with other people, groups or individuals are either increasingly difficult to make, or become ever more extensive and random as remuneration assumes new, subtle and less visible forms.

Inflation has been one of the most obvious of these processes and I will not dwell on that. But so, too, has been the increasing volume and complexity of the tax and benefit system. In 1959 an average family paid 5 per cent of its gross income in income tax but by 1977 that had become 18 per cent. In fact, at all income levels except at the highest and lowest the proportion of income taken in tax has approximately trebled. In 1959 the top tenth of income recipients contributed nearly two thirds of the total Inland Revenues receipts of income tax. By 1976–77 they paid only 40 per cent and the next five tenths were paying over half the total. We know only too well the resentment caused by this increasing burden of taxation which is *very visible* to the individuals concerned. Less visible, however, are the benefits received and financed by these taxes. They vary between families with children and those without: between the young and the old: between those in work and those out of work; and the system has become so complex that it is hard not only for the participants to grasp the net effect, so that resentments grow and myths flourish, but also for the policy makers to judge what the effect of a change in one part of the system will be. There is now a distinction between the distribution of rewards by the albeit imperfect labour market, and the final distribution arrived at after the vast amount of government interventions required to satisfy, in uneasy partnership, the rudimentary requirements of adherence to some precepts of collective social justice – (i.e. that collective provision should be made for the old, the sick and handicapped as well as for education for all as a basic right on the one hand) – and of incentive to productive efficiency (i.e. provision in such a way that the will to work is not weakened, and that education is provided for the needs of the economy). This is a dilemma which faces every capitalist democracy.

Many other features of the system render economic outcomes more like those of a lottery than any rational reward for effort, (by which I mean, of course, the rationality of the capitalist market), need, or deserts. I will give a few examples. The more than proportionate increase in house values results in windfall gains (albeit frequently not realisable) for those who are already owner-occupiers as compared with first-time buyers or renters irrespective of income level or occupations. The value of fringe benefits varies enormously from occupation to occupation (as well as with income level on which I have already commented). Even as basic a provision as an employer's sick pay scheme still shows wide variations both in coverage and levels of benefit between manual and non-manual employment and even more between men and women. But the company car epitomises the system.

The likelihood of a second earner in the family has, as we have seen, increased and in 1976, 60 per cent of all married women were working, but the probability of working is less for those with young children. The percentage of wives contributing to family income rises steadily as one moves up the income scale, except at the very top among the rich. Thus, the standard of living which the family can enjoy is increasingly dependent upon its position in the life cycle and upon the availability of jobs for women. Opportunities for second or supplementary employment (where remuneration may be in cash to evade tax) vary between individuals, occupations and locations and their extent is unknown, but believed to be growing.

Movement into occupations in the intermediate social classes (routine white collar jobs) may be judged by sociologists as constituting upward mobility. But these jobs are valued by their occupants more for the possibility that they will provide marginally more interesting work, congenial working conditions and more fringe benefits, than because 'status' is something valued for itself. But increased social mobility widens the range of occupations in which friends and relations are to be found and hence it will extend the frame of reference within which economic comparisons are made and dissatisfactions formed, but in a partial and random fashion. And the mass media make their own particular contribution here.

The attenuation of the link between 'official' rewards in the labour market and the final control over resources available to ordinary workers has led to a situation where the system of rewards is itself seen to be lacking rationale. But this I believe results, not in any radical challenge to the normative legitimisation of market rewards – such as would arise from an appreciation of the class nature of the inequalities which persist – but to a concentration upon sectional claims and comparisons.

Alongside this fragmented challenge to the lottery of the system, however, there can be discerned a growing consciousness of the shared subordinate position inherent in most employment relationships, as well as an extension of feelings of powerlessness and insecurity to those occupational groups which make up the intermediate classes. This contributes to a greater willingness to take action against the employer in pursuit of justice for one group in particular, but not to any sense of shared class identity.

Whatever movement into this intermediate strata may have meant in the past, it is now no guarantee of stable employment or further upward mobility, as we have seen. The overall probabilities of becoming unemployed may still be higher for manual than for non-manual workers, but the Department of Employment's study of the long-term unemployed showed that the probabilities, once without work, of remaining unemployed for several months are now becoming closer. The two occupational groups most at risk of long-term unemployment are male clerical workers and male general labourers. In the past the internal labour market of a company would have offered some prospect of advancement for non-manual workers but these channels are closing in the present unfavourable economic climate.

But if manual and intermediate non-manual occupational groups are converging in their experience of insecurity and powerlessness, this dose not at the moment provide a sufficient basis of 'solidarity'. Moreover, there is one major sector of employment where the nature of the employment relationship is even more opaque than elsewhere and which certainly deserves more attention than it has hitherto received from sociologists.

## INEQUALITY AND CLASS STRUCTURE

Central and local government together (excluding the nationalised industries) employ *one in five* of the total labour force. But in this sector there is a sharp contradiction in the nature of that employment. For like any other job, the employment contract implies the handing over of the right to use one's labour in return for a package of remuneration. But a crucial difference which has been highlighted and has been demonstrated over and over again in the last few years, is that management in this sector is qualitatively different in that it ultimately lacks the power to determine either levels of remuneration or of employment. (It is the Treasury or the Cabinet). Hence the frustration of organised groups who cannot exert pressure to affect 'profitability' or the effectiveness of an on-going concern, and can, particularly if they take industrial action, only cause discomfort and hardship for clients or the general public including the bureaucratic employees themselves. This dilemma has I believe contributed considerably to the growing view that today's conflicts are *within* classes rather than between them.

I said that my third theme would be to consider whether the reactions to changes in income distribution and to social mobility were such as to have weakened any normative underpinning of the industrial order. My conclusion would be that they have. But it is *not* the nature of structural inequality as revealed by the income and wealth statistics which is rejected. It is not even the gross deprivation of the poor which is as yet not sufficiently appreciated to cause any feelings of great injustice. It is, I believe, that the majority of workers today feel that their treatment by the economic system is unpredictable and lacking in rationale. In the face of this the only strategy open to them is to pursue the interests of their group, as they perceive them. The concern with inequality voiced by government in 1974 has disappeared from political discourse. There are no obvious *general* principles of distributive justice to which appeal can be, or is made by any political party, and which carry conviction. Thus dissent and conflict emerge which are narrowly conceived and defined. At the same time, the growth of trade union membership among the hitherto unorganised, in particular white collar workers, has increased the power of such groups to take collective action and to look after themselves. But the dissatisfactions felt and to which expression is given are not amenable to appeals to a 'greater good', or to the 'best interest' of the community. Groups engaged in these kinds of conflict are calculative. If, on balance, it is not in their interest to push a claim they will not push it. If they anticipate that it serves to do so, they will. And who can complain that rational economic calculation predominates in a market economy except those of us who cling tenaciously to a concept of social justice?

The probabilities are that we face a future where our society will be less capable than it has been at any time in the post war period of giving its members what they want unless there is some radical change. Unemployment generally will grow and long-term unemployment in particular. We have a social security system quite unable to protect those affected from serious and persistent hardship in a no-growth situation. That is a most inauspicious climate in which to encourage the rapid adoption of new technologies and the restructuring of our economy which is so badly required. The problems facing government, then, are not ones deriving simply from the obvious economic difficulties which Britain faces, but from social divisions and antagonisms scarcely denting our political surface compared with Italy or France, but as Miliband suggests, played out in the industrial sphere.

*POSTSCRIPT.* SEPTEMBER, 1993

Mrs. Thatcher had gained her first Conservative victory in 1979 by appealing to the confusion and dissatisfaction of the disparate interest groups described in the closing paragraphs of the lecture. What was not entirely clear at the time was just how much of a watershed that victory represented in terms of British post war history, nor that it would be followed by, to date, fourteen uninterrupted years of Tory rule. Once in power, Margaret Thatcher embarked upon a programme to destroy the post-war political consensus and to implement a wide range of policies based upon 'market' principles. The key elements have been a total rejection of Keyesian economic policies in favour of monetarism; the subordination of the goal of full employment to one of reducing inflation; a dogmatic and ill-thought-out programme of privatization; a frontal attack upon the power and influence of trade unions; a steady retrenchment of the scope of the 'Welfare state' and an erosion of the power of local government and the creation of non-elected elites to control public services. The replacement of Mrs. Thatcher as Prime Minister in 1991 may have softened the rhetoric, but it has not changed the thrust of policy. Some similar trends can be observed developing in other European Countries in the eighties. But the range and scope, as well as the ideological underpinnings, are unique to Britain.

What have been the consequences? In 1979 the remarkable feature of the structure of economic inequality was its stability over the post war period. That is no longer the case. Inequalities of both income and wealth have increased quite dramatically, as a result of growing inequalities in the distribution of original income and of fiscal policies. There are two differences of interest, however. The consequences of fiscal de-regulation, and reduced direct taxation means that more of the wealthiest households owe their position as much to accumulation as to inheritance. At the other end of the scale, the big increase in unemployment, which has affected a wider range of occupations than in earlier periods, has resulted in a more heterogeneous group falling into the lowest income groups.

As for social mobility there has probably been little change in terms of the broad categories of social class used for analysis in 1979. But the steep fall in the proportion of the labour force employed in manufacturing industry has served to weaken further the traditional working class. The new jobs that have been created in the service sector are increasingly associated with insecure employment conditions – part-time and casual working – and legal protection has been weakened or removed. More and more women are to be found in these types of employment. It would not, then, be surprising if the sense of insecurity and powerlessness, of which I spoken in 1979, had increased.

Many leading authorities have described the economic policies of the Conservatives, over the period, as disastrous. Certainly the position of Britain in the national income league table of European countries has fallen. The consequences for ordinary people can be compared to riding a roller coaster from which many have been thrown off. Any buoyancy which might have been experienced during the boom of the eighties quickly evaporated with the onset of the deepest recession for thirty years. Individual economic outcomes resemble a lottery even more closely now than they did in 1979. My conclusion then was that the social antagonisms and divisions, visible at that time, would continue to be played out in the industrial sphere resulting in a continuation of hostile industrial relations. But there I was mistaken. A combination of high unemployment and anti-trade union legislation has weakened beyond measure the power of trade unions and sapped the will of workers to resist. And so far there has been no *political* voice found. The big question for the future is whether it will be.

*Emeritus Professor of Industrial Relations*
*Imperial College of Science Technology and Medicine*
*University of London*

NOTE

\* The Stamp Memorial Lecture delivered before the University of London on 11 December 1979.

REFERENCES

John Goldthorpe and Catriona Llewellyn, 1977, 'Class mobility in Modern Britain', Sociology **II**(2).
John Goldthorpe and Catriona Llewellyn, 1977, 'Class mobility', *British Journal of Sociology* **XXVIII**(3).
R. Miliband, 1978, 'A state of desubordination', *British Journal of Sociology* **XXIX**(4).
J.L. Nicholson, 1979, *The Changing Impact of Taxes and Benefits on Household Income in the UK 1957–76*. Policy Studies Institute.
Sir Henry Phelps Brown, 1977, 'What is the British predicament', *The Three Banks Review*, December, 1977.
*Royal Commission on the Distribution of Income and Wealth*, 1974–79, Reports, No. 1–8. London: HMSO.
R.M. Titmuss, 1962, *Income Distribution and Social Change*, George Allen & Unwin.

KURT H. WOLFF

# FROM 'DUALISM OF HUMAN NATURE' TO 'HUMAN BEING AS A MIXED PHENOMENON'*

It should become clear, even plausible, how it is that our discussion begins with excerpts from a diary.

## I. FROM A DIARY

October 2, 1992

I remember a few days ago I thought I could write, should, had to: I saw the first yellow leaves in the green-leaf sea, a triplet, I was moved, I hadn't written for so long, I couldn't even get myself to start a paper, although I have been fairly clear about it – which may contribute, it occurs to me this moment, to my difficulty in actually starting it – on Durkheim's 'dualism of human nature' and my own 'man as a mixed phenomenon' and the reasons for differences between them – but I must also go back to 'zwei Seelen wohnen, ach! in meiner Brust' ['two souls, alas! are lodged in my breast'].

October 24, 1992

A late friend could no more account for being crazy about *The Bolero* than I can for falling asleep over Windelband's history of philosophy,[1] reading up on dualisms, Pythagoras, Plato, Aristotle, Stoics. My paper: How the decades between Durkheim and me, eight tenths of the 20th century, can account for the differences between his and my approaches to the human being. (In less than eight years, the third millennium, our third millennium, is to start, but nobody talks about anything but the next century.) This last century of the second millennium displays the explosion of society, which Durkheim had embraced, espoused, slipping into his praise the splendor of sociology, fighting against what he thought were windmills only in need of adjustment or, at worst, repair, namely, anomie, but which became totalitarianism, the cancerous growth of science into the nuclear bomb, the poisoning of air, water and earth, medical progress turning into wild population growth, with sociology, meant to be the science of society, lagging behind efforts to rethink science and propose the redistribution of human skills and knowledge, natural resources, territorial units. No wonder, despair, cynicism, isolation, distrust or failures deepening and spreading, and the tiny beginnings of refuge into the single individual. But who is this individual? It is that being who has both exclusively human and shared features. And attention to the ones, the others, and both must guide our endeavors aiming at a society in which we would be proud of our socialization rather than ashamed or deeply embarrassed.

Why must I go further back than Durkheim? To see his dualism as it is preceded by other dualisms, at least in the West. It looks as if most of them for the longest part of Western history see two elements in the cosmos rather than in man; the world is dual; the relocation of dualism in the human being may have come about with the articulation of the concept of society beginning in the 17th century so that the dualism within the human being would be contemporary with that between man and society – no *Faust* without Rousseau, but also no Rousseau without *Faust*.

I still don't know how to write my paper. Its connection with myself is even closer or thicker than it appears here: where do I end and where does the paper begin? It is an old question for me, though in a new formulation. I have been writing for decades but still know nothing but the hope that I have a shimmer of knowledge of my ignorance.

November 6, 1992

It's still coming, it still hasn't come. I have been thinking about the necessary distinction between *dualism*, a name or concept, and *division-into-two*, a fact of the world. Which comes first? One should think the fact, for without it, one should think, there is no concept or name for it. But it may not be that simple. Take the notion of *surrender-and-catch*. Yes, there was the experience, then the name; but then the name has engendered new facts: writings (a fact, facts) on relations among phenomena (e.g., surrender and rebellion: facts). The relation between division-into-two and dualism is between cosmology and epistemology. It is a problem quite different from that of the relation between Durkheim's dualism of human nature and my man as a mixed phenomenon, which is neither a cosmological nor an epistemological but a sociological problem: how to account for the differences between them in the light of changes in the two writers' society or societies that have taken place in the decades between their two conceptions.

Why, then, again, was I driven to look into (some of the) dualisms that have been formulated in the course of (Western) history? It is a search,

# DUALISM OF HUMAN NATURE 237

I realize, for what might be called dominant dualisms, such as Durkheim's (between individual and society) or mine (between exclusively human and shared features or between the transcendental and the empirical subject). But what is a "dominant" dualism?

There should be a paragraph or a page or two on each of the thinkers mentioned before, as well as on Descartes, Kant, the early sociologists (Montesquieu, Condorcet, Rousseau, Saint-Simon, Comte); Nietzsche (Dionysian and Apollonian), and I've reread Faust ('two souls, alas!'), Freud (pleasure vs. reality), and I forgot the Bible, from Abraham to Prophets and Job. All this, and more, a (wild) sketch of dualisms, Durkheim's and mine among them. (See also my 1959 paper on the sociology of knowledge and the paper on Hannah Arendt's *The Human condition*.[2]) I suppose this covers the material for my paper.

November 21, 1992

Something began to get clearer or more in focus when I wrote a letter about my effort with this paper. I mentioned some examples of dualisms, from Pythagoras to Goethe. Now I quote:

Also I-and-Other, and today I got the idea that Other has recently changed. "Other" I mean to refer not only to another person but to whatever is Other – this typewriter, sunset, leaf, dog. It seems to me the boundaries between me and Other have become problematic. Where do I end and where does the Other begin (see, e.g., me and this paper)? Lovers have known this question with respect to each other, but I wonder whether today it is not on the way of being universalized. And if so, the reason might be that for the first time in human and the earth's history we human beings have the knowledge, theoretical and practical (technology), to abolish ourselves and our habitat and all that lives in it (we even have a rich choice of suicides). Is part of the consciousness that this absolutely new fact is bound to be shaping in us that we stick more closely together, quite irrespective of (conscious) strife and murder? And "we," again, means All, not only human beings, since All, to repeat, is facing Nothing.

This last thought, on All and Nothing, experienced in surrender, I have made my own by interpreting Judith Feher's paper on surrender and death.[3] Since I wrote the letter from which I quoted, Auschwitz has come back in again: people in cattle cars on their way to Auschwitz typically were in awful physical proximity so that, like lovers, they may not have known where the boundary was between self and other. And surely so in an air-raid shelter during a bombing attack, in slave-labor camps, in death camps. The historical point is that such situations have become more frequent, more regular,[4] because bureaucratic organization, efficiency, and anonymity, rule by rule alone, yielding of community before social mobility, voluntary and compulsory (refugees), have malignantly increased, all by technological developments.[5] Thus, is I-Other today the dominant dualism? If so, how does this dominance fit with my dualism of transcendental vs. empirical individual or exclusively human and shared features?

I have a hunch now. these dualisms are not only implied, as I thought, by focus on the individual ('vindicating the human subject,' the subtitle of *Survival and Sociology*[6]) and suspension of our ideas about our social institutions and traditions. For such a proclamation of the individual also implies a reconsideration of social organization that recognizes this dual nature of man and accordingly advocates measures that place man's shared features at the service of the uniquely human features: the former are the means of acting on the latter. (This last formulation resembles the distinction between "civilization," the realm of means, and "culture," the realm of ends, formulated in the early 1920s independently, it appears, by Robert M. MacIver in the United States, and Alfred Weber in Germany.[7] What around them made them so divide the world? A subproblem.)

## II. DIVISION-INTO-TWO VS. DUALISM

I leave what was written earlier, in preparation of this paper to start the paper in the more customary sense of the term, emerging from the "subproblem" in parentheses: we look at "culture" and "civilization" as one of an indefinite, perhaps infinite number of divisions-into-two. The expression "division-into-two" is meant to leave open the question of whether the two into which something is divided are so divided in and by themselves or by human agency. "Culture" and "civilization" is of the second type. Examples of the first are apple and pear, man and woman, young and old, light and heavy, heavy and heavier, and an indefinite or infinite number of other divisions-into-two. But a further distinction must be made, between what may be called first-order and second-order conceptualizations. "Apple" and "pear" and the other examples of divisions-into-two "in and by themselves" are first-order concepts, that is, concepts of items which are not themselves concepts. By contrast, "culture" and "civilization," as well as – to name a few others – husband and wife, democracy and totalitarianism, and again indefinitely or infinitely more – are "second-order" concepts, that is, concepts of concepts, of matters conceptualized by human beings. And we must also recognize that there is an indefinite or infinite variety of relations between the two terms of all these types of intrinsic and artificial, first- and second-order concepts. Again, merely to illustrate: juxtaposition (as it may be between apple and pear), contrast (as it may be between man and woman, young and old, light and heavy), conflict, competition, complementation, cooperation, love, mutual exclusion – for all of which all of our examples may serve as instances – and, once again, an indefinite or infinite number more.

The inventory of divisions-into-two of all of these types is so enormous that the suspicion may arise of there being a universal tendency to perceive the world or any part of it as dual. And if so, we may want to ask why, and may come up with the surmise that the root of this tendency is *a priori*, possibly biological, an experience prepredicable, prephenomenal, perhaps most emblematically expressed (only in the West?) as body vs. soul (or any of its variants) – as it is for Durkheim. In contrast to "division-into-two," "dualism" is the concept of it, thus, depending on whether a given division-into-two is a first- or second-order concept, a second- or third-order concept.

### III. DURKHEIM'S 'DUALISM OF HUMAN NATURE'[8]

Both Durkheim's dualism of human nature and my man as a mixed phenomenon are third-order concepts for they conceptualize concepts of concepts. The dualism of human nature is that of the human being as unsocialized (a concept, in fact an "ideal type") and the human being as socialized (equally a concept and ideal type). The conceptualization of these two concepts consists in adding their necessary cooperation: without the unsocialized human being (a variant of Rousseau's natural man, equally an ideal type), there would be nothing to socialize, and without socialization, human beings or people (once again a concept) could not live together in societies (another concept).

The only reason for discussing conceptualization is to stress the distinction between division-into-two, understood as a prephenomenal experience, and dualism, understood as its conceptualization, and this reason elevates the discussion above to a reinvention of the wheel: our context gives rise to the question how it is that this primitive experience was conceptualized by Durkheim (and myself) in the way it was, and how we can account for differences between these two conceptualizations.

What, then, does Durkheim say? He begins by praising civilization, "that ensemble of intellectual and moral goods . . . that has made man what he is . . . [and whose causes and conditions are those] of what is specifically human in man" (325; 206⁹). Durkheim faults the critics of his *The Elementary Forms of the Religious Life*, which had appeared two years earlier than the paper under discussion, for having failed to perceive "the great principle" (326; 207), namely, "the constitutional duality of human nature" (326ˣ; 207), which man has, in fact, keenly felt at all times as the duality of body and soul: while the body is an integral part of the material universe, the home of the soul is elsewhere, in the world of "sacred things." The contrast entails that between sensations, which are individual, and conceptual and moral activity, which is impersonal (326–327; 207–208).

> There is in us a being that represents everything in relation to itself, from its own point of view; in everything that it does, this being has no other object but itself. But there also is another being in us which knows things *sub specie aeternitatis* as if it were participating in some thought other than its own, and which at the same time, in its acts, tends to accomplish ends that surpass its own (327–328ˣ; 209).

These two sides of the human being are in perpetual conflict; both absolute egoism and absolute altruism can never be reached but only approximated. It follows that our

> concepts never succeed in mastering our sensations and in translating them completely into intellectual terms. . . . Doubtless, we sometimes dream of a science that would adequately express all of reality; but this is an ideal that we can only approach ceaselessly, not one that is possible for us to attain (329ˣ; 211).

The conflict between egoism and altruism is the source of both our misery and our grandeur: we are condemned to suffer but we are unique among all beings. Yet, what is the source of our duality?

> The duality of our nature is . . . only a particular case of that division of things into the sacred and the profane that is the foundation of all religions, and it must be explained on the basis of the same principles (335; 217),

as Durkheim has tried to do in *The Elementary Forms of the Religious Life*, where he attempted to show

> that sacred things are simply collective ideals that have fixed themselves on material objects. The ideas and sentiments that are elaborated by a collectivity . . . are invested by reason of their origin with . . . an authority that causes[s] the particular individuals who think them and believe in them to represent them in the form of moral forces that dominate and sustain them. . . . [They] are . . . simply the effects of that singularly creative

and fertile psychic operation – which is scientifically analyzable – by which a plurality of individual consciousnesses enter into communion and are fused into a common consciousness (325ˣ; 217).

The reason why human beings feel themselves double is that they are: they have two kinds of states of consciousness, one strictly individual, the other states

come to us from society; they transfer society into us and connect us with something that surpasses us . . . they turn us toward ends that we hold in common with other men (337; 219).

And because

society surpasses us, it obliges us to surpass ourselves; and to surpass itself, a being must, to some degree, depart from its nature – a departure that does not take place without causing more or less painful tensions. To want to pay heed [*attention*] is, as we know, a faculty awakened in us only by the action of society. To pay heed presupposes an effort; . . . we must suspend the spontaneous course of our representations and prevent our consciousness from pursuing the dispersive movement that is its natural course. We must, in a word, do violence to certain of our strongest inclinations. Therefore, since the role of the social being in the complete being we are will grow ever more important as history moves ahead, it is wholly improbable that there will ever be an era in which man is required to resist himself to a lesser degree, an era in which he can live a life that is easier and less full of tension. To the contrary, all evidence compels us to expect our effort in the struggle between the two beings within us to increase with the growth of civilization (338–339ˣ; 220–221).

Thus ends Durkheim's paper. According to him, we see the human being as a tragic being, not only constitutionally but increasingly in conflict between egoism and altruism, profane and sacred, desire and duty, body and soul, pleasure and reality, as Freud put it; and the fulfillment of any one of these two is as impossible as its elimination. One consequence of this condition is that science, which resides in the second element of these pairs, cannot do justice to the first; it cannot have its home in both. All religions, Durkheim claims, divide the world into sacred and profane and bestow sacredness on certain material objects by making them symbols of collective ideals. Collective ideals means society means civilization: society and civilization thus become one for Durkheim. This identification is experienced by the members of society when their "individual consciousnesses . . . are fused into a common consciousness," and "public festivals, ceremonies, and rites of all kinds" (336; 218) commemorate and perpetuate this experience.

What, then, does it mean, as Durkheim says, that society "surpasses us" and thus "obliges us to surpass ourselves?" In the next sentence, Durkheim introduced "*l'attention*," attention to something, consideration of something, paying heed to something, "a faculty," we know, "that only the action of society" awakens in us. I interpret this at first perhaps puzzling sentence to mean that civilization (or society) keeps on developing, "surpasses itself," in the sense (or at least I cannot think of any other) of moving toward

> a science that would adequately express all of reality ... an ideal that we can only approach ceaselessly, not one that is possible for us to attain.

If my interpretation is correct, Durkheim, though less optimistic than Husserl, is very close to Husserl's 'to the things.'[10] Durkheim's relation to society would thus not be as exclusively, if not blindly, positive as a first reading may suggest, a reading according to which society gives the individual all that is good – altruistic, sacred – in this dual being. And society also sets itself and its members an insoluble, never ending task; everybody is a Sisyphus, we are Sisyphus. Durkheim is more ambivalent than he may initially seem, even though his *Suicide* might have alerted us already 17 years earlier.[11]

## IV. THE HUMAN BEING AS A "MIXED PHENOMENON"

Like Durkheim's, my own conception of the human being is that it is dual. I have gone through several versions of this dualism. First I encountered it as one of the "metaphysical premises" of the sociology of knowledge understood as an "interpretation of the world":

> I want social relatedness but I also want validity. . . . I want maximum intrinsic interpretation and I want maximum extrinsic interpretation. . . . I want to understand phenomena in their own terms and I want to understand them as instances of laws. I want to carry each of these wishes to its extreme in order to do right by both.[12]

The second time, eight years later, I encountered dualism, while analyzing Hannah Arendt's *The Human Condition*, as a characteristic of human nature, although not as Durkheim's sacred vs. profane but as freedom vs. necessity, the two components of the human space:

> ... man's historicity is the changeability inherent in his freedom which, with his unchangeable nonsovereignty [within the world of freedom where the human being is not master of its story and actions], makes up his dual nature. Changeability inheres in his freedom because he lives in two worlds, those of necessity and freedom, and he is free to reduce the limitations from them with knowledge, acceptance, or rejection and thought, speech, or action.[13]

In surrender, the distinction between necessity and freedom vanishes (as do so many other distinctions), and the human being is more itself, more real, as well as more rational than in any other state or activity.[14] Surrender thus is a message of hope absent from Durkheim's dualism. In surrender, the human being is as purely or exclusively human as it can be, changed from empirial (dual) to transcendental subject. This capacity of change is one of the exclusively human characteristics but, because we are and have bodies, we also have shared characteristics: the human being is a mixed phenomenon.

This idea emerged in an analysis of Camus's *The Rebel*:

... we are forever 'divided'; even without slavery, we are ourselves both master and slave, and both master-and-slave and "man against the world of master and slave."[15]

Continuing:

the ineluctably dual consciousness, or nature, of man derives from the fact that he is both part of the cosmos – which is composed of many other parts whose characteristics man shares – and unique in the cosmos. He is a 'mixed phenomenon': he has features that he shares and others that only he has, that are exclusive to him, are essentially human. He is an object, an organism, an animal, has weight, and innumerable other characteristics that also are attributes of other contents of the cosmos, but he also speaks, means, symbolizes, rebels, surrenders, craves freedom, justice, beauty: with these and other elements of his nature he is alone in the cosmos.[16]

It follows that

to study man ... means to study him as the mixed phenomenon he is and to study his products (in the widest sense of the term – social institutions and cultures) as the mixed phenomena they are, being the products of a mixed phenomenon. But to study man and his products as mixed phenomena means to surrender to them. The very act of surrendering to them will determine which of their elements or aspects is done justice to by the usual procedures of science (such as describing, defining, reducing to instances of generalization), and which, instead, by procedures that will emerge from the encounter with them [from surrendering to them[17]]. One reason why surrender to them is the approach called for by mixed phenomena is that it entails suspending received notions as to what is exclusively human and what is not exclusively human in them, inviting us to look afresh at these features and at the line dividing them, thus to test our received notions, consequently to improve our notions.[18]

In the beginning of this paper I mentioned the infinitude of dualisms that have been devised throughout at least Western history. I then presented some of them: Durkheim's altruism-egoism, socialized-non-socialized, society-individual, and my social relatedness (or historicity; origin) vs. validity, intrinsic vs. exterinsic interpretation (and, related, interpretation-explanation). I have not given up any of these, but they have been superseded by what I think is a historically more important and more adequate dualism, the human being's exclusive and shared features. The question arises of how to account for the shift from Durkheim's 'dualism of human nature' to 'man as a mixed phenomenon.'

I think the clue lies in the changes, between the time Durkheim wrote his paper (1914) and today, in the nature and perception of our society and of society. During the eighty years elapsed since Durkheim wrote his paper, there have been two world wars, genocides, the nuclear bomb and its explosions, labor and death camps, pollution of air, earth, and water, increasing gaps between West and East, North and South, rich and poor, and above all, the achievement of the capacity of abolishing ourselves and, as far as we now know, all life whatever.

Since our institutions and traditions have at best not prevented us from arriving at this unprecedented crisis, we cannot trust their capacity to resolve it. Whom, then, can we trust? Ourselves, the endangered species we are on our endangered planet. But how do we define ourselves in this situation? Surely not as the socialized beings we also are, for we must be skeptical, suspicious, and critical of our socialization and seek to transcend it. We must define ourselves, not as empirical subjects but as transcendental subjects, as capable of trying to suspend our received notions, to surrender to our situation, whatever anyone of us might mean by "our," from individual to humanity, that is, all of us, in order to find out what to do, how and what to think and feel. Surely we cannot trust society, as Durkheim could, or thought he could, he who had discovered it as an exciting candidate for a discipline, sociology, and was eager to show, most convincingly in *Suicide*, that sociology had its own subject matter – society or civilization – that was different from that of the discipline closest to it, psychology:

... sociology, which draws on psychology and could not do without it, brings to it, in a just return, a contribution that equals and surpasses in importance the services that it receives from it. It is only by historical analysis that we can discover what makes up man, since it is only in the course of history that he is formed (325; 206).

In the meantime, there have been enormous developments in both disciplines, among the most important, if not the most important, the exploration of consciousness and its psychic (the subconscious, the unconscious) and social bases (sociology of knowledge). In this context, the idea of surrender-and-catch appears as a further step in these developments, released by what has gone on in the world. It is and proclaims a new vindication of the human subject.[19]

## V. RETURNING TO THE DIARY

In one of my diary entries here reproduced I wrote that I don't know where I end and the paper begins – and the fact of my reproducing the diary entries relevant to this paper demonstrates this blurring of the dividing line. But I don't mean my saying so to be a confession; I mean it to be an objective observation, illustrating the blurring of the boundary between self and other, which I also note in my diary. In a way or ways I haven't explored yet, this change in the self-other relation has to do not only with world-historical developments alluded to at several points of both diary and "post-diary" pages, but also with differences between Durkheim's dualism of human nature and the human being as a mixed phenomenon, as well as with the development from one to the other.

This way of writing here practised has grown over many years – and now I am turning quite personal, as appropriate, I hope, to the occasion of this paper. I have been encouraged by Robert S. Cohen's, and Marx W. Wartofsky's, recommendation that my 1976 book, *Surrender and Catch*, be published – and by its actual publication. This book is my first detailed discussion of the idea of surrender-and-catch in several of its bearings, and it also contains a rich description of the various attempts I had made by then at discussing this idea with students. The format of the present writing – I refer above all to the presence of diary notes – is a development of this and subsequent books and thus is quite properly dedicated to Bob Cohen on this significant occasion, with the author's gratitude and warmest congratulations and wishes.

*Brandeis University*

### NOTES

\* Written for a *Festschrift* for Robert S. Cohen.
[1] Wilhelm Windelband, *Geschichte der Philosophie* (1891), 12th ed., Tübingen: J.C.B. Mohr (Paul Siebeck), 1928.
[2] 'The Sociology of Knowledge and Sociological Theory' (1959), in *Beyond the Sociology of Knowledge: An Introduction and a Development*, Lanham, New York, London: University Press of America, 1983; 'Man's Historicity and Dualism: The Significance of Hannah Arendt's *The Human Condition* for Sociology' (1961), *ibid.*, pp. 199–235 and 67–103.
[3] Judith Feher, 'On Surrender, Death, and the Sociology of Knowledge,' *Human Studies* 7 (1984): 211–226.
[4] Cf. *Surrender and Catch*, p. 33: "As our official consciousness is opposed to

surrender as the relation to the world, so it also has all but lost any meaning of 'total' or 'absolute' except as terror. . . . Total experiences [synonymous with surrenders: *ibid.* 22] and their name which proclaims them oppose to terror . . . an image of man for whom the absolute is not only terror but also home, for whom 'extreme situation' calls forth not only his death but also his greatness and his happiness."

[5] Cf. Zygmunt Bauman, *Modernity and the Holocaust*, Cambridge: Polity Press, 1989.

[6] New Brunswick and London: Transaction Publishers, 1991.

[7] See, e.g., Robert M. MacIver, *Society: A Textbook of Sociology*, New York: Farrar and Rinehart, 1937; Alfred Weber, *Ideen zur Staarts- und Kultursoziologie*, Karlsruhe: Braun, 1927; also Robert K. Merton, 'Civilization and Culture,' *Sociology and Social Research* **21** (1936): 103–113.

[8] Emile Durkheim, 'The Dualism of Human Nature and Its Social Conditions,' trans. Charles Blend, in Kurt H. Wolff (ed.), *Emile Durkheim, 1858–1917: A Collection of Essays, with Translations and a Bibliography*. Columbus: Ohio State University Press, 1960, pp. 325–340; 'Le Dualisme de la nature humaine et ses conditions sociales,' *Scientia* **XV**, No. 34 (1914): 206–221.

[9] Here and in the following pages, the first number in parentheses refers to the page(s) of the English translation of Durkheim's paper; the second, to the page(s) of the original. An elevated x following the first number indicates a slight modification of the translation.

[10] Neither Husserl nor phenomenology (or Feuerbach, for that matter), appear in the indexes of the major study of Durkheim, at least in English: Steven Lukes, *Emile Durkheim, His Life and Work: A Critical Study* (1973), Penguin Books, 1975.

[11] Also see 'Anomie and the Sociology of Knowledge, in Durkheim and Today,' in *Survival and Sociology*, pp. 83–96. For a study of Durkheim in which suffering figures as the pivotal element in Durkheim's work, see Stjepan G. Meštrović. *Emile Durkheim and the Reformation of Sociology*, Totowa, NJ: Rowman & Littlefield, 1988.

[12] 'A Preliminary Inquiry into the Sociology of Knowledge from the Standpoint of the Study of Man' (1953), in *Beyond the Sociology of Knowledge*, p. 184. See also 'Sociology and Meaning' (1992), *Philosophy and Social Criticism* **19**, 3 (1993): 287–292.

[13] 'Man's Historicity and Dualism,' in *Beyond the Sociology of Knowledge*, p. 100.

[14] *Surrender and Catch, passim.*

[15] Albert Camus, *The Rebel* (1954), trans. Anthony Bower, New York: Vintage Books, 1960, p. 284; quoted in *Surrender and Catch*, p. 61.

[16] *Ibid.*

[17] *Surrender and Catch*, p. 79.

[18] *Ibid.*, pp. 172–173.

[19] Recall the subtitle of *Survival and Sociology: Vindicating the Human Subject.*

DANILO ZOLO

# THE TRAGEDY OF POLITICAL SCIENCE

> Today the most urgent and demanding task of political science is to analyse and perhaps to question the very ideology of scientific politics.
> – Norberto Bobbio

## 1. 'POLITICAL SCIENCE' VERSUS 'POLITICAL PHILOSOPHY'

The meaning which the term 'political science' has in western culture today is the particular approach to questions of politics which originated in the 'behaviourist revolution' of the United States in the twenty years following the Second World War. Since that time this approach has been widely diffused not only in the United States itself – where its academic practitioners are now reckoned to number not less than two thousand – but also in Europe, above all in England, Germany and the Scandinavian countries.

In contrast to this specific notion of 'political science', the term 'political philosophy' continues to be used in Europe to indicate the more conventional form of study of political matters in the tradition of the classical masters of western political thought, from Plato and Aristotle to Machiavelli, Hobbes, Locke, and Marx. Unlike 'political science', political philosophy is not restricted to the study of the 'observable' behaviour of social agents and the functioning of (contemporary) political systems, but embraces also the much wider problems of the means, ends, and 'meaning' of politics (over and beyond, at a further level of reflexivity, the means, ends, and 'meaning' of the study of politics itself).

The first purpose of the present study is to trace the theoretical components of the debate centred on the comparison of these two disciplines since the 1940s and then, on the basis of this, to give a detailed assessment of the present position of the relations between these two diverse methods of studying and understanding man's political life – the first closely linked to cultural developments in the United States, the second more directly connected with the classical tradition of European philosophy. We shall find that 'political science', and in particular American 'political science', is now at a point of crisis which shows every sign of being about to threaten its very identity as a discipline. The title of a well-received book published in the United States, *The Tragedy of Political Science*[1] may be taken as an eloquent symbol of that crisis.

In the final section I propose to indicate the general reasons by which a radical reconsideration of the methods and contents of contemporary political thought has, in my opinion, now become necessary. And if it is necessary for traditional political philosophy, often inclined to the purely academic revision of the work of past thinkers or to the reframing of archaic metaphysical models, then it is still more necessary, to my mind, for behaviouristic 'political science'. 'Political science' set out fifty years ago with two distinct objectives, first – its acknowledged aim – of attaining *certain and objective knowledge* of political facts, in as much as its own approach was based, unlike Marxist idealism and historicism, on the empirical analysis of social phenomena; and second – an unacknowledged aim, but one deeply rooted in the subjective motivations of its practitioners – of demonstrating the optimality of (American) democratic institutions, as forming the realisation of liberty, pluralism, and equality of opportunity.[2] Paradoxically 'political science' is the study which finds itself in greater crisis today. For this there are three main reasons: first, the current general uncertainty that exists over the bases of scientific knowledge and, more especially, over the epistemological status of the 'social' sciences; second, the rapid and continuing increase in complexity of the social phenomena which 'political science' aims to explain and to predict empirically; third, and above all, the growing 'evolutionary risks' which threaten democratic institutions within post-industrial societies, including the United States, where the processes of democracy are undergoing the unsettling, even alarming, transformation into 'televisual democracy'.[3]

## 2. FROM THE 'BEHAVIOURISTIC REVOLUTION' TO POST-EMPIRICISM.

In this section a summary division of the relationships of the two theoretical approaches will be attempted, taking account almost entirely of developments within the area covered by the English language. According to this division there are four main phases to be identified, though they naturally overlap to some extent chronologically:

1. The behaviouristic program as set out and established between 1945 and 1965, the most important expositors, especially in the initial period, being Gabriel Almond,[4] David Easton,[5] Heinz Eulau,[6] Robert Dahl,[7] K.W. Deutsch,[8] David B. Truman.[9]

2. The debate over the so-called 'decline of political theory', in the course of which appeared the first reaction, predominantly defensive in character, against behaviouristic political science. Notable authors in this debate at the end of the Fifties and the beginning of the Sixties were P.H. Partridge,[10] I. Berlin,[11] and J.P. Plamenatz.[12] A position of great importance, clearly marked out by its openly anti-modern and conservative stance, was taken by Leo Strauss in his 'ontological' critique, developed in his famous work, 'What is Political Philosophy?'.[13] The introductory pages by Eric Voegelin to his *The New Science of Politics*[14] may also be taken to belong to this category.

3. The crisis of behaviourism, accompanied by the fading of the scientific optimism which had been characteristic of the initial period, and the emergence of a growing uneasiness in the discipline manifested first in attempts at methodological reform inspired by Popperian 'falsificationism', subsequently in the internal criticism voiced from the left by members of the 'Caucus for a New Political Science,[15] and finally in the overt self-criticism which has come from some of the most influential American political scientists, such as Gabriel A. Almond[16] and Charles E. Lindblom.[17] This phase gained particular intensity during the so-called 'decade of disillusion' from 1965 to 1975 and ends with the publication of David Maria Ricci's *The Tragedy of Political Science*,[18] followed a year later by David Easton's article 'Political Science in the United States. Past and Present',[19] both of these works being severe criticisms by political scientists of the entire development of their discipline.

4. The Anglo-American revival of political philosophy in the Seventies by such writers as John Rawls, Robert Nozick, Ronald Dworkin, and Bruce A. Ackerman. This revival abruptly broke with the tradition of Anglo-Saxon analytical philosophy which had pronounced political philosophy dead, and realigned itself with the great evaluative, ethical, and prescriptive themes of classical political philosophy. In conjunction with this, there emerged a more mature form of epistemological writing, found in the works of a number of writers committed both to political philosophy and the philosophy of the social sciences, notably Alasdair MacIntyre,[20] Alan Rvan,[21] Charles Taylor,[22] Sheldon

S. Wolin,[23] and John Dunn.[24] By and large the epistemology which unites these authors may be defined as 'post-empiricist'. They have, however, been deeply influenced by the 'revolt against positivism' of the Seventies, and in some cases have been directly inspired by the epistemology of Thomas Kuhn. In their critique of prevailing political science they start therefore not from a reproposition of the traditional concerns of European political philosophy, but from a general criticism of the empiricist 'received view'. However, unlike the majority of critics belonging to the first phase, they do not dogmatically deny the importance of the contribution offered to political philosophy by sociological investigation into political agents and systems.

In the succeeding pages, rather than provide further illustration of these four phases in the debate between the supporters of political science and their adversaries, I shall attempt to reduce to their essential points the theoretical reasoning of both sides, introducing one single, basic, diachronic dimension – the distinction to be drawn between the behaviouristic program as first formulated and the very much more uncertain and moderate terms in which political science is presented today, in the middle of the Nineties. In the same way, in dealing with the arguments of the critics of 'political science', I shall make a distinction between those arguments adduced in the first defensive reaction in the debate on the 'decline of political theory' and the epistemologically more advanced arguments put forward by present-day supporters of the 'post-empiricist' approach.

### 3. THE ORIGINAL PROGRAMME OF POLITICAL SCIENCE

A useful starting-point for consideration of the original program of behaviouristic political science exists in the precise formulation of it given by David Easton,[25] to which elements may also be added from the more recent analysis contained in Jürgen Falter's historical account of the development of the discipline.[26] Adherence to the 'behaviouristic revolution' implied, according to the belief of its founding fathers, at least the following five assumptions, attached to each of which was a goal which had to be met in order for the results of the research to be considered scientific:

1. *Explanation and prediction on the basis of general laws.* In both the behaviour of political agents and the functioning of political systems there were regular recurrences to be observed. The basic job of the

political scientist was to identify these recurrences and to express them in the form of general laws of a causal or statistical nature which would permit the explanation or future prediction of political phenomena. To this end it was thought that the political scientist should not merely amass data and restrict himself to generalisations about them within limited bounds of space and time. Instead he should select and organise empirical data in the light of wide-ranging theories, in just the same way as was done in natural sciences such as physics and biology.

2. *Empirical verifiability and objectivity.* The validity of nomological generalisations in political science could be ascertained at root by means of empirical verification taking the behaviour of political agents as its reference point. Only by adopting this type of procedure would political scientists be able to claim the status of certain and objective knowledge for their theories and statements, possessing the same degree of (intersubjective) cogency as discoveries in the natural sciences.

3. *Quantification and measurement.* Rigour was also necessary in the recording of data, formulation of results, and carrying-out of tests on political behaviour. The political scientist had therefore to adopt the techniques of quantification and precise measurement which were already in use in the 'exact' sciences and which produced results also in such social sciences as economics and psychology.

4. *Systematicity and cumulativity.* The research carried out by political scientists could be shaped in forms analogous to those already established in the praxis of longer-existing scientific communities. It would have to be performed 'systematically'; that is, there would be constant interaction between a logically structured and consistent language of theory and empirical research governed by a rigorous inductive method. The progressive accumulation of empirical data would permit the gradual development of theories and the formation would thus be achieved of a nucleus of discoveries shared by the whole community of political scientists. In this way it would be possible to engender genuinely professional organisation of political research, overcoming the subjectivism of traditional 'political philosophers' with their never-ending and boundless disagreements.

5. *Avaluativity.* The explanation and empirical prediction of political phenomena should be kept strictly separate from evaluations and prescriptions of an ethical or ideological kind. This was an essential condition of science and one which would provide intersubjective binding of propositions in political science. The political scientist was there-

fore under an intellectual obligation to abstain from any kind of ethical or ideological evaluation in the course of his studies and ought always to make clear what values he was employing every time he abandoned his scientific stance by thinking it opportune to express moral or ideological evaluations of the aims of his research or by attempting to extract prescriptive indications from them. Viewed in this way, political science was diametrically opposed to conventional political philosophy, which had never developed the distinction between factual propositions and value judgments and had been conceived predominantly as a form of 'prudent' and prescriptive thought rather than of objective knowledge.

It will be obvious that the elements of this methodological list, which is intended to express the core of behaviouristic theory, implies the series of generalised philosophical and epistemological options which the standard empiricist view has inherited from Viennese logical positivism and has joined with other elements from the American tradition, such as operationism, pragmatism, and Watson and Skinner's behaviouristic psychology. Central to such options is the decision to take political questions as belonging to the sphere of the empirical sciences on the grounds that no difference exists in principle, at least so far as our ability to know and predict them is concerned, between the 'behaviour' of natural objects and the individual and collective behaviour of human beings.

### 4. THE POLITICAL PHILOSOPHERS' ARGUMENTS

First reactions from the supporters of political philosophy took the form, as we have seen above, of a debate on the 'decline of political theory'. The debate was set in motion by Isaiah Berlin's famous study, 'Does Political Theory Still Exist?', whose principle 'defence' was the claim that a philosophical dimension existed in political thought which no 'science' of a logico-deductive or empirical kind would ever be able to eradicate or disguise, since the problems it dealt with were of neither a logical nor an empirical order; on the contrary these problems involved ideological and philosophical options of a very general kind, as well as ever-present value-choices, such as the question of the justification of political obligation.

This argument was accompanied by the charge that political science had proved itself quite incapable of constructing a 'theory' of any practical significance from the standpoint of what actually happened in 'politics' or of any relevance for those personally involved as politicians. Behaviourists had failed to come up with a theory capable of 'replacing' political philosophy or, even less radically, of relegating it to the metalinguistic position of providing analysis and clarification of the language used in the study of politics.

In addition, the preoccupation of political science with the analysis of empirical behaviour and facts (which it appeared to assume as the only domain of political study) left wholly out of account any consideration of what politics were for, as well as ignoring the reasons that made legitimate (or illegitimate) the exercise of power – topics which, from Aristotle onwards, had been at the very centre of Western political thought. A 'science' which sacrificed any discussion of the 'values' of politics on the altar of methodological rigour and involved itself exclusively with the 'facts' abandoned all claim to be able to state, let alone advance or solve, *the problems of politics*, in as much as these problems could hardly not involve some decision on the aims, limitations, or meaning of politics. In times of crisis or of accelerated change in political systems or of turbulence in the ideologies which are active upon them, the 'neutral' political scientist could only find himself condemned to a position of intellectual impotence. The ambitious attempt to imitate the model of the natural sciences imposed on political science such elevated standards of procedural rigour that they were themselves the simple cause of its methodological obsession and, at the same time, of its frustrations over the paucity or minimal importance of the results achieved.

Considerably different arguments have been advanced by the critics of political science referred to above as 'post-empiricist' and influenced by the crisis of the standard empiricist view in England and America. They have no qualms in referring to political science as a 'putative science' and contest not so much its results as the epistemological assumptions themselves by which it is claimed to be a 'science' within the framework of modern 'social sciences' and which contrast it, in epistemological terms, with political philosophy. Whatever judgment one wishes to pass on the results of 'political science' – and this may even, hypothetically, be largely positive – it remains, according to their view, quite untenable to claim that 'political science' attains its results by being a 'science' and by adhering to its own epistemological premises, when in fact the results are attained precisely in proportion to the extent in which it either departs from those premises or applies them in a purely metaphorical, even rhetorical, fashion. The *de facto* paradigm of political science is not the hyper-rationalistic one claimed for it by its methodology but is rather what Lindblom has called 'muddling through', or the wriggling out of difficulties with the least damage sustained, through the application of pragmatic techniques to the solution of problems step by step and case by case without any overall cognitive strategy being applied at all.[27]

If this may be considered as the essence of the new polemic against 'political science', various closely-linked arguments develop from it. They may be expressed schematically by the following five points which, either directly or indirectly, challenge the five original assumptions behind behaviouristic political science which we examined above.

I. The possibility of making a sufficiently full record (and over a sufficiently long period) of recurrences in the behaviour of political agents and in the functioning of political systems simply does not obtain. Political science does not have the capacity to develop general laws of a causal or statistical kind which permit explanations, still less predictions of a nomological and deductive type, to be made. The reason why it can neither explain nor predict is not some temporary state of immaturity of the discipline or of a present lack of technical development, but for reasons basic to its theory. They are the same reasons why nomological and deductive explanation and the prediction of 'single events' become so very difficult in the physical, chemical, and biological sciences.[28] In addition, the social sciences suffer from such other more specific problems as the high degree of unpredictability in the behaviour of individuals, the increasing complexity of social relations, and the non-linear, reflexive, nature of societal functions, especially those to do with power.[29] Even in the physical sciences, according to post-empiricist epistemology, the formulation of universal and invariable laws which take no regard of historical and evolutionary considerations, is simply not possible.

II. The validity of nomological generalisations in political science – no less and perhaps even to a greater extent than in all other social or 'natural' sciences – is not susceptible to empirical verification (or, *à la* Popper, falsification), provided that these terms are to be used in a more than purely metaphorical sense. For the 'facts' against which the explanations and predictions have to be rigorously verified (or 'falsified') are in practice themselves the result of choices determined by the methodological dictates of some preceding philosophy or other. The confirmations are only as good as the theories behind them, and stand or fall in conjunction with those theories, as has often been seen in physics with theories supported at first by empirical tests but then falling into disuse, such as the theory of phlogiston or the theory of ether. In other words there is no such thing as an 'observation language' which can be kept rigorously distinct from the language of the theories. Theories are always, to some degree, linked with general philosophies or *Weltanschauungen*, with all the historical and sociological conditioning that these imply. Accordingly there can be no firm base whatsoever for the idea that the testing of theories, in political as in other sciences, consists in the checking of their 'correspondence' to the 'facts'.

Nor, on the other hand, can the so-called 'comparative method' often claimed by political scientists, such as Giovanni Sartori,[30] as specific to scientific research on politics be taken in any way to be a 'method of testing' or even, more generally, as a method at all. It is merely the operation of evaluation and selection of data which is bound to be involved in a theory's initial development under even the most elementary inductive technique.[31]

III. Very little scope exists in the sociology of political behaviour for measurement and quantification. The sole exception to this is perhaps the analysis of election results (which has come wrongly to be described as 'observation' of electoral 'behaviour', when in practice it involves no 'observable' social behaviour but only quantitative aspects of ritualised social procedures). The use of quantitative techniques or of any form of measurement worthy of the name is made totally redundant by the impossibility of our attributing political significance to social behaviour while paying no attention to the 'motivation' of the agents, with all their ideologies and with the purposes – whether declared, latent, or pretended – of their 'political action'.[32]

IV. Political science has never succeeded in its purpose of 'accumulating' a nucleus of unanimously agreed theories, in order to form an undisputed stock-in-trade of the discipline. The original (ingenuously inductive) aim of accumulating cognitive data by advancing empirical research into highly individualised or marginal areas of politics ('case studies') has given rise to the famous 'hyperfactual' distortions so typical of the provincialism of American political science. The strictures of C. Wright Mills on this score are well known, and similar criticism has arisen also within the discipline, especially in some of the more important works by David Easton.

Equally ingenuous is the attempt to give conceptual unification to the vocabulary of political theory, as has been attempted for some time now by Giovanni Sartori. For this purpose he has founded at the University of Pittsburgh the controversial 'Committee on conceptual and terminological analysis' (COCTA). But, as the members of this committee in fact admit, the semantics of modern political science resemble nothing so much as the 'tower of Babel'.[33] What these attempts always seem to leave out of account is the fact that the (necessarily imprecise, subjective, and conventional) metaphorical element of the language of theory can never be eliminated and that it is precisely in this element that, to a large degree, the representative and informative capacity – as well as the heuristic richness – of concepts and theories resides.

V. The aim of avaluativity generally turns out to be impracticable in the social sciences and especially in the study of politics. As soon as a researcher passes from the elementary level of classification of data to the development of non-trivial theories, i.e. theories of sufficient complexity to be applied effectively to political experience, it becomes inevitable that he orients himself, whether consciously or not, in accordance with value-choices of a political, ethical, or ideological nature.[34] Analysis, in particular of the relationships of power, cannot separate itself from the influence which existing relationships of power exercise reflexively on the social, economic, and cognitive premises of the research itself. The element of evaluation cannot in practical terms be identified and removed from theories, so long as premises of value lie concealed behind them and so long as such premises influence the perception itself of phenomena and the very selection and posing of problems. In all these cases there is no criterion for applying to the language of theory the healing balm of Weber's *Wertfreiheit*.

These positions all clearly point to the conclusion that no form of 'political science' can be significantly distinguished either, at one end, from the sociology of politics or, at the other, from political philosophy. The difference is a simple one of degree and thematic preference.[35] It remains that a 'post-empiricist' political theory would need to include all of the following elements: analysis of present conditions; reconstruction of past political thought; consideration of the values and goals of politics; epistemological meta-argument on the methods and procedures of political research.

## 5. THE 'TRAGEDY' OF AMERICAN POLITICAL SCIENCE

In using the somewhat emotive term 'tragedy', my purpose, in common with David Ricci, is to call attention to the situation of acute unease prevailing today in American political science after many of its expositors, including some of the most influential, such as Gabrie A. Almond and David Easton, have severely criticised not only the original program of political behaviourism but also the subsequent development of the discipline. American political science, Ricci observes, seems incapable of producing an effective form of 'political knowledge' because of the obligation it places itself under to reach a fixed and absolutely precise – in a word, 'scientific' – knowledge of political life. This obligation to reach an unreachable 'scientific' knowledge drives the political scientist away from the political topics, such as the crisis of democratic institutions, which are crucial to the society in which he lives. These topics cannot be properly handled by one whose profession commits him to political neutrality. Political science runs the risk therefore of a 'tragic' self-negation by virtue of being a 'politically disinterested' science.

Almond and Easton both recognise not only the unsuitability but also the basic impossibility of remaining faithful to the requirements of the behaviouristic program. Almond rejects the idea that political science should follow the path of imitation of the natural sciences, which he calls a 'flirtation with erroneous metaphors'. He finds that the nomological and deductive model, with its implicit assumption of determinism and causalism, has no value whatsoever for the explanation and prediction of social phenomena, while logico-mathematical axiomatisations are also, to his mind, of little effect, for the reason that their formal rigour brings with it a self-defeating simplicity which renders them inadequate in the face of the complexity of political phenomena. In place of these he advises the use of 'weak' heuristic theories which make no claim to legitimise themselves on the basis of their explanatory and predictive power, but instead confine themselves to the 'interpretation' and 'understanding' of politics as a 'process of adaptation and attainment of goals' within bounded decisional frameworks.[36]

Easton is still more radical. In his thorough review of the history of political science in the United States, he links the outcome of the discipline (with its assertion of the ideological neutrality of the political scientist) to its adherence to the myth of the 'end of ideologies', a myth which, in his view, has served in practice to conceal the domination of the democratic-conservative ideology. He maintains forthrightly that political science in the United States has in fact shaped itself to the climate of persecution of liberals and dissidents initiated by McCarthyism in the Fifties, because this has led the discipline to legitimise at the theoretical level a disinterest in actual social problems and political criticism, and so to offer to students of politics a haven in which they could shelter themselves from the dangers of actual political and ideological engagement. He attributes the failure of behaviouristic political science to such things as an underestimation of the very real transformations which have occurred in American society, an inability to make social prediction, scant regard for the historical dimension, reliance on a dogmatic conception of the 'scientific method' drawn from neopositivism, and a touching faith in the evaluative neutrality of science.

In Easton's view, American political science, following the crisis of behaviorism, now lacks any common aim or standpoint, and is devoid of cognitive tautness and imagination, having reached a particularly crucial point of uncertainty over its own identity. As a solution to this, Easton, like Almond, proposes abandonment of the original premises of behaviourism on account of their connection with the positivistic view of science which has proved itself to be unsupportable. Viewed from the position of epistemological standards, political analysis should be satisfied with achieving the formulation of plausible, even if not 'rigorous', reasons for political behaviour, together with a success at 'comprehending' phenomena as a result of careful attention to empirical data but not as a result of some pretence that this comprehension is based on the crucial element of verification or falsification of such data.[37]

## 6. FOR NEW RELATIONSHIPS BETWEEN POLITICAL SCIENCE AND POLITICAL PHILOSOPHY: AN ITALIAN WAY?

The difficulties identified above in North American 'political science' are to be seen also in the epistemologically rigid form of political science introduced to Italy by Giovanni Sartori in the Sixties. The aim of Sartori and his disciples has been to present political science as the only testable and trustworthy form of political knowledge. This has been accompanied also by strong ideological polemic on their part against all 'holistic' conceptions of politics. This polemic too they have naturally claimed to be purely scientific. But it is important to recognise that, behind the 'cognitive' motivation of such a form of 'political science', there lies the straight decision they have made in favour of an elitist conception of democracy and their violent aversion to any properly democratic (and not strictly liberal) conceptions of it, such as those represented by radical democratic, socialist, and Marxist doctrines.

Today the time seems ripe in Italy also for taking stock of the bases and the 'achievement' of 'political science' and above all for reopening the question about what has in truth become for it a shirt of Nessus – the positivist dogma of the separation of 'judgments of fact' from 'judgments of value' and the associated principle of the ethical and ideological 'avaluativity' (*Wertfreiheit*) of scientific theories. As Norberto Bobbio has pointed out, this is a doctrine which implies a particular ideology, that of 'scientific politics', i.e. political and social relations rationalised along technocratic lines so that facts form the sanction for the 'end of ideologies'.[38]

On the other hand Italy has also seen in these years a notable revival in political theory, of which the following indications may be cited: the journal *Teoria politica*, edited by members of the school of Norberto Bobbio; the appearance of the first Italian journal actually to bear the title *Filosofia politica*, edited, under the direction of Nicola Matteucci, by a group of historians of political philosophy who see their task as 'critical and hermeneutic investigation into the tradition of western political thought'; the spread of political-philosophical and political-historical writing which aims to challenge not only the claims to hegemony but even the legitimacy itself of 'political science' practised in the mode of Sartori. Here the theses of contemporary German neo-Aristotelianism (the so-called *Rehabilitierung der praktischen Philosophie*) are well represented, and re-examination is taking place of the western democratic tradition in the light of such authors as Carl Schmitt, Eric Voegelin, Leo Strauss, and Hannah Arendt.

How, then, are Italian students of political science reacting to the situation of crisis which is currently affecting their subject in its country of origin and to the re-evaluation now being carried out, especially in Europe, of the cognitive aims of political philosophy?

Satori maintains that political science in Italy has always been immune to the failings and excesses of American political science, that it has never been either properly behaviouristic or properly positivist, with the result that it is today in a position of advantage over the United States, especially with regard to comparative political studies. But he adheres all the same to the idea that political science, unlike political philosophy (which he condemns for producing no 'testable' knowledge), should respect 'the methodological canons of empirical knowledge'. And these he identifies, even now, and with no further qualification, as logical rigour in definitions, observability of phenomena, empirical verifiability of theories, and the cumulativity of scientific knowledge.[39]

Sartori's opinion, however, influential though it has been, appears now to be in something of a minority. Recent views expressed by Luigi Graziano,[40] Domenico Fisichella,[41] and above all Gianfranco Pasquino[42] show, when compared with Sartori's, a quite different reaction to the situation of crisis surrounding the 'bases' of their subject, and far less optimism over the cognitive results so far achieved. This reaction is often indirect and not consciously stated, but it may be perceived for the most part in their attempts to reach epistemological compromises, in a number of uncertain formulations, and in their decision to put to one side many of the most pressing questions, as may be seen especially in the *Manuale di scienza della politica*, edited by Pasquino himself with the help of several of the most respected members of the second generation of Italian political scientists, who now seem at some pains to distance themselves from the position maintained by Sartori.

The *Manuale* gives an important indication of the present state of the discipline in Italy. What emerges from it more than anything else is, despite the promise of its title, the weakness of its methodological approach and the glaring deficiencies it reveals in the discipline. It amounts in effect to no more than a collection of sometimes excellent but hardly homogeneous essays on individual topics, linked only by their adherence – sometimes implicit, sometimes 'obvious' – to the general outline of western democratic values. The trend of the articles is predominantly historical-political and philosophical-political, and their 'factual' analyses are constantly intermingled with a great many value judgments. Scarcely any of the articles makes mention of, still less applies, the classical canons of behaviourism, once the semantically rhetorical use of such terms as 'scientificity', 'observation', 'measurement', 'empirical test' etc. is excluded.

In this respect it is significant to find Pasquino, in the final pages of his Introduction, strongly expressing the need for political science to redefine itself afresh against political philosophy, whose rich complexity of themes needs to be taken into account in order for any genuine comparison of two disciplines to take place above the level of academic rivalry and territorial expansion. Pasquino's hope is that this interaction between political scientists and political philosophers will result in a new 'theory of politics', one able to cope with the accelerated and more complex demands of politics in the world today.

The views expressed by Pasquino form, to my mind, an interesting commentary on the crisis of 'political science' in Europe. For, in common with the post-empiricists, I find it difficult to draw a strict dividing-line between the theoretical and linguistic natures of 'political science' and 'political philosophy'. There is no definite, and still less definitive, epistemological stance to be seen in the social sciences, and especially in political science today. Our understanding of society lacks any precise bounds or foundations. Consequently we all find ourselves in the position of the sailors in Neurath's metaphor, who are forced to repair and restructure their ship in mid-ocean, while supporting themselves on the old structure without the ability to put in to port and rebuild it anew. We can none of us escape this situation of circularity.

It is important to remember that the dialogue between political scientists and political philosophers need not proceed in purely academic forms but can produce results also from the political standpoint. For this reason it seems necessary to me that each of the two disciplines should have the courage to face up to its history and release itself from the hold of its respective traditions. Both need to concern themselves more with the 'problems' than with the 'facts' of politics, not to mention mere questions of method or the paying of ritualised obeisance to the classical masters of political thought. Rather than restrict themselves to the problem of defining the boundaries between themselves, they should recover an appreciation of, and a concern for, the great social and political questions of our time. Among these I would include: the fate of democracy in post-industrial societies now dominated by modern forms of technology; the increasing power man is reflexively gaining over his environment and even over his own genetic and anthropological identity; the unrestrained violence of international relations; the economic chasm separating the peoples of the post-industrial societies from the remainder of the world; the problems of ecology, energy, and demography now emerging as a danger of epic proportions to the world in the immediately approaching decades.

As for political philosophy, there are certain leading aspects of its 'Old European' tradition on which it now needs to turn its back. These include its generic humanism, its moralism, its predilection for designing models of the 'ideal stage', its propensity towards the simplications of Messianic politics, its distaste for precise and unbiased analysis of phenomena. There is no room today, amid the complexities of modern society, for a political philosophy which sets out to 'rehabilitate' or retraverse the old paths of Aristotelian metaphysics or dogmatic theology. This is something not yet fully realised by those Italian and German political philosophers who now, having brilliantly exposed the failings of the 'modern' code of popular-democratic certainties, return nostalgically, in the company of Carl Schmitt, Leo Strauss, and Eric Voegelin, to the old tradition of theology and metaphysics with its assemblage of ontological ingenuities, moral dogmatism, and hierarchical conceptions of politics. Nor does it seem at all likely that we can recover a morality based on natural law, in either its utilitarian (Harsanyi) or contractualistic (Rawls) forms, which now seem little more than elementary schemes for the justification of existing economic and political conditions. The growing complexity of social phenomena, if nothing else, has rendered defence of such schemes impracticable.

Political science, for its part, needs now to set about freeing itself from its obsession with methodology, the pretensions it entertains to scientific standards, its unrealisable aspirations towards evaluative neutrality, and its disdain for consideration of social history and change. But it should by no means forsake the lesson of rigour and conceptual clarity which it has taught nor fail in its function of pursuing 'empirical' research on politics, so long as this signifies the provision, without any positivist bias, of information on, and the comparative study of, contemporary political systems. For without this there will never be constructed any 'political theory' worthy of the name.

[*Translated from the Italian by David McKie*]

*University of Florence,*
*Dept. of Theory and History of Law*

### NOTES

[1] D.M. Ricci, 1984, *The Tragedy of Political Science*, New Haven: Yale University Press.
[2] R.A. Dahl, 1956, *A Preface to Democratic Theory*, Chicago: The University of Chicago Press.
[3] T. Luke, 1986–87, 'Televisual Democracy and the Politics of Charisma', *Telos* **70**: 59–79.

[4] G.A. Almond, 1966, 'Political Theory and Political Science', *American Political Science Review* **4**.
[5] D. Easton, 1962, 'The Current Meaning of "Behavioralism"', in G.C. Charlesworth (ed.), *The Limits of Behavioralism in Political Science*, Philadelphia: The American Academy of Political and Social Sciences.
[6] H. Eulau, 1963, *The Behavioral Persuasion in Politics*, New York: Random House.
[7] R.A. Dahl, 1961, 'The Behavioral Approach in Political Science: Epitaph for a Monument to a Successful Protest', *American Political Science Review*, **55**.
[8] K. Deutsch, 1966, 'Recent Trends in Research Methods', in J. Charlesworth (ed.), *A Design for Political Science: Scope, Objectives, and Methods*, Philadelphia: The American Academy of Political and Social Sciences.
[9] D.B. Truman, 1951, 'The Implications of Political Behavior Research', *Items*, December.
[10] P.H. Partridge, 1961, 'Politics, Philosophy, Ideology', *Political Studies* **9**: 217–235.
[11] I. Berlin, 1969, 'Does Political Theory Still Exist?', in P. Laslett and W.G. Runciman (eds.), *Philosophy, Politics and Society*, 2nd Series, Oxford: Blackwell.
[12] J. Plamenatz, 1984, 'The Use of Political Theory', in A. Quinton (ed.), *Political Philosophy*, Oxford: Oxford University Press, pp. 19–31.
[13] L. Strauss, 1959, 'What is Political Philosophy?', in L. Strauss, *What is Political Philosophy and Other Studies*, Glencoe, Ill.: The Free Press.
[14] E. Voegelin, 1952, *The New Science of Politics*, Chicago: The University of Chicago Press.
[15] J.F. Falter, *Der 'Positivismusstreit' in der amerikanischen Politikwissenschaft* (Opladen: Westdeutscher Verlag, 1982), pp. 53–62; P.J. Euben, 'Political Science and Political Silence', in P. Green and S. Levinson (eds.), *Power and Community. Dissenting Essays in Political Science* (New York: Random Housem, 1970), pp. 3–58.
[16] G.A. Almond and S.J. Genco, 1977, 'Clouds, Clocks, and the Study of Politics', *World Politics*.
[17] C.E. Lindblom, 1979, 'Still Muddling, not yet Trough', *Public Administration Review*, November–December.
[18] D.M. Ricci, op. cit.
[19] D. Easton, 1985, 'Political Science in the United States. Past and Present', *International Political Science Review* **6**: 1.
[20] A. MacIntyre, 'Is a Science of Comparative Politics Possible?', in P. Laslett, W.G. Runciman and Q. Skinner (eds.), *Philosophy, Politics and Society*, 4th Series (Oxford: Basil Blackwell, 1972), pp. 8–26; A. MacIntyre, 'The Indispensability of Political Theory', in D. Miller and L. Siedentop (eds.), *The Nature of Political Theory* (Oxford: Clarendon Press, 1983).
[21] A. Ryan, 1972, ' "Normal" Science or Ideology?', in P. Laslett, W. G. Runciman and Q. Skinner (eds.), *Philosophy, Politics and Society*, Oxford: Basil Blackwell, pp. 86–100.
[22] C. Taylor, 'Neutrality in Political Science', in P. Laslett and W.G. Runciman (eds.), *Philosophy, Politics and Society*, 3rd Series (Oxford: Basil Blackwell, 1967), pp. 25–57; C. Taylor, 'Political Theory and Practice', in C. Lloyd (ed.), *Social Theory and Political Practice* (Oxford: Clarendon Press, 1983).
[23] S.S. Wolin, 1969, 'Political Theory as a Vocation', *American Political Science Review* **63**.

[24] J. Dunn, 1985, *Rethinking Modern Political Theory*, Cambridge: Cambridge University Press.
[25] D. Easton, 'The Current Meaning of 'Behavioralism', op. cit.
[26] J.F. Falter, op. cit.
[27] J. Hayward, 1986, 'The Political Science of Muddling Through: The *de facto* Paradigm?', in J. Hayward and P. Norton (eds.), *The Political Science of British Politics*, Brighton: Wheatsheaf Books, pp. 3–20.
[28] D. Zolo, *Reflexive Epistemology* (Dordrecht-Boston-London: Kluwer Publishers, 1989).
[29] N. Luhmann, *Macht*, (Stuttgart: Enke Verlag, 1975), English trans. in N. Luhmann, *Trust and Power* (Chichester: Wileym, 1979).
[30] G. Sartori, 1985, 'La scienza politica', *Mondoperaio* **38** (11): 114.
[31] A. MacIntyre, 1983, 'The Indispensability of Political Theory', op. cit., pp. 8–26; N. Bobbio, 'Scienza politica', in N. Bobbio, N. Matteucci and G. Pasquino (eds.), *Dizionario di politica*, Torino: Utet, p. 1023.
[32] N. Bobbio, op. cit., p. 1025.
[33] G. Sartori, 1975, 'The Tower of Babel', in G. Sartori, F. Riggs and H. Tuene, *Tower of Babel*, Pittsburgh: International Study Association.
[34] C. Taylor, 'Neutrality in Political Science', op. cit., pp. 25–57.
[35] D. Zolo, 1987, *Complessità e democrazia*, Torino: Giappichelli, pp. 157–183.
[36] G. A. Almond and S.J. Genco S.J., op. cit.
[37] D. Easton, 'Political Science in the United States. Past and Present', op. cit.
[38] N. Bobbio, op. cit., pp. 1025–1026.
[39] G. Sartori, op. cit., p. 118.
[40] L. Graziano, 1984, 'Vecchi e nuovi concetti e paradigmi', in *La scienza politica in Italia: materiali per un bilancio*, Milano: Angeli, pp. 7–13.
[41] D. Fisichella (ed.), 1985, *Metodo scientifico e ricerca politica*, Roma: La Nuova Italia Editrice, pp. 11–80.
[42] G. Pasquino, 1986, 'Natura ed evoluzione della disciplina', in G. Pasquino (ed.), *Manuale di scienza della politica*, Bologna: Il Mulino, pp. 13–37.

SELECT BIBLIOGRAPHY

G.A. Almond and S.J. Genco, 1977, 'Clouds, Clocks, and the Study of Politics', *World Politics* **29**: 4.
I. Berlin, 1962, 'Does Political Theory Still Exist?', in P. Laslett and W.G. Runciman (eds.), *Philosophy, Politics and Society*. Oxford: Blackwell.
N. Bobbio, 1983, 'Scienza politica', in N. Bobbio, N. Matteucci and G. Pasquino (eds.), *Dizionario di politica*. Torino: Utet.
J. Buchanan and G. Tullock, 1962, *The Calculus of Consent. Logical foundations of Constitutional Democracy*. Ann Arbor: University of Michigan Press.
R.A. Dahl, 1961, 'The Behavioral Approach in Political Science: Epitaph for a Monument to a Successful Protest', *American Political Science Review* **55**: 4.
K. Deutsch, 1966, 'Recent Trends in Research Methods', in J. Charlesworth (ed.), *A Design for Political Science: Scope, Objectives, and Methods*, Philadelphia: The American Academy of Political and Social Science.
A. Downs., 1957, *An Economic Theory of Democracy*, New York: Harper & Row.
D. Easton, 1962, 'The Current Meaning of "Behavioralism"', in G.C. Charlesworth (ed.),

*The Limits of Behavioralism in Political Science*. Philadelphia: The American Academy of Political and Social Science.

———, 1985, 'Political Science in the United States. Past and Present', *International Political Science Review* **6**: 1.

P.J. Euben, 1970, 'Political Science and Political Silence', in P. Green and S. Levinson (eds.), *Power and Community. Dissenting Essays in Political Science*. New York: Random House.

H. Eulau, 1963, *The Behavioral Persuasion in Politics*: New York: Random House.

J.F. Falter, 1962, *Der 'Positivismusstreit' in der amerikanischen Politikwissenschaft*, Opladen: Westdeutscher Verlag.

J. Hayward, 1986, 'The Political Science of Muddling Through: The *de facto* Paradigm?', in J. Hayward and P. Norton (eds.), *The Political Science of British Politics*. Brighton: Wheatsheaf Books.

P. Laslett, 1956, 'Introduction', in P. Laslett (ed.), *Philosophy, Politics and Society* Oxford: Basil Blackwell.

D. Lerner and H.D. Lasswell, 1951, (eds.), *The Policy Sciences: Recent Developments in Scope and Method*. Stanford: Stanford University Press.

C.E. Lindblom, 1979, 'Still Muddling, not yet Through', *Public Administration Review* **39**: 6.

A. MacIntyre, 1972, 'Is a Science of Comparative Politics Possible?', in P. Laslett, W.G. Runciman and Q. Skinner (eds.), *Philosophy, Politics and Society*. Oxford: Basil Blackwell.

———, 1983, 'The Indispensability of Political Theory', in D. Miller and L. Siedentop (eds.), *The Nature of Political Theory*. Oxford: Clarendon Press.

P.H. Partridge, 1961, 'Politics, Philosophy, Ideology', *Political Studies* **9**: 3.

G. Pasquino, 1986, 'Natura ed evoluzione della disciplina', in G. Pasquino (ed.), *Manuale di scienza della politica*. Bologna: Il Mulino.

J. Plamenatz, 1967, 'The Use of Political Theory', in A. Quinton (ed.), *Political philosophy*. Oxford: Oxford University Press.

D.M. Ricci, 1984, *The Tragedy of Political Science*. New Haven: Yale University Press.

A. Ryan, 1972, '"Normal" Science or Ideology?', in P. Laslett, W.G. Runciman and Q. Skinner (eds.), *Philosophy, Politics and Society*. Oxford: Basil Blackwell.

G. Sartori, 1975, 'The Tower of Babel', in G. Sartori, F.W. Riggs and H. Tuene, *Tower of Babel*. Pittsburgh: International Study Association.

L. Strauss, 1959, 'What is Political Philosophy?', in L. Strauss (ed.), *What is Political Philosophy and Other Studies*. Glencoe, Il.: The Free Press.

C. Taylor, 1967, 'Neutrality in Political Science', in P. Laslett and W.G. Runciman (eds.), *Philosophy, Politics and Society*. New York: Barnes & Noble.

———, 1983, 'Political Theory and Practice', in C. Lloyd (ed.), *Social Theory and Political Practice*. Oxford: Clarendon Press.

E. Voegelin, 1952, *The New Science of Politics*. Chicago: The University of Chicago Press.

S.S. Wolin, 1969, 'Political Theory as a Vocation', *American Political Science Review* **63**: 4.

D. Zolo, 1988, 'Theoretical Language, Evaluations and Prescriptions: A Post-Empiricist Approach', in E. Pattaro (ed.), *Reason in Law*. Milano: Guiffrè.

———, 1989, *Reflexive Epistemology*. Dordrecht, Boston, London: Kluwer Academic Publishers.

———, 1992, *Democracy and Complexity*, Cambridge: Polity Press.

RICCA EDMONDSON

# REASONING IN THE SOCIAL WORLD: PROLEGOMENON TO A SOCIOLOGY OF ARGUMENT

## 1. THE CAPACITY TO REASON VERSUS THE VARIABILITY OF OPINIONS: WORLDS OF ARGUING IN DIFFERENT PARTS OF THE SOCIAL SPHERE

Argumentation is, fortunately, endemic to much human interaction, and this essay introduces an attempt to regard this practice both as a reasonable activity and as a social activity. This approach is intended to allow us to deal creatively with variously stated problems raised by the practical difficulty of co-operating and agreeing in the social world with people who may be formally classified as reasonable beings, but among whom reasonable consensus can be hard or impossible to reach.

Writing in the seventeenth century, the Anglican divine Jeremy Taylor poignantly expresses some of the frustration which can be connected with arguing for one's point of view:

> That which will demonstrate a truth to one person, possibly will never move another. Because our reason does not consist in a mathematical point: and the heart of reason, that vital and most sensible part, in which only it can be conquered fairly, is an ambulatory essence, and not fixed; it wanders up and down like a floating island, or like that which we call the life-blood; and it is not often very easy to hit that white by which only our reason is brought to perfect assent: and this needs no other proof but our daily experience, and common notices of things.[1]

Experiences similar to this have made it fashionable in our own time to embrace relativism, to become disgusted with what are perceived as attempts to impose conformity on myriad cultures and traditions, to conclude that there is no hope of convincing everyone of anything, and quite possibly no need either. Yet if we look at some of the potentialities for misunderstanding which daily life offers, we can still contend (as Taylor would have done) that these potentialities do not allow us to infer that there is no real possibility of reasonable argument, or that 'all reasoning is relative'. I shall suggest that discovering more about the social roots of practical reasoning can strengthen rather than curtail recognition of the power of reason. Rather than discussing reasonableness or rationality prima facie, that is, it is instructive to look at what people are actually doing when they successfully or unsuccessfully attempt reasonable debate. We need to discover more about practices and mechanisms connected with how reasoning functions, or fails to function, in different social environments. Such an approach offers the possibility of reconciling the notion of reasonableness with the recognition that there are various methods of seeking it and no once-and-for-all discovery of reasonable answers. It can allow us to acquire grounded knowledge about what sorts of reasonable position can be communicated between social settings, and how. Much later in the investigation, we may be able to say more about what reason actually 'is'.

In order to examine reasoning in practice I should like to make two initial claims. First, in any situation of practical human action, reasoning is partly constituted by phenomena which have only rarely been conceptualised as belonging to sensible argument at all. For example, many constructive sources of common understanding are of a socio-emotional nature and have been neglected as constituents of reasonable discourse (see below). Secondly, it is important to attend to the social and cultural contexts of argumentative interactions, which take place within more or less tightly defined interactive settings. Within these settings, particular combinations of elements are accepted as conventional, comprising separate argumentative 'worlds' – between which it may or may not be difficult to move, according to case.[2] Crucially, these worlds are distinguished in terms of underlying legitimative forms which make arguments appear to function and interact reasonably from the points of view of inhabitants – who are nonetheless seldom able to put these legitimative forms into words. In order to understand such factors, we need a sociology of argument, able not only to account for misunderstanding but to analyse understanding when it does occur, and chart the reasons for it. Such a sociology of argument could make the modes of arguing within such worlds more transparent to the inhabitants of others, and help to clarify different types of justification for what are claimed to be reasonable positions. The possibility of achieving some such understanding has the potential of contributing to diminishing the disharmony involved in daily interaction.[3] It will at least offer a more realistic account of what reasoned argument is.

In the following sections, I shall examine, first, some aspects of misunderstanding between contemporary worlds; secondly, attitudes to arguing in other historical periods, in which the existence of different worlds of argument was also experienced; thirdly, a case which offers some hope for negotiation between worlds. We can then go on to investigate some of the components of arguments which can contribute to their use as reasonable interventions in the settings in which they are used.

### 1a. *Worlds of Argument: Possibilities for Misunderstanding*

Travelling from place to place, and engaging in conversation and other forms of life with different groups of people, it is hard not to be struck by the fluidity of what it is accepted as reasonable to do, say or feel in different sociopolitical situations. It must be partly for this reason that many people fear all new social settings as a matter of course, and that their anxieties may be worth taking seriously. It is not just that identifiable items of social behaviour, such as table manners or queuing habits, are different in new social worlds. In new circumstances – new jobs, new towns, new parties even – not only are some conventions of thought, discourse and behaviour likely to contrast with those one knows, but one cannot know in advance which will differ, or where to look for the patterns which will make the new conventions seem coherent.

This difficulty is compounded by the fact that one cannot generally discover local conventions structuring argument just by asking. In a new culture, not only may you be ignorant of what tentative arguments, questions and suggestions, about education, say, are regarded as acceptable by the teacher of your child; no-one will be able to tell you. There simply will not exist a worked-out meta-account of the local interpretations attached to interaction between parents and teachers. It will not be possible for any person, indigenous or not, to say explicitly and exactly how far to go before barriers of respect, intimacy and so forth are reached or broken. It is not just that principles defining local attitudes do not need to be put into words by their users; it is not only principles which are involved. Practice runs along lines of partially submerged habits, assumptions, feelings about what is appropriate and what is not. Such heterogeneous configurations can precisely not be termed 'social rules', although their force is sometimes analogous with that of rules. They are not particularly clear or consistent, they are composed of attractions, aversions and tendencies too various to class as injunctions, and they cannot be plainly stated.[4] Here we should recall the memorable insight[5] that the most effective source of information about the parameters of new social arenas comes when you offend against them, and it is often only possible to tell that you have offended when you are duly punished.

This punishment may take extreme form, which illustrates the seriousness of argumentative breaks between social 'worlds'. The 'worlds' which I claim are made up of conventions, habits, emotional attitudes, points of view, some of which I shall try to explicate below, make certain positions seem reasonable and others not in a way which may have heavy significance – in courts of law, for instance. A case in point is the *cause célèbre* of the Birmingham Six, men who were imprisoned in England for sixteen years for a bombing which it was eventually accepted that they had not committed. The defendants were unable to convince the court that their own accounts of how they had spent the day in question were reasonable at all. It was hence their different 'worlds' of reasoning which helped to convict them, as a newspaper article on one of their appeals may be read to indicate.

Of the 17 people imprisoned on foot of the Birmingham, the Guildford, and the Maguire cases, 11 went to the same school on the Lower Falls.
   This . . . indicates how specific and local their culture is. This culture comes into conflict with other cultures when the prisoners stand in the dock of English courts.
   . . . Last Thursday, Mr Judge, QC, was scrutinising the men's way of life quite directly, as he traced their movements through their last day of freedom, the day on the night of which the bombs went off.
   . . . Again and again he invited the bench to look at the men's own accounts of how they had spent that day, and to join him in finding them frankly incredible.
   Mr Judge is a graduate of Magdalene College, Cambridge. His distinguished career includes some years as prosecuting counsel to the Inland Revenue, which the precision of his manner no doubt reflects. His recreations include music and cricket, and he lives in a very beautiful part of rural England.
   He was addressing the Lord Chief Justice, educated at Shrewsbury and Trinity College, Cambridge, the Right Honourable Sir Patrick O'Connor, Downside and Merton College, Oxford, recreation, golf, and the Right Honourable Lord Justice Stephen Brown, Queen's College, Cambridge, recreation, sailing, clubs, the Garrick and the Naval . . .
   Put before their lordships was a day in the life of six Irishmen, five of whom spent it somewhat vaguely trying to arrange to go to McDade's funeral in Belfast. The sixth, Hughie Callaghan, was just hanging around.
   Take Callaghan. He went into town that morning to collect £17 social security money, and then he met his wife, who needed that money for the household and knew his ways of old.
   He told her a lie. He said the social security hadn't paid out, and she went home. In his own mind, everything was going to be all right, because he had plans to meet a fellow who owed him money that night. This meant he could drink all day and still go home with the right money.
   He went to a pub and a club and then called in on Richard McIlkenny to repay £1, and then wandered off with McIlkenny and the others to the station, where he saw them off. He could not, after all, go home.

... Mr Judge found all this behaviour about trains highly suspicious. Why should they be hanging around a railway station at that time of night, he asked? And why should Callaghan be there, when he was not even travelling?

... These are all facts about which very different cultural judgments can be made. It is possible that to an Irish court it would seem as surprising as all this does to Mr Judge, had all the men had the fare, all had luggage, and all been on time.[6]

This is a dramatic instance of the effects of the clash of two worlds: of the way in which reasonable arguing in one context can fail, with far-reaching results, to make any impact in another. No wonder, then, that social strangeness is viewed by many people with trepidation, if what they thought was plain sense on their own parts can be transformed into absurdity overnight by no will of their own. No wonder, either, that the process of trying, not to conform with other people's expectations but to change them, is risky to undertake.

### 1b. *Worlds of Argument which Differ Through Time*

The above instance shows the existence of worlds of arguing which co-exist in time but not in social setting. Before going on to try to describe some of the components of the worlds in question, we should notice that although different attitudes *towards* reasoning alter between historical periods, this does not mean that the problems we are dealing with here are specific to the twentieth century. Even so, they are problems which have been regarded with greater or lesser degrees of hopelessness. In order to absorb something of the ambience of a social situation in which the risks associated with trying to argue between social worlds were embraced in some danger but nonetheless with enthusiasm, it is instructive to consider aspects of seventeenth-century England. This was a period in which arguing was an activity of considerable social prominence, moreover an activity which seems to have been experienced as a social one whether or not it was overtly conceptualised as such. Cognitive diversity was recognised as a fact of life, even if some people thought it regrettable. Nonetheless it was not generally assumed that the existence of different points of view entails that no reasoned movement can be made between them.

Then, as now, a person who attempted seriously to communicate with others of different opinions could easily feel, as Jeremy Taylor did, that

... reason is such a box of quicksilver that it abides nowhere; it dwells in no settled mansion; it is like a dove's neck, or a changeable taffeta; it looks to me otherwise than to you, who do not stand in the same light that I do.[7]

Still, in this quotation there is no implication that there is no such thing as reason, only that its perception depends on one's point of vantage. (I hope to show that this claim is not as relativistic as from a modern standpoint it sounds.) Taylor, a theologian who was for a period expelled from his own parish, had experience of failing to convince people with arguments which had seemed perfectly respectable to other audiences; of finding, as he moved from one to another argumentative 'world', that what had elicited approval in one context seemed absurd or pernicious in the next. Nonetheless he was not tempted to conclude that reason in human affairs is necessarily defective, or argumentation pointless. However, his working definition of the reasoned position he was defending did not contain only cognitions or principles. McAdoo's account of Taylor's approach points to its characteristic practices, attitudes, habits of mind and passionate feeling.[8] I shall try to show that features such as these are legitimate parts of reasoned arguing and that their analysis can explicate what Taylor means when he remarks that what looks reasonable from one vantage-point often cannot be *perceived* as such from another. If we admit that there are socioemotional aspects of argumentation (see below), we can see how arguments and perspectives can change even when 'facts' do not.

### 1c. *Worlds of Argument: Possibilities for Understanding*

To conclude this section, I want to give one example of a clash between worlds which could, at least in retrospect, be successfully understood. The anthropologist Lawrence Taylor reports that in the 1970s, concern about economic neglect and decay in his rural area of northwestern Ireland led Fr. John McDyer to initiate a number of often highly productive plans for invigorating his district.[9] Besides schemes for promoting tourism or infrastructural development, a number included attempts at co-operative ventures, some of which met with surprising initial success. Not so, however, the following case. McDyer observed that the men of the area were able to cooperate smoothly, enforcing their own unofficial norms, when carrying out the joint operations involved in salmon-fishing. Since they did not own rights over the waters concerned, they were technically poaching, and their work remained circumscribed by its clandestine nature. It therefore seemed reasonable to McDyer to suggest that the local people should come together to buy the fishing rights on the river in order to set up a regular salmon-based industry in the area. To his astonishment the fishermen were horrified and wanted nothing to do with the notion: it was an insane scheme which could lead to nothing but confusion and even, possibly, bloodshed. How could their reaction be explained?

Lawrence Taylor, after considerable examination of the view, history and habits of the fishermen concerned, offers the conclusion that while McDyer's suggestion seemed reasonable enough to anyone who accepts that what works under illegal circumstances is unlikely to be threatened by making those circumstances legal, the fishermen precisely did not accept this inference. For them, Taylor says, poaching was associated with forms of neighbourly assistance perceived as natural as between private individuals owning their own separate property. Common ownership they associated, in contrast, with the unnatural strife exemplified by children at odds over an inheritance; and the business world they connected with competing self-interests defended ruthlessly, as by bailiffs. For the fishermen, therefore, McDyer's plan involved an irresponsible confusion of several distinct social and conceptual areas; an anything but sensible proposal.

The cheering aspect of this case resides in the fact that although neither McDyer on the one side nor the fishermen on the other were aware of the argumentative underpinnings of their positions, Lawrence Taylor was able to use his knowledge of the social and cultural contexts of both parties to reconstruct them. For each set of participants, the other side appeared dangerously irrational; nonetheless, their respective habits, conduct, and fragmentary assertions could be connected to yield themes which revealed what underlay the argumentative connections they made. In this case, rather than both parties being insane, each might claim a reasonableness intact within the worlds of practice they inhabited. The habits and inferences structuring each world were incompatible, it is true, but not in any sense entailing the principle of relativism, for both were reasonable views though both could not be followed at the same time. In what follows I shall try to reconstruct some components of social worlds which show what sources of reasonableness they may incorporate.

## 2. SOME SOCIO-EMOTIONAL COMPONENTS OF REASONING

I have claimed that unless we recognise those aspects of argumentative practice which are socio-emotional in nature, much of what is in effect reasonable will escape us. These components of arguing may take the form of, first, conventions and preconceptions about who is permitted to argue, under what circumstances, in what style; and, secondly, ingredients which are more directly constitutive of reasoning itself. The existence of conventions of the former type appears most obvious when we confront them in societies at some distance from our own daily interaction. England in the seventeenth century, for example, displayed an intriguing mixture of habits and views about arguing. This was a period somewhat distinctive in argumentative terms because at that time the world of thought was ceasing, for a significant number of people, to be divided up like a chess board, in which a man was only entitled to jurisdiction over the arguments appropriate to his own square (the partial release of women as a group from this type of confinement came some two or three centuries later).[10] In no society is entitlement to argue on a particular topic simply a question of intellectual competence,[11] but in relatively static societies, the boundaries of perceived impertinence are rigidly drawn. The priest is held to be the man whose job it is to argue about religion, and the teacher is the person entitled to adjudicate on questions of education, even if cowherds or mothers think of more interesting things to say about these matters. Such attitudes to argumentation can still be found in rural areas of contemporary Europe, and profoundly instructive misunderstandings result if newcomers to these areas attempt to argue as if they were entitled simply to say what they thought relevant to the case in hand. Such settled social conventions surrounding arguing are, in times of revolution, apt to be discomposed.

Despite the political upheavals in Civil War England it does not appear to have been supposed by many people that there were no social criteria at all for entitlement to argue; this is demonstrated by the Army Debates defining in terms of property ownership what category of person could be trusted to reflect reasonably on the state of the country and hence could be entitled to a vote.[12] Nonetheless the participants in those debates were themselves people from a social group not formerly thought an appropriate source of opinions on public matters. Argumentation was therefore becoming *more* fluid, open to more men (and a few women, especially where religious matters were concerned). The connection was less tight between entitlement to argue about a particular subject-matter and occupying a social status directly connected with it. This tie was eventually displaced, in some areas of society, by the intellectualist view that since all arguments are subject to the criteria of reason, anyone theoretically capable of applying these criteria is entitled to think and say exactly what he or she is convinced is the case. It is doubtful if this is even now a truly popular opinion; in most work situations, for example, what it is possible to argue is strongly influenced by lines of social power and control. Nonetheless, in written texts the view endorsing freedom of argument is often treated as the only one. This is a position which has been intensified during an era of mass production and employment, where physical rootedness during one's lifetime has come to be regarded as insignificant compared with the more abstract but more pressing imperatives of work. Now, in place of clearly perceived social criteria, abstract ones are taking their place.[13]

How is it, therefore, that a belief in the ubiquity and invariability of right reason has come to fruition in parallel not only with a diminution of interest in arguing as an activity, but also with frequent expression of the opinion that as far as human questions, at any rate, are concerned, there is no really effective reasoning at all, and no grounds to treat arguing too seriously, for in moral, political, and practical questions you can virtually take your pick?

I have suggested that, apparently paradoxically, it is a lack of attention to the social constituents of arguing itself that brings about this situation. If we insist that arguments are verbalisations of purely cognitive events, then there is no way to interpret differences in opinion shown by apparently reasonable people who agree about 'the facts' except by concluding that reason itself must be somehow a random affair. Now, conventions such as those just mentioned form part of the *praxis* of arguing; and they certainly influence the manner, duration and content of arguments. There are, though, other socioemotional elements which form even more clearly intrinsic parts of arguments, and whose consideration also contributes to a more flexible account of what arguing can be.

In practice, though not in most versions of theory, it is readily possible to notice the relevance of personal feeling, experience and involvement to arguing. A British Minister of the Environment in the 1980s became an exponent of conservationism only after he had been taken to view dying Scandinavian woods and lakes at first hand. This is an instance of the functioning of 'pathos' in a sense I shall be using here: a socioemotional experience made the minister open to the force of arguments which had left him comparatively cold when the details of the effects of pollution did not form vividly felt parts of his experience. This is by no means to claim that this gentleman became unaccountably irrational after his tour. He merely grew to accord more social and political significance to environmental effects, and thus became (reasonably) inclined to accept different arguments from those he had supported while assuming that the situation gave no special cause for concern. Different situations of feeling bestowed reasonableness upon different sets of arguments.

It is argumentative phenomena such as this which have traditionally been dealt with under the Aristotelian rhetorical triad of ethos, pathos, and logos.[14] 'Pathos' I shall take to refer to some blend of sociopolitical or socioemotional factors which can enable recipients of an argument, given their own (partly socially derived) predispositions and needs, to comprehend and take it in – whether or not they finally accept it. There is no implication here that such factors are necessarily to be impugned; indeed, ethos and pathos form, on the view I am putting forward, legitimate and indispensable parts of arguing. The idea of ethos refers to a speaker's or author's need to establish some reliable form of self-presentation, and to establish the same for key methods and witnesses. Rightly, we do not give the same credence to malicious fools as to well-informed, well disposed people. The specifically sociopolitical element of ethos derives from the fact that we often do not know from personal experience whether a given arguer is of the latter type; therefore, social markers – such as rank in a particular profession – are taken (however fallibly) as substitutes for personal knowledge. It follows that attempting to establish the ethos of particular social categories of person is itself a highly political as well as an intellectual process: the acquisition of a positive ethos is a practical path to political power. Lastly, while the third element of arguing, logos, is that most similar to what tends to be thought of as constituting an argument today (its cognitive aspect), it is by no means an asocial, non-fluctuating element, as I shall claim in the next section.

For our present purposes, one of the most significant elements of these three components taken together is that they make up a picture of argumentation which is no longer rigidly disconnected from human social affairs. They provide part of the instrumentarium we need in order to understand arguing as a social phenomenon. Moreover, taking them into account allows us to acknowledge that arguments may fluctuate on socioemotional grounds without automatically forfeiting their claim to reasonableness.

### 3. STRUCTURAL FEATURES WHICH UNDERLIE 'WORLDS': CLUSTERS OF TOPOI AND LOCAL POSTULATES

The components of pathos and ethos, important as they are, are unable by themselves to render the whole structural outline of what I am calling social 'worlds' of reasoning. This term, 'world', is intended to retain the three-dimensionality of the metaphor needed to envisage the structured and setting-related locations of the argumentative practices we are dealing with. Although these practices are conventionally divided neatly into logical or conceptual components on the one hand and historical or social influences on the other, I shall concentrate here on one special type of 'logos' whose composition more closely resembles a combination of the two. Reasonable discourse in any area is in fact heavily underdetermined by the components of agreed-upon facts, logical inferences, and (historically influenced) clusters of concepts – and even adding the socioemotional components mentioned in the previous section does not complete the analysis of arguing as it actually takes place. Arguing is composed also of background attitudes, general or particular perspectives, evaluations, suppositions – some of which sculpt acceptable argument at so subterranean a level as to be imperceptible in most practical situations. As I have tried to stress, what looks entirely reasonable in Catholic working-class Northern Ireland looks absurd in upper-class England, say, to such an extent that it is not easy to mediate between the inhabitants of these contexts by pointing to obvious elements of disagreement. Instead we need to start by asking about the means by which people negotiate their own worlds, or sets of worlds, in such a way as to produce pieces of reasoning acceptable to themselves and other inhabitants. How do the other inhabitants know how to contradict them if they wish to do so? How do they know what type of rejoinder will function as a contradiction? What inferential structures yield arguments which 'make sense' to participants in one argumentative setting but not in another?

It appears to me that neglected aspects of the answers to such questions can be dealt with by buttressing 'ethos' and 'pathos' with an explication of our third element of arguing, 'logos', and in doing so emphasising the ancient tradition of describing 'commonplaces'. These commonplaces make up much of the 'intellectual' content of arguing; they have traditionally been held to constitute locations to which arguers resort when searching for intelligible structures round which to build arguments.[15] I would claim that although these structures have generally been analysed (if at all) insofar as they contribute to individual arguments, they also *underlie* whole *patterns* of argument accepted as reasonable in specific social settings. The store of acceptable legitimations for arguing is not invariant between settings; on the contrary, settings are partially defined by the patterns of argument which tend to be regarded in them as effective or for some other reason preferable. Patterns of arguing may thus be given by social settings as people enter them. It is this setting-relatedness, this blending in to other aspects of setting, which makes it easy for argumentative forms to subsist on strata below conscious or reportable attention. It is this which compounds the difficulties of translating the views of people who are accepted as reasonable from one context to another.

The field broadly classed under the term 'commonplace' is described by Aristotle in terms of 'topoi' and reputable opinions.[16] According to his analysis, a general topos falls between a logical and a substantive rule. General topoi are thought-structures of what at first appear to be of heterogeneous kinds – they range from 'If a quality does not in fact exist where it is *more* likely to exist, it clearly does not exist where it is *less* likely' to, say, 'The things people approve of openly are not those which they approve of secretly.'[17] If regarded in their functions as providing recipes for arguing, these all have the capacity both to generate and to legitimise arguments or interpolations in argument. It therefore seems to be this argumentative or sense-making *capacity* which distinguishes the type. General topoi all function at a level broad enough to make them independent of context – even though it would seem that particular social or intellectual contexts develop traditions of favouring arguments based on some general topoi rather than others. In contrast, 'special' topoi are concerned with the basic values of a society's discourse: 'Whatever is x is good', 'beautiful', 'fair' or 'true'. It is clear that these two categories between them can shape a considerable extent of argument. Other people's social worlds can be made more intelligible by discerning unstated special topoi which may form recurrent themes linking (and making intelligible) their habits or judgments: 'Whatever is respectable is good,' for instance.[18]

I suggest, though, that there exists at least one more category to be appended to this account of special and general topoi. Trains of argumentation are often constructed on the basis of general assumptions which resemble topoi in that they function to delineate some arena in which connections are sought, legitimise connections made within that area, and mark out further areas in which connections are *not* sought, in which the territory of relevance ends. Yet they are more specific than general topoi, and less overtly evaluative than special topoi. An instance might be, 'Whatever shows tolerance is socially constructive.' This is a *Hinweis* for arguing or indicator of what sort of argument to make, and there are some social contexts in which such a generalisation is accepted as bestowing reasonableness and can be used to fashion well-functioning arguments. It can be classed as a topos in that it has dual functions, both selective and legitimising, in constructing arguments. This category, which may be termed a directed topos or local postulate, can be distinguished from what Aristotle terms a 'reputable opinion', a claim about the world which either simply happens to be widely accepted or is accepted by experts in a field, and which also provides a starting-point for argument. For Aristotle, a statement such as, 'Those who love are loved,' is a reputable opinion because it is (or was) a commonly accepted account of what can normally be taken to be the case. It indicates nothing about the contexts in which it might occur, or about what other inferences to make. A reputable opinion ('Children love their parents,' say) is a subject-related claim suitable for insertion into some context, whereas local postulates as I propose them are not substantive in the same way. It seems to me that they function as procedural indicators which *define* contexts as well as directing arguers what to select or emphasise within them. As such they help separate one style of interpreting the world from others – 'Whatever shows class conflict is socially relevant,' for example, versus 'Whatever shows harmonious interconnectedness is socially relevant.'

Thus a local postulate defining and legitimating an area of discourse tells us what to look for and what not to look for; and it confers relevance and coherence on a text or piece of behaviour containing some sequence of reported interconnections – which otherwise would not actually *be* a sequence at all. This concept can, then, account for a range of argumentative components. It appears, for example, that what we now call an 'attitude' is constituted partly by a tendency to think or feel in terms of one of these, or a cluster of these, local postulates.

Attitudes are therefore not necessarily deeply personal, exclusively emotional, internal phenomena; they can be analysed in terms of socially-located tendencies to argue in particular ways.

The fact that received ideas and attitudes can differ from place to place has been conceptualised recurrently in terms of landscape-related images referring to 'maps' or 'territory'. These have strong intuitive and pictorial appeal but a distorting effect can arise from their two-dimensionality, their concentration on the surface content of atomised concepts and their omission of attention to the structures of arguing in which ideas are used. In the social sciences this can result in 'attitudinal maps' which class together items which are apparently alike or look conceptually similar but which may in fact have very different social origins and locations. Thus, apparently conceptually related positions may in social reality fit within different clusters of opinion and behaviour, and their adherents may draw from them different consequences. The connection is only one of surface resemblance, and will by no means tell us everything about the social 'worlds' concerned. For instance, to locate the Irish political parties Fianna Fáil and Fine Gael together on such a map, or the Roman Catholic and Anglican Churches, or the Labour and Communist parties, might have heuristic value for some purposes but would be misleading if taken to show genuinely shared practical worldviews. First, this would ignore the potentially transforming effects of socioemotional and political factors on opinion. Secondly, it would ignore the fact that some arguments which share surface similarity can derive from different deep structures of topoi, which may yield significantly different correlates and consequences for those who hold them. Lastly, it would ignore the effects of different combinations of argumentative legitimators. We may both hold that political corruption, say, is abhorrent. But this opinion in itself is compatible with radically varying structures of opinion about political intervention: quietism versus activism, say. The world of interrelated argumentative habits in which I live can transform the effects of a given single component, according as it combines with others.

When we try to see how people choose between such different worlds of arguing, the notion of different levels of topos demonstrates how one argument can be countered by another. The social world is not made up of contexts all governed by topoi in logical contradiction to each other; rather, claims of appropriateness appealing to topoi tend to be offset by competitors rather than trounced by disproofs. Searching for the most appropriate topos in a given setting involves evaluating, in the light of a particular background of attitudes, beliefs, customs and so forth, what fits the present circumstances – whether to plump for an argument assuming that two similar results are likely to proceed from similar antecedents,[19] perhaps, or whether to take a sceptical line about searching for varieties of cause. If we are arguing with each other from different positions or different entire worlds, where my world is characterised by a tendency to display topoi advocating timidity and yours is more adventurous, what I think of as a highly apposite topical argument – let us cut our losses and not venture further – is liable to be dismissed as feebly inappropriate by you, who are more inclined to see reason in urging that now we have come so far, it is better not to waste our efforts and to press ahead.[20]

The selective, legitimating, sense-giving functions of dominant topoi enable them to regulate even what we regard as a relevant and established fact and what we dismiss as uncertain or unconnected. A society which is organised with respect for the special topos, 'Whatever is part of military life is excellent,' say, will perceive only faintly and formulate only vaguely states of affairs which are associated with more tentative and vulnerable states of living. This has a huge impact on the existence of public languages – on the fact that, as mentioned above, there are some social contexts in which it simply does not make sense to try to make certain sorts of claim or engage in certain sorts of behaviour. This may make up a large part of what Durkheim meant by the impossibility of controverting 'social facts'. In a related fashion, 'themes' in society or literature, 'narcissism' and so forth,[21] are discerned when behaviour is read as if it resulted from a particular topos or reputable opinion. We may be urged to interpret the social behaviour we witness as if it were based on such a topos, on the inducement of understanding more about it if we do.

It is the case, then, that the 'logos' components of arguments can differ between times and places: they differ between worlds. In a parallel fashion, selections and combinations of all three varieties of topos can themselves constitute a particular emotional or sociopolitical tone in argumentation. Although not necessarily explicitly value-laden, nor necessarily equipped with identifiably evaluative 'concepts', they convey the choices in attention and relevance which a speaker or writer has made. In this respect the gulf between description and evaluation can be seen to disappear. The sympathetic tone of some particular account need not derive from 'bias' as commonly understood, but can originate, say, in a tendency to argue from topoi which expect co-operation rather than confrontation. Take a local postulate such as, say, 'The daily details of interaction are important for understanding social life.' The supposition that small events do not occur without a pattern, and are important rather than unimportant, is as much a personal ethic as a theoretical conviction. At the same time, it is as much a theoretical conviction as it is a personal ethic. Hence the element of logos in an argumentative sequence is as capable of conveying personal and evaluative positions as are ethos and pathos. It is clear also that such elements play a crucial role in informing not just the patterns of reasoning inherent in a single discourse but those characteristic of whole social approaches or *mentalités*. Collections of topoi help to explain how it is that people moving within social worlds find it reasonable to *select* certain details for notice and then *connect* them together in patterns which are experienced as intelligible. This constitution of intelligibility is thus a major role for topoi, and is both a social and an intellectual process. The range of topoi considered appropriate in some given situation is governed partly by social-historical criteria, and social-historical situations are partly constituted by the patterns of topoi used within them. But in their actual use, topoi direct the selection and inferential structure of arguing and so must be categorised as intrinsic to 'the argument itself'.

## 4. SOCIAL WORLDS AS FORMS OF DISCLOSURE

We have until now been emphasising separate components of worlds of arguing. Putting these together, we can consider in this section how worlds taken as wholes are actually felt by people moving within and between them. Worlds of arguing seem normally to be experienced chiefly in terms of social phenomena, camouflaged, so to speak, in the patterns of everyday living; on the other hand, aspects of these social phenomena can function as argumentative disclosures in themselves. Aspects of social worlds, that is, unobtrusively govern arguing by means of the argumentative forms they incorporate; and, often, social actions are interpreted as indicators of positions which could be expressed in terms of arguing. This helps to delineate the global impression which worlds of arguing give from within. In the following, final section I shall examine some of the ways in which people negotiate between these worlds and what makes some of them better at doing it than others.

## 4a. *Arguing as a Social Experience*

To see how it is that argumentative patterns are woven into the social experience of their users rather than experienced as independent structures between which one might simply choose, we can look again at the ways in which their components blend into their settings. 'Ethos' and 'pathos' are clearly qualities with heavy social components. According to different social situations, different categories of person are endowed with the ethos of arguers and reasoners – only men, perhaps, or only older people, or only younger people. The criteria for ethos – how much do we need to know about people to approve their status as arguers?[22] – tend to be established via social developments, rather than as the result of abstract decision-making. It may be only the people who are active 'producers' in a society who are allowed to function as competent arguers; in other situations, the vulnerable inhabitants of private spheres may be taken to have something to contribute; but these are states of affairs which develop rather than being settled upon after overt debate. On the side of pathos, not only the question what feelings are considered legitimate influences on arguments, but feelings themselves are partly socially produced and learned.[23] Historical work on variations in family form, for instance, makes clear that people actually feel love and friendship differently, for different categories of person, in different times and places.[24] Even the capacity to notice that one has feelings at all depends on social setting, to the extent that when there is a public language available in which to frame some emotion or impression, it attains much more perceived significance than when it could not be framed in verbal arguing at all.[25] Providing a language in which to describe some set of phenomena, then, as political movements standardly attempt to do, is among other things an activity connected with pathos. It is an attempt to influence felt interaction so that people can open themselves to arguments whose force they formerly might not have felt.

If ethos and pathos are formed according to patterns and conventions which are partly of sociopolitical origin, this is just as true of phenomena connected with logos. We have seen that given topoi are at home in some settings rather than others, and also that these settings are characterised by praxis-regulating conventions. Socio-political conventions regulate when and where arguments are started and finished, at committee meetings, in parliaments, in chance meetings on the street; this influences their content and structure.[26] Topoi and praxis may interact: social perceptions of time, for example, play a variety of roles in argument. Arguments may be initiated slowly, emphasising the incorporation of all participants, as in rural areas, or speed may be of the essence, as in metropolitan life. Rural argumentative strategies are often consensual and organic, advancing in wave-like fashions appropriate for moving the participants along together; in urban environments, confrontational styles may be approved which would give offence in rural settings. In general, moreover, arguments are not always, or perhaps seldom are, pursued to their logical conclusions. Matters such as the acceptable style and length of an argument inevitably affect its content, and cut-off points are accepted where an argument rubs up against another of more social power, or where it is cut short by topical conventions – for instance by respect for those in authority – or by pragmatic, institutional imperatives such as the need to reach a conclusion within some given period.

The perspectives, emphases and criteria of relevance which I have briefly touched on all help to constitute argumentative links which are seen as acceptable in some social places and not others: they all go towards forming criteria of effective meaningfulness which make operative certain inferences and indeed whole patterns of discourse, and rule out others. For example, in a relatively new society where not a great deal is commonly known about its history, it 'make sense' to lecture for an hour about an obscure seventeenth-century annalist, recounting in detail what he did from day to day, without any obligation to show why the audience should take an interest in this person. The social situation of the group in question bestows reasonableness on such conduct – it helps to reconstruct a common past. In another culture, where everyone is confident of the existence of a good deal of knowledge about the same historical period, it would scarcely 'make sense' to give the same talk. A speaker would need to show what general point it illustrated or how it fitted into debates of contemporary relevance in order to give sense to the account.

The prevalence of particular languages and conventions of arguing in particular social settings both defines those settings and regulates behaviour within them. We know that we are in a different setting when people not only act differently but conceptualise and justify their conduct differently.[27] Phenomena classifiable under the concept of logos are particularly potent here, for the capacity to determine what makes sense, where and when, has strong sociopolitical effects as well as causes. Many people have had the experience of noticing in themselves that they moderate their conduct in new circumstances, but this is not just a question of social tact. It is not only that it seems unreasonable to swear when visiting elderly relatives in convent nursing homes; it is difficult, for instance, to try to argue for the virtues of departmental democracy when joining a more authoritarian university. This is not merely a matter of social pusillanimity in the face of some powerful organisation; rather, it is part of the power of that organisation that it defines the discourses which 'make sense' within its ambits. Social contexts can bestow, or refrain from bestowing, on one's actions the intended type of intelligibility – so that what is accepted as intelligible on one side of the Pyrennees may not make sense on the other.

### 4b. *Social Behaviour Interpreted as Arguing*

Since arguing is experienced as intrinsic to social setting, showing apparent allegiance to some aspect of a situation may be interpreted, or misinterpreted, as endorsing positions taken as indigenous to that setting. Thus, small conventions, such as refraining from lounging against walls or putting one's hands in one's pockets, may be taken to speak for whole complexes of what is accepted as pragmatically reasonable in the same context.[28] This is part of the reason why such an emphasis is laid on apparent trivialities of conduct in institutions; conventions are scarcely ever just conventions. They are composed of details which are seen to indicate discourse-governing values expressible in the form of topoi: seeing control rather than spontaneity as good, for example. Hence people may quarrel or even kill about apparently minute details of social conduct, as happened in the royalist versus Puritan disputes of the Civil War period mentioned above. The 'world'-representing or symbolising capacity of these details are in practical reality are taken to indicate acceptance of whole ways of life.

Social performances, such as giving a lecture in a certain way or getting more coal in before you go to bed at night, thus follow patterns for which commonly accepted topoi can be produced to give sense. Fetching coal now rather than waiting until it is needed may be in effect held to show assent to a web of generalisations such as, 'It is advisable to prepare early for the future,' or 'It is better to take control over details than to allow oneself to be controlled by them.' It is for this reason that ways of life are forms of disclosure in themselves. They follow patterns which could in principle be reconstructed according to explicit topoi, though it is more usual to show understanding by going along with them than to try to express them in words, which are clumsy for the purpose.[29] This has the corollary that the striking practical wisdom of certain individuals, or even certain cultures, must depend in part on the reconstructive creativity of those interpreting actions. In some traditions, in particular contemplative religious ones, wise individuals may actually refuse to 'put words on' their conduct, but permit their followers to reconstruct the topoi with which to interpret their ways of life.

Conversely, the finesse required in order to show by one's actions that one has understood the world of topoi in which one lives goes towards explaining why it is so hard to live permanently in a new culture. The newcomer is constantly performing residual gestures which are read as exuding a non-local grasp of everyday interaction. Social ambience itself is made up of small-scale interactive details which are interpreted in practical life in argumentative terms – in terms of what they allow people to think, say and do in public and be taken seriously. The very existence of ambience witnesses to the capacities of social beings to infer from parts of social worlds to their wholes: part-conveyors of argumentative attitudes are readily taken up in terms of the entire worlds they are hold to disclose.

Particular social moves or argumentative conventions, then, are not otherwise empty agreements about what to do and how, but are capable of being understood as standing for particular social worlds, on either a symbolic or a part-for-whole basis. We can carry this observation further by suggesting that particular worlds are marked by characteristic enthymemes and combinations of enthymemes. Enthymemes, inferences which are frequently not only incompletely stated by also rely heavily on socioemotional components,[30] are indicative of particular worlds for at least two reasons. First, particular socioemotional constituents of ethos and pathos, as well as some accepted patterns of logos, are accepted as parts of legitimate inferences in given settings but not in others. Secondly, there are, according to setting, items which are so much taken for granted that they do not need to be stated, and may seldom or never be reflected upon at all.[31] The fact that something need not be said is itself a potent cement of understanding – but it causes clashes when two worlds meet. It may in fact be unclear to all the participants in a misunderstanding either that the unstated items exist, or what they are. Here, social worlds may function as forms of disclosure to their inhabitants but as forms of veiling and disguise to strangers.

This does not entail, however, that we should assume the existence of a hermetically sealed sequence of worlds within which reasonableness is only internal. If at least some aspects of some worlds did not partially overlap, we should find even less mutual intelligibility in social life than we do. It is to the forms and products of this understanding that we shall now turn.

### 5. GETTING FROM ONE WORLD TO ANOTHER, OR MELTING HORIZONS: CONTRIBUTIONS FROM THE SOCIOLOGY OF ARGUING

What is involved, then, in understanding views, or defending views, as between worlds? This essay has argued that it is easy and common to meet with *mis*understanding of more and less limited extents when trying to communicate between worlds – and there clearly are some people and groups with whom it is hard or impossible to communicate.[32] Often, particular claims cannot be intelligibly made to certain people at certain times at all.[33] Thus the dependence of an argument's acceptability on contextual features of a particular space and time can make the communication of certain messages extremely risky. Hence we need to know more, not only about the structure and function of different worlds of arguing, but also how to navigate between these argumentative situations.

In daily life we are surrounded by such attempts at negotiation, some of which appear at least partly successful. Rather than confining ourselves to examining practices such as anthropology and historiography,[34] it is instructive to observe more mundane adventures, such as taking package holidays or trying to accommodate oneself to one's parents-in-law. The pragmatic acceptance of 'good enough' communication in some of these cases indicates that the beginnings and ends of worlds need not be impermeable, fixed or wholly mutually exclusive.[35] For example, habits in relation to ethos, pathos or logos may overlap, which furnishes a bridgehead for attempting communication. Nonetheless, what overlaps between one world and the next may be only part of some cluster of topoi or some set of assumptions about the place of ethos or pathos, say; and it must be borne in mind that even the same configuration in a new setting can take on a new appearance. New settings alter the imports of arguments entering them, by providing new contexts of interpretation and use.[36] Understanding other worlds cannot be completely accounted for, then, only in terms of literal coincidence between parts of those worlds.

Given the caveats just mentioned, let us consider one approach to comprehending action in a new 'world'. Suppose I find myself in a new culture where a centrist politician gives a job to someone I regard as a right-wing fanatic. Before concluding that the politician must therefore be a fascist in disguise, I might survey other possible explanations for the appointment; perhaps in terms of other topoi I have experienced or heard about. The special topos, 'Whatever shows loyalty is good' might be such a candidate; if I ask about past associations between the two protagonists I may discover an alternative explanation in terms of loyalty. Whether I find this acceptable in terms of my own standards is a separate issue. The point is that the topos concerned is perceptible *as* a topos even if one does not share it. This also appears to have been the process followed by Lawrence Taylor in the article discussed above (section 1c). He may *himself* not adhere to the view that, say, 'Co-operation is natural providing it takes place between people whose private spheres are secure,' but he can *recognise* it as a potentially coherent view and discern its role in legitimising inferences. It seems possible, therefore, to reconstruct patterns underlying unfamiliar forms of arguing even if one does not use them within one's own world.

The capacity to take such an hermeneutic approach to understanding is one which varies between individuals. It is necessary to be diligent enough to seek for alternative patterns from which behaviour might ensue, for example, and to feel sufficient personal interest in others' views to wish to do so; these are matters connected with pathos. The ethos one attributes to the person whose actions one is trying to explain will also play a part in the investigation: is this a person whose motives one expects to involve some form of integrity, or not? The recognition of patterns in new behaviour is thus not an exclusively cognitive process any more than the 'worlds' we are describing are exclusively cognitive 'worlds'. The willingness and the creative capacity to discern patterns which render strange conduct intelligible are partly constituted by moral and socio-emotional proclivities. These proclivities, which differ in frequency and degree between worlds as well as among individuals, may furnish their bearers with more or less sensitivity and competence in developing new interpretations. Cultural backgrounds which allow some relaxation in rigidity of expectation may predispose their inhabitants to this, as may emotional states which encourage in their bearers tolerance rather than its opposite. (One of the reasons for which people enjoy falling in love is the ease in communication it facilitates, by conferring on the protagonists a respite from insisting on the exclusive naturalness of their own cultural prejudices.) Among other requirements for negotiation through new worlds are traits such as sufficient sensitivity to recognise unfamiliar signs as indicators of argumentative position; flexibility in adapting one's own patterns of expectation; and sufficient courage not to be cowed by sanctions meted out in response to error. The interpretation of new worlds is, as the components of rhetorical arguing should lead us to expect, a moral and personal matter as well as simply an intellectual one.

Finding one's way through a new world will seldom stop at interpreting one action, but is likely to be a long-drawn-out, more and less conscious process. If an initial interpretation involves imputing a topos unfamiliar to the interpreter, it may cautiously be tried as the basis of an explication for other forms of behaviour by the same, then other inhabitants of the culture, and even tested in relation to future events – with whom will people co-operate in public? With which members of their own culture are they likely to disagree? Confirmation or otherwise of relevant projections will show over how wide an area the particular topos is used in this 'world'.[37] The consistency of a new world need not, then, be the kind of consistency one expects and its boundaries may be different from the familiar ones; but once the stranger has learned to function inside it, it eventually ceases to seem bizarre that relevant criteria should begin and end where they do. At this stage it begins to be difficult to notice and describe the negotiation of new situations. Developments which, when they were new, might have seemed too unfamiliar to describe, can give way to such habituation that the newcomer can no longer recall what appeared strange in the first place.[38]

Those entering the worlds of others by no means always assent to learning the ways of those worlds, of course; often they may, deliberately or otherwise, dominate or alter them. There are several ways in which parts of one world can be deliberately transported, into others. This happens during the development of hegemonic ideologies and during cultural domination; or else – in a potentially more positive form of constructing understanding – artistic forms may stimulate new attitudes through methods which may or may not be verbal. Outside spheres such as the academic one, in which a somewhat artificial value is placed on verbalisation, communication often takes place in a rich mixture of forms

– as in painting, for instance, when a new style – impressionism or cubism – may form an attempt to force audiences to respond to the world differently. Even when no precise claims and counter-claims are being made, this is less unusual than it may seem, given that even in verbal arguing, topoi are often *counterposed* by other topoi rather than explicitly contradicted. Some second topos is produced with the implicit aim of showing that it fits more productively into the audience's world than did the original one. Implicit argumentation of this kind may now be gaining in significance with the advancing power of mass media and their need to communicate with millions, or hundreds of millions, of people at once; its implicitness allows it enough flexibility to be incorporated into a huge variety of different 'worlds'.[39] Words, in contrast, may be highly culturally specific, and when they are not, they degenerate into a 'newspeak' which also fails to communicate.

What I have outlined here attempts to supply a realistic account of arguing, one which is not misleadingly over-cognitive but which recognises the role in arguing of human social processes. I have claimed that the social embeddedness of these processes cannot be taken to imply that there is in the last analysis no consistency of reasonableness worth maintaining between one world and another. Quite the contrary, it appears to me that we need to know how reasoning works in practice in order that this and other values can be effectively defended. Recognising the influence of social contexts on people's views alerts us to the facts that if we wish to defend, say, the injunction, treat equal ceases equally, we have to find out what counts as equal for the interlocutor to hand and why. In rare cases, it is true, we may have hit upon a world in which the injunction cannot be made to make sense at all. If there is such a complete clash among worlds, criteria will need to be sought, according to the audience concerned, for choosing between entire worlds rather than elements of their contents; this is a radical form of argument but one which sometimes cannot be avoided.[40]

Here I have tried to incorporate more into what we understand as reasonableness without detracting from the general validity of the concept. Far from suggesting that it is only the local standpoints of arguers which allow assessments of their arguments, I have asserted that in empirical experience there often are bridgeheads of some sort between social settings, even if their discovery requires assiduous searching in non-obvious places. Acknowledging patterns of inference whose contents one would not use oneself does, admittedly, demand recognition that reasonableness may inhere in argumentative form rather than substantive agreement: a pattern may fall within the ambit of the humane even if its contents seem odd. But this is exactly what tolerant, creative and flexible individuals are used to conceding in the course of their everyday interaction. The practice of arguing reasonably as they find themselves called upon to do it therefore yields crucial insights which can be built upon in understanding arguing – to positive or negative ends, as the case may be.

*Department of Political Science and Sociology*
*University College, Galway*

## NOTES

[1] Jeremy Taylor, 1660; *Doctor Dubitantium or The Rule of Conscience*; Bk. I, ch. 2, Rule VI; p. 485 in vol. XI of the *Works*, R. Heber (ed.). London: Rivington, 1828.
[2] Some people are particularly gifted at moving among worlds of arguing and mediating between them, which is why this essay is dedicated to Dr R.S. Cohen.
[3] No student of politics could deny that much daily interaction is dedicated to *avoiding* transparency in order to further interests of various kinds; I am not offering to reform this aspect of the social world but to open up the option of increased mutual comprehension for occasions when it is both practicable and desired.
[4] This point is argued at more length in Edmondson, 1987, *Rules and Norms in the Sociology of Organisations*; Berlin, Max-Planck-Institut. This examines cases where empirical sociologists claim to be observing 'rule-governed behaviour' and finds that it is only in a minority of instances that this claim appears to be accurate as it stands.
[5] This position is most notably associated with the name of Harold Garfinkel, whose maxim was to test the existence of hypothesised social assumptions by deliberately disregarding them. Garfinkel's students had to struggle against powerful inhibitions when bringing themselves to perform activities they knew other people would perceive as irrational, even though they themselves recognised the reasonableness of these experiments in terms of their own social and intellectual worlds. This shows the power exercised by one's current setting, whether or not it is one's most significant world of residence.
[6] Nuala O'Faolain in *The Irish Times*, Dec. 1 1987, p. 9.
[7] Taylor, 1660, op. cit., vol. XII; Bk. II, ch. 1, Rule I; p. 209.
[8] Vid. *Anglican Heritage: Theology and Spirituality*; H.R. McAdoo, Norwich: The Canterbury Press, 1991.
[9] This is a case analysed by Lawrence Taylor in '"The River would run Red with Blood": Community and common Property in an Irish fishing Settlement', in B. McCay and J. Acheson (eds.), *The Question of the Commons*. Tucson: University of Arizona Press, 1987; pp. 290–307.
[10] There have been other periods in which profound social change and a propensity to take pleasure in arguing accompanied each other, as in the salons of the Enlightenment.

Here too, alterations in social permission to argue seem to have played a significant role in developing the practice.

[11] Cf. the dispute between Marianne and Elinor which gives the theme of Jane Austen's *Sense and Sensibility*. Elinor's objection to Marianne's endorsement of speaking 'without attention to persons or circumstances' (1811; ch. 10 and passim) is not connected with social propriety merely, but also with the conditions of establishing mutual intelligibility between people. Marianne's misunderstanding with Willoughby demonstrates that it is not sufficient to say everything that occurs to one in order to achieve comprehensibility.

[12] Cf. the argument between Ireton and Petty about whether qualification to vote should be based on abstract consideration of the reasons for government, or on assessment of what social arrangements are as a matter of fact likely to conduce to arguments and decisions based on a sober assessment of the nation's interests (cf. *Puritanism and Liberty*, A.S.P. Wodehouse's edition of the Clarke Manuscripts; London: Dent & Sons, 1938; pp. 61–62). Many of the debates represented in the Leveller Documents (North Library, British Museum) centre on related questions.

[13] Since physical mobility during one's lifetime is becoming so frequent in the industrialised world, it is a serious question how the ensuing mixes of people from different social 'worlds' succeed in understanding each other on more than a superficial level.

[14] Here I am starting from the treatments given in Edmondson, *Rhetoric in Sociology* (London: Macmillan, 1984), and Markus H. Wörner, *Das Ethische in der Rhetorik des Aristoteles* (Freiburg/Br., Alber Verlag, 1990).

[15] The editor of the French translation of Aristotle's *Topica*, J. Brunschwig (*Aristote – Topiques*, Paris, Société d'Edition 'Les Belles Lettres', 1967; introduction) points out that the work covers a field which is by no means understood today. We should bear in mind that not only Aristotle but authorities such as Cicero and Erasmus took an interest in the area. This essay is intended to function partly as an approach to explicating the reason for their concern.

[16] I am here building on interpretations suggested in Wörner (1990, op. cit.), myself (1984 op. cit.), also W. der Pater ('La Fonction du Lieu et de l'Instrument dans les *Topiques*', in G.E.L. Owen (ed.), *Aristotle on Dialectic*, Oxford University Press, 1968; and O. Bird, 'The Rediscovery of the "Topics": Professor Toulmin's 'Inference-Warrants'; *Proceedings of the American Catholic Philosophical Association* 34, 1960. The work of J.L. Austin, Stephen Toulmin and Chaim Perelman is crucial for this field.

[17] Ar. *Rhet.* 1397b 13ff. and Ar. *Rhet.* 1399a 30f. respectively. Compare the translation of the *Rhetoric* by W. Rhys Roberts (Oxford University Press, 1946).

[18] This is not an Aristotelian topos, needless to say, but is intended to summarise an instance of ethnographic reconstruction reported in 'A Labouring Man's Daughter' by Marilyn Silverman (in C. Curtin and J. Wilson (eds.), *Ireland from Below*, University of Galway Press 1989).

[19] Compare Ar. *Rhet.* 1399b, 5f.

[20] This is equally an Aristotelian topos and underlies many governmental activities, for example when huge financial sacrifices are made to protect the status of local currency.

[21] Recent proponents of what they allege are dominant themes in American society are Christopher Lasch and Tom Wolfe. Commentators of this type are effective when they offer fruitful topoi and postulates to try out in interpreting the society around us.

[22] Hence the question whether the public has a 'right to know' about the private lives

of politicians, on the basis that their intimate conduct is one indicator of how they might be expected to form decisions affecting the public. This question is answered differently in different cultures, as recent controversies show.

[23] Cf. especially A.R. Hochschild, 'Emotion work, feeling rules, and social structure', *American Journal of Sociology* **85**, 551–575.

[24] The work of Laslett and Ariès on family life is often quoted; for different frames in which children and childhood are perceived, see also C. Curtin and T. Varley, 'Childhood in Ireland, in Curtin *et al.* (eds.), *Culture and Ideology in Ireland*, Galway University Press, 1987.

[25] Cf. Edmondson, 1992, 'Empirical Approaches to Studying the Elderly: Some Comparisons', paper to Budapest conference of the I.S.A. Research Committee for the History of Sociology; also 'Rhetoric as Method in Sociology', paper to section on the Sociology of Language of the German Sociological Association, Trier 1988.

[26] See Edmondson, 1989, 'The Argument and the Subject of Social Action'; paper for Symposium on Rhetoric and Social Action, University College, Galway, 1989.

[27] This seems to me to underlie the significance attributed by J.L. Austin to the different justifications people use in discourse ('A Plea for Excuses', in J.O. Urmson and G.J. Warnock (eds.), *Philosophical Papers*, Oxford University Press, 1970).

[28] Part of the art of acting is being able to discern and reproduce the small gestures which represent a way of life or cast of mind. Actors can sometimes be detected practising following the ways of life of people whose roles they are studying; in my view this aids them to work out, at least unconsciously, which activities give the clue to distinctive life-worlds. This raises a question about ingenuousness for newcomers to particular social worlds, since following residual social conventions is usually expected to be done spontaneously. The 'fitting in' required of newcomers must be more conscious than that of the indigenous inhabitants, but that does not (necessarily) mean that it is in some way dishonest.

[29] Hence, presumably, Winch's emphasis, following Wittgenstein, on telling that someone knows a social rule by watching that person fall in with it. This is a point illustrated by the film *Terminator 2*. At the start of the film the little boy who is the hero cannot explain to the Terminator why he should not kill people. After a period of living closely with the child, learning to talk slang and to joke and to engage in all the daily activities normal to him, the Terminator ceases to behave as if killing were a taken-for-granted option.

[30] Cf. Edmondson 1984, Wörner 1990. op. cit.

[31] Often, underlying enthymemes may not be stated as such but expressed in terms of stories. The regular but not usually deliberate patterning produced by dominant topoi in a social world results in characteristic enthymematical narratives; often, they form stories which may be lifted from their original sequences and applied, as it were allegorically, to new situations. In a given social world, people will tend to use one sort of template, to tell one sort of story (or set of stories) rather than others: these templates and stories will be made up of intertwined sets of dominant topoi, of patterns of expected development, of admissible and forbidden feelings taken to underlie action.

[32] Vid. Edmondson 1984 – it is an error to suppose that for every possible piece of communication, there must be some way or ways in which to frame it so that it will make sense to anyone. Admittedly, in heavily biased situations which are governed by extreme anxiety, such as Cold War and now Muslim-Western dialogues, protagonists may give up looking for means to communicate far too soon.

³³ It is a frequent theme in novels to deal with people who feel alienated from a particular world but attempt to communicate with those within it – H.G. Wells's *Ann Veronica* for example, or Gaskell's *North and South*. In the latter, Margaret's personal involvements are a means of assisting her to come to understand the world of mill-owners.

³⁴ Candid anthropologists such as Nigel Barley reveal that they do not know exactly how they accomplish what they do (*The Innocent Anthropologist*, Harmondsworth: Penguin 1987, passim). Other anthropological texts show a methodological opacity which makes it virtually impossible to reconstruct from them the practical techniques of *Verstehen*, but an effort is being made in this article to contribute to such a task.

³⁵ The boundaries of what one defines as one 'world' will depend on the purposes of the analysis – clearly there exist overarching worlds comprising whole cultures, within which there are subgroups, some of which are subcategories of others. The degree of difference regarded as significant enough to define the boundaries of a 'world' depends partly on the task in hand.

³⁶ Thus, one step in effective conversation must involve taking account of how one's own arguments might appear when lodged in an interlocutor's world. This is also a type of social sensitivity which is more easily practised by some people than others.

³⁷ Many of the quotidian hypotheses referred to by Alfred Schutz (*Collected Papers*, M. Natanson (ed.), The Hague: Nijhoff, 1973 edn.; vol. 2, 73) are in my view likely to be of this nature. The question what type and degree of comprehension counts as enough, however, depends on the aim of the operation in hand.

³⁸ This scarcely implies that all attempts at hermeneutic understanding need be accurate as they stand; the virtues of this approach may even generate their own forms of error. For instance, people who become bewitched by whole cultures may feel that they are merely observing them, when what they entranced by is rather the combination between their worlds of origin and the new one. Such confrontations enable newcomers to emphasise items not necessarily significant to the indigenous inhabitants, and to become transfixed by the continual challenge and delight of transforming incomprehension into meaningful patterns. People entering rural cultures from industrial ones, for instance, often enthuse over the tendency of inhabitants to relax regulations on behalf of individuals – to forget the rules of the road in favour of halting in the middle of the highway to converse with neighbours. Part of the charm ascribed to such conduct may derive from the impression that the indigenous person is rejecting the values of industrial culture in order to embrace those of personal communication. But, if the indigenous person is really indigenous, this is not exactly so – that person sees his or her conduct as the natural way to behave, not as an act of rebellion against anonymous bureaucracy. It is the observer who conceptualises it as the one thing *rather than* the other, and this understanding is in part, therefore, illusory.

³⁹ This may of course produce more efficacy than genuine unity of understanding and has concomitant dangers as well as advantages.

⁴⁰ Hence the relevance of the mechanisms of missionary activity, Cold War argumentation, and so forth. Note that conversion in such cases is often a process of choosing a way of life, or group of people, rather than of espousing particular doctrines for cognitive reasons – a point made by the late Laurence Bright, O.P. As far as the 'incommensurability' argument is concerned, some elements of some worlds are likely to turn out to be incommensurable, and others not – this is partly an empirical question.

CHARLES L. GRISWOLD, JR.

# RHETORIC AND ETHICS: ADAM SMITH ON THEORIZING ABOUT THE MORAL SENTIMENTS*

I

Adam Smith began lecturing in Edinburgh in 1748, when he was 25 years old, and with great success. In 1751 he accepted the Chair in Logic at Glasgow, and about a year later he occupied the chair of Moral Philosophy at Glasgow, a position he held until 1764. The activity of teaching thus consumed much of his energy for 16 years, until he was forty-one years of age.[1] Every indication we have suggests that Smith was an effective, dedicated teacher. He seems to have understood the rhetorical dimension of that enterprise.[2] Indeed, rhetoric was among the topics that interested Smith deeply from the very start of his teaching career. While at Edinburgh, he lectured on rhetoric, belles lettres, and eventually on jurisprudence.[3] Clearly there existed lively interest in the issues of rhetoric and belles lettres, and just as clearly Smith's dedicated teaching and imaginative approach to the subject helped fan the flames.[4] During his occupancy of the Chair in Logic, he again taught rhetoric and belles lettres in addition to logic and metaphysics, as well as part of the moral philosophy course usually taught by the occupant of the Chair in moral philosophy (evidently Smith discussed "Natural Jurisprudence and Politics").[5] Smith redefined "logic" so as to radically deemphasize Aristotelian logic and to emphasize rhetoric and belles lettres instead. A contemporary report informs us that Smith thought the changes warranted by the fact that "the best method of explaining and illustrating the various powers of the human mind, the most useful part of metaphysics, arises from an examination of the several ways of communicating our thoughts by speech, and from an attention to the principles of those literary compositions which contribute to persuasion or entertainment."[6] Students of Smith will detect in this statement an indication of the important role psychology plays in Smith's work on ethics and economics. Smith did not give up lecturing on rhetoric during his tenure as Professor of Moral Philosophy; the reports of his lectures on rhetoric and belles lettres in our possession evidently date from 1762–63.[7]

It seems reasonably clear that Smith's work in ethics and economics had its source in his lectures at Glasgow University and in Edinburgh. The first edition of the *Theory of Moral Sentiments (TMS)* was published in 1759, and the second edition in 1761.[8] Of course, Smith's views on ethics may well have developed in significant ways during his stay at Glasgow.[9] What seems to remain constant, however, is his interest in rhetoric and in the effective communication of ideas – in rhetoric and pedagogy, in effect. His written work ought to be counted as a species of pedagogy. He clearly cared a great deal about precision and effectiveness when communicating his views, whether in the classroom or through the written word. Smith's extreme care with the manner in which his work reached the public is well attested. He seems to have worried about student note-taking on the no doubt justified basis that students might garble his words;[10] he did his best to ensure that nothing he had not prepared for publication would survive his death; he worried a great deal about the publication of Hume's *Dialogues* (though there the issue primarily concerned their reception); and he took his work on ethics through six editions, 'combing and curling' (to use Dionysius of Halicarnassus' characterization of Plato's tinkering with his texts) over and over.[11] Smith was a careful, self-conscious writer. Any connections between his rhetoric and the views he wished to communicate should be taken as deliberately crafted on his part.[12] While this 'principle of charity' is generally defensible when interpreting philosophical writing, we have particular reason for holding to it when reading Smith.

Smith's early abiding interest in rhetoric signals at the very least an awareness that what one wishes to say or write to others is shaped by the demands of the audience one envisions and by the constraints of the medium in question. But to leave the matter there would be to accord rhetoric a merely instrumental role in the communication of ideas. We have reason to believe that Smith understood that the subject matter itself may require expression of a certain sort if it is to be represented accurately. Smith's work evinces a sophisticated awareness of the problem of the relationship between "form" and "content" in discourse. The problem is an old one in the history of rhetoric, and Smith was acquainted with the discussions of rhetoric, composition, and style going back through Cicero to Aristotle. Smith might have known from Hume too that the interpretation of a text is linked to the way in which the text is written, as well as to what the author wishes to covey.[13] As both Hume and Berkeley wrote dialogues, Smith had before him contemporary examples of alternative ways to write philosophy (of course, there were numerous classical examples as well, including Cicero, Lucian, Augustine, and Boethius).[14]

In the present essay, I propose to reflect on the "rhetoric" of Smith's theorizing in the area of moral philosophy.[15] I shall be concerned with what we might call the philosophical rationale for the rhetorical dimension of the *TMS*. The issue I wish to focus on concerns not so much how the categories of rhetorical discourse set out in the *Lectures on Rhetoric and Belles Lettres* might apply to the *TMS*,[16] but rather Smith's understanding of the rhetorical demands generated by his conception of ethics. It is obvious at a glance that, say, Spinoza's way of writing an ethics would for Smith completely distort the subject matter. This is because Smith's conception of ethics is radically different from Spinoza's; for Smith, ethics just does not lend itself to *that* kind of articulation. To overstate slightly; as Spinoza's *Ethics* is modeled on geometrical deduction, so Smith's is modeled on literary, indeed "dramatic," representation. As Spinoza mathematicizes ethics, so Smith aestheticizes ethics. In neither case is the rhetoric of theorizing merely window dressing. In neither case is the rhetoric in question determined by the intended audience alone. Rather, an understanding of what it means to "do" ethics heavily influences the form the rhetoric takes.

The question as to the philosophical rationale for the rhetorical dimension of the *TMS* is a special case of the general problem of the connection between "form" and "content" in philosophical writing. In recent years the general issue has been subjected to extensive discussion. It is being pursued with vigor all the way from Greek philosophy (where Plato's use of the dialogue form has earned particular attention) to modern philosophy (one thinks of the discussions of the form of Descartes' *Meditations*) to contemporary, post-structuralist discussions of the ways in which rhetoric and argument influence each other.[17] A good deal of attention has also been devoted to the specific issue of writing about ethics.[18] The present essay attempts to enroll Smith in these broader discussions.

I turn first (part II) to discussion of some of the rhetorical features of the *TMS*. In part III, I shall focus on Smith's reflections on what it means to do ethics, and shall then reconstruct Smith's rationale for writing about the 'moral sentiments' in the ways specified in part II. In part IV, I discuss Smith's scepticism about the role of reason in ethics, I conclude in part V by pointing to three questions raised by my account of Smith. The issues are complex; the present essay should be taken as a preface to a more detailed treatment.

## II

As noted above, it seems a reasonably well-attested fact that much of the *TMS* originated in the Edinburgh and Glasgow lectures. There are interesting traces of that lecture-hall origin in the published text. I refer not to the occasional use of "upon a former occasion" where "in chapter Y above" would have been more appropriate,[19] so much as to the strong sense of audience and ethical community evident throughout the *TMS*. The sense of audience and ethical community are signalled by Smith's use of what I shall call the "protreptic 'we'." The pronoun first occurs in the second sentence of the *TMS*.[20] His use of the pronoun is "protreptic" in that it is intended to persuade us to view things in a certain light, and at least occasionally to act in a certain way. Smith uses the first person plural pronoun both to adduce evidence – in the form of concrete observations about ethical situations and "our" reactions to them – for propositions in moral psychology, and to pass ethical judgments on all manner of issues.[21] The *TMS* is full of unabashed moralizing. Evidently Smith does not think that moralizing and theorizing are, in a work on ethics, at odds with each other. Smith's theorizing about the expression of moral sentiments relies on his self-understanding as a moral actor. The premises from which the writing of ethics proceeds are drawn not from the nature of practical reason and of the will, as in Kant, but from shared insights into what the virtues are and into the reasons for which certain dispositions are denominated virtues.

The dual function of the "we" just adumbrated reflects the duality of questions that Smith is examining in the *TMS*. We are told in part VII of the book that in "treating of the principles of morals" one is to discuss, first, what it is that makes a character praiseworthy. This comes to understanding "virtue." Second, we are also to examine "by what power or faculty in the mind is it, that this character, whatever it be, is recommended to us?" This comes to understanding moral psychology (p. 265). While in pursuing answers to both questions Smith appeals to "our" predilections, there is a crucial difference between the two. Answers to the "what is virtue" question can be expected to affect our ethical behaviour while, Smith says, answers to the moral psychology question will not do so. The latter question is a purely theoretical one (p. 315). Consequently Smith's protrepticism will be double in intent, with both 'practical' and 'theoretical' purposes, and may assume a double "we". In the course of his discussion (excepting part VII), however, Smith does not specify systematically which of the two questions he is dealing with, and in any event he appeals to "our" ordinary self-understanding throughout. Thus while the two questions, and the two "we's," are distinct, the answers to them may be interdependent.[22]

Two other features of the rhetoric of the *TMS* should be noted before we progress further. The first is the remarkable role played by examples throughout Smith's discussion. We are asked over and over again to consider this or that situation and this or that reaction to a situation, and to draw the appropriate moral. The examples are sometimes elaborated into little stories about a human life. As Smith can write elegantly, the stories are sometimes quite powerful (e.g., I think of the sketch of the life of the "prudent man" on pp. 214–215, of the lover of wealth on pp. 182–183, and Smith's imagined dialogue between King Pyrrhus and Cineas on pp. 149–150). The examples and stories are often invented by Smith; but, secondly, literature also plays an important and variegated role in the *TMS*. Dramas, novels, and poems offer Smith a source of examples and stories; tragedies in particular seem to intrigue Smith. The notion that we are to understand literature as a source of moral theory is clearly, and strikingly, evident in Smith.[23] So permeated with examples, stories, literary references and allusions, and images is the *TMS* that at times it presents the character of a novel; narrative and analysis are interwoven throughout.

Especially with respect to the appeals to our ordinary self-understanding, the *TMS* exhibits a distinct and interesting rhetoric that would find a natural home in the living context of oral, classroom discussion. As written, the book is addressed to a much broader audience, and one is entitled to ask whether Smith had reflected on the constraints that writing, as distinguished from speaking, imposes on rhetoric. There exists at least one important indication that the answer is affirmative, namely the organization of the work. The sequence of topics of the *TMS* deviates significantly from what we would have expected in a classroom presentation of the same material – and indeed from the order Smith apparently followed when presenting the material orally. I refer to the placement at the *end* of the work of the discussion of the history of moral philosophy.[24] I shall argue in the following section of the paper that there are good methodological reasons for the placement of the historical material at the end of the written work.

We can begin to understand these features of Smith's ethical writing by looking first at his reflections, presented in part VII of the *TMS*, as to how ethics has been understood, and then by turning to his views as to how ethics ought to be understood.

### III

In part VII of the *TMS*, Smith reviews the history of moral philosophy, focusing first on the "what is virtue?" question, and second on the moral psychology question. I note, first, that there are some remarkable silences in part VII. The discussion is interestingly selective. Thematically, Smith entirely omits Plato's metaphysics, even though the theory of Forms is intrinsically connected, in the *Republic*, with the ethics (to which Smith approvingly refers). A similar pattern occurs with the treatment of Aristotle and the Stoics. In general, Smith shows scarcely any interest in the *TMS* in the epistemological and metaphysical views of his predecessors. He proceeds as though ethics were a discipline that has been undertaken in isolation from these other fields. That this is, to say the least, a limited take on the history of moral philosophy cannot have escaped his attention.[25]

Moreover, he himself proceeds as though the isolation is warranted: the *TMS* is not part of an articulated metaphysical or epistemological scheme (say, in the way that Aristotle's *Nicomachean Ethics* has ties to the *Metaphysics, De Anima*, and so forth), and Smith did not publish a book on metaphysics or epistemology. There is no Smithean analogue to, say, book I of Hume's *Treatise*, to Locke's *Essay*, and so forth. A hypothesis (to be further explored below, part IV) for explaining Smith's selectivity suggests itself, namely that in one way or another, Plato and Aristotle (among others) allow a link between virtue and reason that goes far beyond anything Smith will allow. Smith's self-understanding is that, while reason figures in morality in terms of the inductive derivation of general rules, morality is basically the affair of "immediate sense and feeling" (p. 320).[26] It is often said that Smith's ethics is anti-rationalist, and with proper qualifications this is no doubt true (e.g., see pp. 318–320). Smith's anti-rationalism – or better, his scepticism about claims to objectivity in ethics – suggests that he thinks he can size up what is useful in the history of ethics while just dropping from the discussion an immensely long and complex dimension of that history. Smith's remarks in the *Essays on Philosophical Subjects* about the psychological value of economy in theoretical explanation suggests a further motive for wishing not to 'burden the imagination' with more elaborate, metaphysical, explanations of the phenomena.[27]

In terms of the figures discussed in *TMS* VII, the most striking omission concerns the classical Sceptics. Indeed, Smith says not a word about Scepticism in any of his writings. Yet Scepticism and issues concerning Sceptical ethics were alive and well in Smith's time, thanks in good part to Bayle, Berkeley, and above all, Hume.[28] As in the case just examined, the omission is best understood as deliberate. Before considering the rationale for this omission, let me take note of another feature of *TMS* VII.

When Smith argues against the views of others, both in part VII and elsewhere in the *TMS*, the argument ultimately proceeds something as follows. Such and such a view is incorrect because it does not do justice to the self-understanding of ordinary moral actors (among whom Smith includes himself). This approach is applied to Hume's view of the centrality of utility to virtue (p. 188); to Epicurean reductionism (p. 299); to Chrysippus' reduction of morality to a system of rules (p. 291); to Hutcheson's reduction of virtue to benevolence (p. 304); to Hobbes' reduction of approbation to self-love (pp. 315–317); to the tendency of philosophers "of late years" to overlook the relevance to ethical analysis of the cause that excites the affections (p. 18).[29] For Smith, these various reductions are the product of the philosopher listening to the demands – at times, aesthetic demands – of his or her own reason, rather than to the only evidence that could verify or falsify a philosophical analysis of morality, namely the pre-philosophical self-understanding of ordinary moral actors (cf. p. 299).

How does Smith think it best to proceed in ethics? I begin by noting three senses in which Smith's recommended procedure is Aristotelian.[30] The first we have already signalled; it consists in Smith's insistence – reminiscent of Aristotle's anti-Platonic view of the relevance of the "endoxa" for the moral theorist – that we orient ourselves by the self-understanding of moral actors. The nature of the evidence is pre-philosophical.[31] Broadly speaking, the Aristotelian assumption is that human nature is such as to be suited to understand *arete*. Not everyone fully understands his or her own moral beliefs, of course, and for Smith as for Aristotle entire cultures can be misled about crucial moral questions (such as, for Smith, infanticide; pp. 209–211). The appeal to the standards of the ethically well-developed actor is not just an appeal to *any* actor. Ordinary experience must be refined through reflection. To be sure, Smith's and Aristotle's reasons for making philosophy answerable to the *endoxa* differs, thanks to Smith's avoidance of metaphysics.[32] Yet their general procedures are strikingly similar.

Second, just as Smith agrees that we cannot understand ethics from an external perspective, so he would seem to agree that lectures on ethics cannot be expected to make the vicious into good persons.[33] Such lectures can, however, reinforce ethical life (*E.N.* 1103b26–29), but those "young" in age or spirit will find them useless. Aristotle (like other "ancient moralists") describes the various virtues and vices vividly, presenting us "with agreeable and lively pictures of manners," and this in part to educate our capacity to judge ethically – thereby indirectly assisting us to be ethical. Smith declares: "In treating of the rules of morality, in this manner, consists the science which is properly called Ethics, a science which, though like criticism it does not admit of the most accurate precision, is, however, both highly useful and agreeable." Smith goes on to note that ethics is "of all others the most susceptible of the embellishments of eloquence" such that "whatever precept and exhortation can do to animate us to the practice of virtue, is done by this science delivered in this manner" (p. 329). Aristotle's approach to ethics, in other words, is commendable in part because it is protreptic.

As noted above, Smith writes as pedagogue, in part to improve the morally educable, in part to satisfy "philosophical curiosity" (p. 315). Presumably Smith thinks of the *TMS* as at least in part a work in "Ethics," and to that extent as falling under his own description of that "science." Insofar as he seeks to improve his readers – to convey a teaching about virtue that may influence our behavior – his protrepticism is of ethical intent. Like Aristotle, Smith is not merely reflecting the intuitions and moral vocabulary of his time, as in a mirror. He seeks to encourage an intelligent, discerning pursuit of given virtues (understood as excellences or refinements of character).

Third, Smith's procedure is Aristotelian in its rejection of standards of extreme exactitude in ethics. With the exception, for Smith, of justice, he does not think there are any exact rules for judging virtue or duty. In general, he is strongly critical of moral absolutism. Nor does he think that there is a decision procedure (we would think of Kant's categorical imperative) for determining how to act well. Right ethical action and reaction require good judgment. Smith offers us an analogy. Casuists, he tells us, proceed as though they were grammarians, attempting to specify exact rules, taking as their model the rules of justice, which do admit of precise specification (pp. 175–176, 226–227, 327 ff.). Smith associates that approach with the Christian theological tradition, as well as with fanaticism.[34] By contrast, Smith suggests that we should recognize that the virtues at best admit of rules that are "loose, vague, and indeterminate," rules comparable "to those which critics lay down for the attainment of what is sublime and elegant in composition, and which present us rather with a general idea of the perfection we ought to aim at, than afford us any certain and infallible directions for acquiring it" (p. 327).[35] Smith implies that Aristotle writes like a critic rather than a grammarian (pp. 327–329).

The analogy illustrates not just varying standards of exactitude, but also three other points. First, one task of the critic is to describe the spectacle he or she observes, to convey it as vividly as possible to others who in turn will "see" the original through an act of imagination. Second, the critic analyzes the performance, considering the coherence of its parts, the principles which connect it into a unified act, the intentions of the event. In the realm of Smithean ethics, this would correspond to the moral psychology that provides the underlying explanation of moral judgments. Finally, the critic will evaluate the performance; her or his activity is in this regard inescapably value-laden. Some standard as to what would constitute a good performance is indispensable. Smith mentions two possible standards, the one of "complete propriety and perfection" that no human being could achieve; the second "the idea of that degree of proximity or distance from this complete perfection, which the actions of the greater part of men commonly arrive at" (p. 26). He indicates that he has chosen the latter standard.[36] When doing so he claims that "the present inquiry [presumably the whole of the *TMS*] is not concerning a matter of right, if I may say so, but concerning a matter of fact" (p. 77). If Smith writes as a critic, however, he cannot but choose a standard by which to judge the performance, even if it is the lower of two standards. As judging critic, Smith will pass evaluative judgments, and this he does constantly throughout the *TMS*, just as Aristotle does in the *E.N.* This is not a violation of a theorist's objectivity; rather, it is an extension of Smith's understanding of what it means to do ethics objectively. Smith writes as a critic, but the critic is both a certain kind of moral actor and moral theorist. Let me explain further.

While Smith leaves it unclear what species of critic is the best model for the doing of ethics, the controlling metaphors of the *TMS* strongly suggest that the theater critic is the appropriate model.[37] The very terms "spectator" and "actor" are of that provenance. The striking "theatricality" of the moral world for Smith is worthy of separate treatment.[38] I restrict myself to noting that critics are spectators of a certain sort; for they too sit in the audience, and indeed in some portion of the audience that allows them to observe both the *dramatis personae* and the audience. In writing, they act in the theater. It is more important for my purposes that every critic is a moral actor; one criticizes, for Smith, in the *persona* of a moral actor reflecting. And, like those one criticizes, the critic is in the theater too. The critic's objectivity is not that of an entirely external spectator outside the theater (one on the outside looking in through the windows, as it were). Objectivity would seem to lie instead in a certain impartiality (defined so as to include detachment but not so as to imply a completely external vantage point), to borrow another term from Smith's account of ethical judgment.[39]

Smith uses the simile of the critic not just to describe how moral actors render judgments, but how ethicists should go about doing ethics. On other occasions too Smith uses the terms appropriate to his moral psychology to describe his own position as moral theorist relative to actors on the stage. He casts the philosopher/scientist as a spectator of the "theatre of nature" and the moral theorist as a spectator of the "spectacle of human life" (p. 59).[40] Provoked to wonder by various aspects of the drama he or she observes, the philosopher's imagination develops a systematic explanation. We do not go too far in suggesting that the *TMS* is in part a "theory" in one Greek sense of the term; it is in part a viewing, a looking-at.[41] With respect to both the theorist and moral actor, successful reflection on the phenomena requires "sympathy" in Smith's sense of the term. The key notions of Smith's moral psychology – imagination, spectator, theater, sympathy – are applicable to his stance as theorist. The transposition of vocabulary from "practical" to "theoretical" realms reflects Smith's rejection of philosophical constructions that ignore the pre-philosophical evidence, as well as Smith's propensity to treat the theorist – at least the moral theorist – as a reflective moral actor.

Not only does Smith's theoretical vocabulary reflect our ordinary moral vocabulary, his rhetoric reflects the ways in which he thinks we actually go about forming moral judgments.[42] It is a key tenet of Smith's theory that we arrive at moral judgments by considering the details of particular situations. As already noted, moral "rules" are inductively arrived at 'later', on the basis of experience of the particulars. Hence his constant use of examples, stories about this or that case or event, and images, is grounded in the primacy of particulars over rules. Smith takes it that we pass judgments from within social contexts; we are spectators of each other, but spectators aware of being actors in the eyes of other spectators. Smith's use of the first person plural replicates what he takes to be a fact of moral psychology, namely that one's moral judgments are mediated by community.

Further, Smith thinks that in making moral judgments about another we put ourselves into that other's situation. This requires an act of the imagination. The imagination is the glue that holds society together; without it, we can share no moral connections with one another. Smith's constant invocations of examples and stories, and his frequent allusions and references to various dramas (particularly tragedies), elicit the work of the moral imagination. Smith's use of the "we" is both protreptic and consistent with his conviction that we can talk philosophically about moral judgments only with people who can imaginatively enter into the particulars of a situation and who are capable (with sufficient reflection) of rendering a judgment that is impartial. To understand morality we need to get "inside" people's intentions and responses, to see a situation from their standpoint. It is not enough to observe patterns of social behavior and cooperation, for those we share with bees and ants. The reader of a Smithean work in ethics needs to be drawn into the drama, his or her emotions and imagination need to be put to work. For the "evidence" in ethics, once again, derives from "our" moral self-understanding. The activity of articulating the subject matter already implies a community. Since Smith writes as a Smithean, he cannot "prove" that his normative judgments are "right" except by persuading other, potentially impartial, persons to agree. And that is how we proceed (according to Smith) in the course of passing judgments as moral actors. The rhetorical dimension of the *TMS*, then, is not mere literary adornment. It reflects Smith's conception of the subject matter.

We may now see better why Smith puts his discussion of the history of moral philosophy at the end of the book. In so doing, Smith wished to avoid casting his argument as an academic one, and as one dependent on a grasp of the history of the subject matter. The arrangement fits Smith's view that ethics is a subject that a reflective pre-philosophical moral actor can enter into, as well as the view that the subject matter of ethics is oriented by the judgments of the ordinary moral actor (not by the judgments of a professor about the history of the "science").[43] A *written* philosophical work runs particular risks of encouraging an "academic" detachment from ordinary life and of reducing ethical debate to a merely theoretical, perhaps casuistic, enterprise. The placement of the historical material in part VII helps indicate that ethics is more an exercise of 'practical reason' than 'theoretical reason.'

It should be evident by this point that Smith assumes that reason will have a restricted role to play in moral theorizing. In the final section of this paper, I would like to reflect briefly on this assumption.

IV

I referred above to Smith's moral theory as (with qualifications) anti-rationalist. There are two sides to this anti-rationalism worth mentioning here. First, and most obvious, Smith's account earns the title by restricting the role that reason plays in moral judgment. Reason can formulate general rules of morality, and deliberate about means to an end, but it does not provide us with "the first perceptions of right and wrong" (pp. 319–320). The limited role Smith assigns to "reason" represents the other side of the coin of the emphasis on the pre-philosophical world as the source of moral intuitions. It does not follow that Smith's moral theory is simply non-cognitivist or emotivist. Second, Smith's account is sceptical about objectivity in ethics. Strictly speaking, a moral theory could be anti-rationalist in the first sense but still be objectivist. However, Smith associates moral rationalism with moral realism (pp. 318–319), and in light of the history of moral philosophy, the association is not unnatural. Reason is easily seen as grasping objective "moral facts" while sentiment expresses subjective dispositions. Smith's dispositionalist account of the meaning of moral terms by reference to moral sentiment clearly subjectivizes them; he states that the meaning of moral terms consists in what pleases or displeases the "moral faculties" (p. 165). Smith writes:

Reason may show that this object [any particular object] is the means of obtaining some other which is naturally either pleasing or displeasing, and in this manner may render it either agreeable or disagreeable for the sake of something else. But nothing can be agreeable or disagreeable for its own sake, which is not rendered such by immediate sense and feeling. If virtue, therefore, in every particular instance, necessarily pleases for its own sake, and if vice as certainly displeases the mind, it cannot be reason, but immediate sense and feeling, which, in this manner, reconciles us to the one, and alienates us from the other. (p. 320)

It is clear that, for Smith, not only are there no answers to questions about what "really" is good or bad independent of the responses of "our" passions or sentiments, but also that moral qualities are "rendered" such by the moral sentiments.[44]

Smith's anti-rationalism and anti-realism permit us to characterize his position as Sceptical. The precise meaning of this term as applied to Smith, and the relation of Smithean Scepticism to the Sceptical tradition, are very complicated issues. I shall say something about each issue by way of illuminating my central theme concerning the rhetoric of Smith's theorizing. Given my limited purposes, my comments should be taken as programmatic.

Smith does not hold, to begin with, that as moral actors we treat morality as a Sceptic might. Rather, we act as though moral realism were true, as though moral qualities existed objectively in the nature of things. As the sentences just quoted suggest, virtue is taken to please "for its own sake." Smith does not say whether people act as though some form of moral rationalism (understood as the view that ethical qualities are grasped by reason rather than sentiment) were also true, but nothing in his account prevents it, and his assumptions about the connection between reason and objectivity support it. In practice, then, we are not moral Sceptics. Nor, secondly, does Smith think that his moral psychology will in practice undermine our self-understanding as moral actors. As noted, Smith explicitly declares that debates about moral psychology have no effects in practice (p. 315), and he no doubt takes that to be all for the better.[45] Hence Smith's Scepticism is limited to the realm of theory. His Scepticism seems to be of the Humean rather than Pyrrhonist variety for, like Hume, he seems to hold that no amount of Sceptical theorizing can budge our everyday beliefs.[46] We do not live by philosophical theory. Hume's account of moral judgment in terms of the passions runs along the same general lines as Smith's, and indeed the remarks about the moral sentiments that Hume puts into the mouth of the Sceptic sound rather like Smith's remarks.[47]

Presumably Smith does not think, however, that his discussion will have the effect of transforming us into Sceptics even on the level of theory. In his discussion of the relationship between theory and practice early in the book, Smith declares that "we approve of another man's judgment [about "the subjects of science and taste"] not as something useful, but as right, as accurate, as agreeable to truth and reality: and it is evident we attribute those qualities to it for no other reason but because we find that it agrees with our own" (p. 20). That is, while in fact another person's judgment appeals to us because of subjective psychological properties – our intellectual sentiments, as it were, of surprise, wonder, and admiration (p. 20) – we talk as though the reason had to do with the correspondence between a judgment and reality.[48] In his discussion of Newton, Smith suggests that the more satisfying to the intellectual imagination a proposed system is, the more given are we to taking the system as describing how things are objectively.[49] As the *TMS* represents – at least in its discussion of moral psychology – examples of Smithean theorizing, he presumably expects it to be treated for the most part as an attempt to represent how things are. The TMS is protreptic with respect to that portion of the audience interested in moral psychology in that it seeks to persuade that a certain interpretation of morality satisfies the mind and soothes the intellectual imagination. If my account is right, it also seeks to persuade us – indirectly – that discussion of metaphysical and epistemological issues can largely be dropped, and this for Sceptical motives.

If Smith's theorizing is meant to articulate how things really are, objectively, then it is both "insulated"[50] from practice and amounts to "negative Scepticism." A pure Sceptic will avoid making any claims as to whether reality can or cannot be known, whether moral judgments really reflect "human nature." A pure Sceptic will not have a "philosophical position" on these matters; he or she will "suspend judgment" on the entire controversy, adopting a sort of passive openness towards the debate. In practice this Sceptic will be guided by various non-philosophical sources, including natural impulses, custom, and feelings. He will be guided by how things "appear" to him but will not assent to anything that is non-evident to him (anything that goes beyond the appearances). Within these constraints, the Sceptic can have conceptions about things.[51] This is the sort of Scepticism that Sextus Empiricus wished to affirm. Sextus talks about things appearing to reason and about objects of thought as appearances.[52] When engaging in philosophical argument, this sort of Sceptic will adopt the linguistic practices of his opponent and will argue *ad hominem*, but without affirming these practices as his own.[53] By contrast, a negative Sceptic has a position – say, that nothing in morals can be known objectively – which he or she will defend as being the case.

This position would be perfectly compatible with holding that, say, in the realm of non-moral facts we may possess "positive" knowledge about how things are (in the natural sciences, for example).[54]

The question as to whether or not Smith's "theory" of moral sentiments is "dogmatic" is not easy to settle. I hazard the judgment that Smith did not think his theory "dogmatic," even while he indicates that it will be read as though it were. First, Smith is Sceptical across the board; his account of theorizing in science, philosophy, and ethics is constant, and it appeals throughout to the "intellectual sentiments." Second, what we might provisionally refer to as Smith's meta-theory to the effect that theorizing is of such-and-such a sort is not tied by Smith to any metaphysics or epistemology that would warrant him to speak about "reality" or "human essence" or "how things are objectively." Smith nowhere argues for his own presentation in those sorts of terms. I noted above Smith's striking silence with respect to metaphysical and epistemological issues. I am pointing out that he does not argue for his own views, if arguing for a view means showing it to be objectively the case. I add that he hardly ever *argues* against rationalist and realist views. In general, he simply drops metaphysics and epistemology from the discussion. He proceeds as though he has suspended judgment about such matters. He instead presents us with a view of the matter that makes an appeal to "our" sensibilities and singles out various principles (such as that of "sympathy") which explain the ways in which we observe moral sentiments operating.

If Smith takes himself to be a non-dogmatic Sceptic with respect to theorizing, there is good reason for him to have avoided arguing for the truth of non-dogmatic Scepticism. He may have believed that to do so would be to contradict the doctrine in question. That is, I suggest that Smith grasped the self-referential problem – a problem long discussed in the history of philosophy, starting with Plato's *Theaetetus*, then taken up at the start of Sextus' "Outlines of Pyrrhonism" (I 14–15, in a chapter titled "Does the Sceptic Dogmatize?"), and also sketched by Hume – inherent in a Sceptic's assertion that Scepticism is "true."[55] For such an assertion would seem to belie the Sceptic's view that no statements about the "truth" of philosophical doctrines can be sustained in face of arguments to the contrary. I suggest, then, that in theorizing about ethics Smiths *enacts* Scepticism. He may therefore be interpreted as following out Hume's Sceptical program to its limit, and perhaps as doing so more consistently than Hume did.[56]

And here, finally, we again see why rhetoric would be so important for Smith. For if as a non-dogmatic Sceptic, Smith does not have a philosophical meta-theory (our provisional use of the term may now be dropped), it is natural for him to frame his theorizing in non-philosophical, rhetorical terms. Specifically, the rhetoric of the theater lends itself to the sort of framing in question. That rhetoric provides a way of articulating Smith's role as spectator viewing a peculiar "appearance," namely the appearance that is the "spectacle of human life." If we go along with Smith's picture of the theorist as a critic within the theater, as suggested above, we have a way of articulating the distance that permits him to offer an analysis of the drama, without pretending that he is somewhere outside the theater proposing a 'view from nowhere.' The picture does not commit itself (at least initially) to a meta-theory and attendant epistemological and metaphysical assumptions – all of which would compromise Smith's non-dogmatic Scepticism.[57] The vocabulary of drama and criticism thus seems fundamental to Smith's articulation of his own position as theorist. My explanation of Smith's sceptical view of "theory" helps us understand why it is that in the *TMS* the curtain just goes up, as it were, on a view of the moral sentiments, without the philosophical commentary on issues epistemological and metaphysical that one would expect from a philosophical theorist.

That view would seem to require a rather odd detachment of Smith from himself as theorist. While others will commonly take him as proposing a philosophical theory "agreeable to truth and reality," Smith will not (as author) view himself in that light (at least not if "truth and reality" mean more than "how things appear to a reflective and well informed moral actor"). Nor will he view his moral psychology as a practical guide. It seems that he will contemplate his own reflecting self as though – to borrow Hume's phrase – his self were a kind of theater.[58] A similar detachment from self has been thought to be a feature of ancient Scepticism.[59] One expects that in contemplating his theory of moral sentiments, Smith finds the tranquillity or repose of the imagination that is (according to Smith himself) the goal of theorizing.[60]

This is not a repose that many are likely to share. The other theatergoers will, presumably, find little time for pursuing the intricacies of Smith's theories; they will not possess the self-knowledge those theories can offer. A bit like the cave dwellers in Plato's simile, they see without realizing why they praise or blame. Unlike Plato, Smith does not take this as a cause for unmixed lament; with the Stoics, Smith finds redemption in this earthly drama.[61] And if people were to read Smith, they would probably (as Smith's own observations suggest) take him as articulating a view "agreeable to reality." Tranquillity might be theirs, but in missing the Sceptical dimension, they would forgo the full measure of detachment mentioned in the preceding paragraph.

V

I conclude by taking note of three questions to which my discussion gives rise, only the first of which I shall attempt to answer here.

To begin with, Smith stresses in the lectures on rhetoric that the "Newtonian" method of explanation is considerably more satisfying than the Aristotelian. The former seeks to provide a minimal set of principles in light of which the variegated phenomena can be explained, while the latter is considerably less systematic in intent.[62] I have stressed the "Aristotelian" elements of Smith's theory, such as the insistence on the importance of examples to the theorist, and the theorist's reliance on an understanding of particulars. Are the Newtonian and Aristotelian sides of the argument at odds with each other? I noted above that Smith accuses other philosophers of having succumbed to the philosopher's propensity to place more importance on accounting for the phenomena from as few principles as possible than on preserving the complexity of the phenomena. Assuming that Smith is not accounting for the phenomena from too few principles, it does not seem that his effort to account for them in light of several principles (such as the principle of "sympathy") must conflict with his Aristotelianism. The 'Newtonian' method is recommended by Smith solely because of its aesthetic, imagination-soothing qualities. In so far as it leads the theorist to overlook the phenomena, its appeal would diminish. Moreover, the explanatory principles are not prescriptive but descriptive. They are arrived at by careful reflection on the particulars, and must always remain open to revision in light of examples. A good deal of Smith's writing and lecturing in political economy and law seeks to test his explanatory principles in light of examples drawn from other periods and cultures.

A more difficult question is whether Smith's moral psychology conflicts with his view that the ordinary self-understanding of moral actors should guide theorizing; for Smith's moral psychology has implications concerning the "objectivity" of ethical judgments that, as we have seen, differ – on Smith's own account – from "our" self-understanding.

Is the moral psychology implicitly reductive in that respect? The critic of the play need not simply repeat the actors' lines; how far may the Smithean critic depart from the actors' self-understanding before being liable to the criticism of having rewritten the play?

Finally, we may ask whether the Smithean "we" is unavoidably parochial. Is Smith's "we" reducible, for example, to a community with which only eighteenth and nineteenth century urbanized Westerners may be expected to sympathize? Similar questions have been raised about Aristotle's and Hume's appeals to communities: MacIntyre, for example, argues that both communities are reasonably well defined, and relatively local.[63] The question can be posed not just from without – with reference to Smith's system as a whole – but also from within that system. In the *TMS*, Smith clearly passes various judgments that are not universally shared (by his own admission, the judgment about infanticide is a case in point). Does Smith's view of ethics grant the irreducibly historical character of ethical judgments, such that no rational mediation between competing judgments is possible? For Smith, is there anything like 'rational mediation' within the moral drama? I expect that Smith would see the "outside" and "inside" questions as connected. We are free to see them as distinct.[64]

*Philosophy Department*
*Boston University*

## NOTES

\* This article is reprinted (with light emendations) from *Philosophy and Rhetoric* Vol. 24, No. 3 (1991): 213–237. Copyright 1991 by The Pennsylvania State University. Reproduced by permission of The Pennsylvania State University Press.

*Author's Note*: One of my first substantive philosophical discussions with Bob Cohen, some four years ago, concerned Adam Smith and Marx. Bob's sympathetic and critical grasp of both thinkers impressed me deeply. I offer him this essay as a contribution to our ongoing dialogue and friendship.

[1] While Smith's departure from Glasgow terminated his formal teaching career, he then served as tutor to Henry Scott (3rd Duke of Buccleuch) for two years.

[2] Concerning Smith's success as teacher at Glasgow, see Raphael, *Adam Smith* (Oxford: Oxford University Press, 1985), p. 12; and Campbell and Skinner's *Adam Smith* (London: Croom Helm, 1985), pp. 40–42. The latter quote a report from John Millar about Smith's teaching, in the course of which Millar comments on Smith's plentiful use of examples "calculated to seize the attention of the audience, and to afford them pleasure, as well as instruction" (p. 41).

³ See D. Stewart's comments on Smith in *Biographical Memoirs of Adam Smith, William Robertson, and Thomas Reid*, W. Hamilton (ed.) (1858; rpt. New York: A.M. Kelley, 1966), p. 10; and W. R. Scott's *Adam Smith as Student and Professor* (1937; rpt. New York: A. M. Kelley, 1965), pp. 50, 54–55.
⁴ See *Adam Smith*, by R.H. Campbell and A.S. Skinner, pp. 29–30.
⁵ See Smith's letter to W. Cullen of Sept. 3, 1751: in the *Correspondence of Adam Smith*, E.C. Mossner and I.S. Ross (eds.) (Indianapolis, IN: Liberty Press, 1987), p. 5. D.D. Raphael notes that while at Glasgow "Smith's lectures on moral philosophy were divided into three sections, natural theology, ethics, and jurisprudence." "This followed the practice," Raphael continues, "of his predecessors and indeed the general tradition of Scottish moral philosophy of the period." D.D. Raphael, *Adam Smith*, p. 13. "Jurisprudence" included, for Smith, political economy.
⁶ D. Stewart quoting from a report by J. Millar, in *Memoirs*. p. 11. Millar goes on to report that for Smith there was "no branch of literature more suited to youth at their first entrance upon philosophy than this, which lays hold of their taste and their feelings."
⁷ These are now available in *Lectures on Rhetoric and Belles Lettres (LRBL)*, J.C. Bryce (ed.) (Indianapolis, IN: Liberty Press, 1985). In his introduction to the volume, Bryce notes that the substance of the lectures seems to have remained more or less constant since Smith began lecturing on rhetoric in 1748 (p. 12).
⁸ Smith published other editions of the *TMS* in 1767, 1774, 1781, 1790. He published editions of the *Wealth of Nations* in 1776, 1778, 1784, 1786, 1789. In some cases the variations between editions is extremely minor, in other cases great. Stewart indicates that the *Wealth of Nations*, like the *Theory of Moral Sentiments*, had its origins in the lectures at Glasgow (*Memoirs*, pp. 10, 66).
⁹ "There is also clear evidence that Smith's ethical theory developed significantly in the course of his twelve years as Professor of Moral Philosophy, both before and after the publication of the first edition of *The Theory of Moral Sentiments* in 1759." Raphael, *Adam Smith*, p. 14.
¹⁰ Raphael, *Adam Smith*, pp. 15–16.
¹¹ See Dionysius of Halicarnassus, *On Literary Composition*, R. Roberts (ed.) (London: Macmillan, 1910), pp. 264–265. As Adam Smith himself put it (with reference to his revising the *Theory of Moral Sentiments* in preparation for the sixth edition): "I am a slow a very slow workman, who do and undo everything I write at least half a dozen of times before I can be tolerably pleased with it. . . ." Letter to T. Cadell of March 15, 1788, *Correspondence*, p. 311.
¹² We ought not to rest content with noting, say, that Smith's use of examples in the *TMS* is to be explained by the origins of the book in his oral lectures and by the young age of his auditors at Glasgow (auditors who, as mere schoolboys, would require plenty of rhetorical language if they were to stay interested).
¹³ I refer to Hume's "Of Essay Writing" (1742) and "Of Simplicity and Refinement in Writing" (1742). Smith might also have known of Hume's comments about Plutarch's use of the dialogue form in a note to "Of the Populousness of Ancient Nations" (1754); see *David Hume: Essays Moral, Political, and Literary*, E.F. Miller (ed.) (Indianapolis, IN: Liberty Press, 1985), p. 463, n. 278.
¹⁴ D. Marshall notes that Shaftesbury advocated writing in dialogue form: *The Figure of the Theater* (New York: Columbia University Press, 1986), p. 3.

[15] Consequently I shall not here discuss the questions as to whether the *Wealth of Nations* has its own distinctive rhetoric, whether the form/content issue arises in that book as well, and whether the rhetorical frameworks of the *WN* and *TMS* cohere with one another. However, my comments at the end of section IV of this essay about the constant sense of "theorizing" for Smith point to a unitary interpretation of the rhetoric of the two works. I note that in the *TMS* Smith discusses ways in which political theory can persuade and animate, and concludes "Nothing tends so much to promote public spirit as the study of politics. . . . Upon this account political disquisitions, if just, and reasonable, and practicable, are of all the works of speculation the most useful. . . . They serve at least to animate the public passions of men, and rouse them to seek out the means of promoting the happiness of the society" (pp. 186–187). D. Stewart tells us that the *WN* was intended to influence the "actual legislators" who craft the laws composing a nation's political economy; *Memoirs,* pp. 55–56. D. Winch cites Smith's comments on the American Revolution as examples of didactic and rhetorical discourse. *Adam Smith's Politics* (Cambridge: Cambridge University Press, 1978). p. 171.

[16] For discussion of those lectures see W.S. Howell's "Adam Smith's 'Lectures on Rhetoric': an Historical Assessment," in *Essays on Adam Smith*, A.S. Skinner and T. Wilson (eds.) (Oxford: Clarendon Press, 1975). pp. 11–43; and J.M. Hogan, "Historiography and Ethics in Adam Smith's Lectures on Rhetoric. 1762–1763." *Rhetorica* **2** (1984): 75–91.

[17] For the discussion with respect to Plato, see the Introduction and chapter 6 of my *Self-knowledge in Plato's Phaedrus* (New Haven, CT: Yale University Press, 1986).

[18] For a start, see the essays in *Literature and the Question of Philosophy*, A.J. Cascardi (ed.) (Baltimore, MD: Johns Hopkins University Press, 1987), particularly the essay by M. Nussbaum entitled "'Finely Aware and Richly Responsible': Literature and the Moral Imagination."

[19] As in the *TMS*, A.L. Macfie and D.D. Raphael (eds.) (Indianapolis, IN: Liberty Press, 1982), p. 190. The editors note there that "The word 'occasion' again shows the original lecture form of the material. . . ." (As the Glasgow edition of Smith's works is now standard, when citing from the *TMS*, I shall cite by page number rather than by section and chapter numbers. The Liberty Press editions are exact reproductions of the Glasgow editions published originally by Oxford University Press. Page numbers in my text refer to the *TMS*.)

[20] For example, Smith opens the *TMS* with these sentences: "How selfish soever man may be supposed, there are evidently some principles in his nature, which interest him in the fortune of others, and render their happiness necessary to him, though he derives nothing from it except the pleasure of seeing it. Of this kind is pity or compassion, the emotion which we feel for the misery of others, when we either see it, or are made to conceive it in a very lively manner. That we often derive sorrow from the sorrow of others, is a matter of fact too obvious to require any instances to prove it. . . ." The last sentence of the second to last paragraph of part VI (added in ed. 6) runs "In our approbation of all these virtues, our sense of the agreeable effects, of their utility, either to the person who exercises them, or to some other persons, joins with our sense of the propriety, and constitutes always a considerable, frequently the greater part of that approbation" (p. 264). For an infrequent first-person singular statement consider *TMS*, p. 41: "If the chief part of human happiness arises from the consciousness of being beloved, as I believe it does, those sudden changes of fortune seldom contribute much to happiness."

[21] I add that Smith's use of the "we" helps to overcome the barrier presented by the written word; it helps to put author and reader onto the same stage, as it were.

[22] In a letter to Thomas Cadell of March 31, 1789, Smith remarks that in preparing the sixth edition of the *TMS* he is supplying a "new sixth part containing a practical system of Morality, under the title of the Character of Virtue" (*Correspondence*, p. 320). That description seems roughly accurate.

[23] Smith also sees literature, particularly drama, as a source of moral education. E.g., see the *Wealth of Nations (WN)*, R.H. Campbell and A.S. Skinner (eds.), 2 vols. (Indianapolis, IN: Liberty Press. 1981), vol 2, bk V, pp. 796–797.

[24] See Raphael. *Adam Smith*, p. 14: "There is reason to think that his [Smith's] lectures on ethics in their earliest form began with a historical survey of moral philosophy from Plato to Hume." The same point is proposed in Macfie's and Raphael's note at the start of *TMS* part VII (p. 265. n. 1).

[25] Smith discusses aspects of Plato's and Aristotle's metaphysics in "The Principles which lead and direct Philosophical Enquiries: illustrated by the History of the Ancient Logics and Metaphysics," in *Essays on Philosophical Subjects (EPS)*, W.P.D. Wightman (ed.) (Indianapolis, IN: Liberty Press, 1982), pp. 119–129. On p. 125 Smith states that "It [the theory of Ideas] is a doctrine, which, like many of the other doctrines of abstract Philosophy, is more coherent in the expression than in the idea; and which seems to have arisen, more from the nature of language, than from the nature of things." The linguistic trap seems to amount to the use of the same word (e.g., "beauty") in many different contexts, thus suggesting a self-same "essence" as referent. When Smith comments on metaphysics, it is by way of criticism. See also Smith's "Considerations Concerning the First formation of Languages, and the Different Genius of Original and Compounded Languages," in *Lectures on Rhetoric and Belles Lettres (LRBL)*, J.C. Bryce (ed.) (Indianapolis, IN: Liberty Press, 1985), pp. 214, 219, 221, *et passim*. Cf. *LRBL*, pp. 10–11 on the "metaphysical" use of prepositions.

[26] Likewise, "the general maxims of morality are formed, like all other general maxims, from experience and induction" (p. 319). Experience and induction are guided by the moral sentiments.

[27] Cf. the view articulated in the "History of Astronomy" essay to the effect that new systems of astronomy are adopted because they evince a superior degree of coherence, simplicity and beauty, thus surprising the imagination with their novelty and relieving it from burdensome complexities the earlier systems had invented to account for the phenomena (*EPS*, p. 75 *et passim*).

[28] On the reintroduction of Scepticism at the dawn of the Enlightenment, see R. Popkin's "The Revival of Greek Scepticism in the Sixteenth Century," in his *The History of Scepticism from Erasmus to Spinoza* (Berkeley CA: University of California Press, 1979), pp. 18–41: C.B. Schmitt's "The Rediscovery of Ancient Scepticism in Modern Times," in *The Sceptical Tradition*, M. Burnyeat (ed.) (Berkeley, CA: University of California Press, 1983), pp. 225–251; and M.A. Stewart's "The Stoic Legacy in the Early Scottish Enlightenment," in *Atoms, Pneuma, and Tranquillity*, M.J. Osler (ed.) (Cambridge: Cambridge University Press, 1991). Stewart begins by remarking that "In Scottish thought in the eighteenth century we find something of a reenactment of the ancient debates between Stoics and Skeptics." Hume is a key player in the reenactment. On p. 11 Stewart remarks: "The seventeenth century, then, far from rejecting the ancients as Bacon had commended, restored to the public curriculum the three schools of thought [Stoic,

Epicurean, Sceptic] which had competed against Aristotelianism in the ancient world." Smith possessed Sextus' writings; see *Adam Smith's Library: A Supplement to Bonar's Catalogue*, by H. Mizuta (Cambridge: University Press, 1967), p. 55.

[29] It is interesting for my purposes that Hume puts into the mouth of "the Sceptic" a very similar complaint about the tendency of philosophers to force their "favourite principle" to account for the phenomena. See "The Sceptic," in *David Hume: Essays Moral, Political, and Literary*, p. 159.

[30] Smith discusses Aristotle in *TMS* pp. 270–272. He remarks that "It is unnecessary to observe that this [Aristotle's] account of virtue corresponds too pretty exactly with what has been said above concerning the propriety and impropriety of conduct" (p. 271). Smith approves of Aristotle's notion that virtue lies in the mean (pp. 27. 244–245, 271; cf. p. 204).

[31] See pp. 313–314: "A system of natural philosophy may appear very plausible, and be for a long time very generally received in the world, and yet have no foundation in nature, nor any sort of resemblance to the truth. . . . But it is otherwise with systems of moral philosophy, and an author who pretends to account for the origin of our moral sentiments cannot deceive us so grossly, nor depart so very far from all resemblance to the truth. . . . But when he [an author] proposes to explain the origin of our desires and affections, of our sentiments of approbation and disapprobation, he pretends to give an account, not only of the affairs of the very parish that we live in, but of our own domestic concerns." I note that Smith here is referring not so much to the "what is virtue" question as to the "principle of approbation" by which we denominate one thing virtue and another not. The passage is located at the very end of the discussion of the first question.

[32] Cf. J. Barnes' "Aristotle and the Methods of Ethics," *Revue Internationale de Philosophie* **34** (1980): 509: for Aristotle. "Human nature is so constituted that we possess a faculty for grasping truth – even if that faculty must be refined by experience. There is a gap between the premises that men have a natural aptitude for knowledge, and the conclusion that *ta endoxa* constitute a deep well of truth. But for Aristotle the gap is easily bridged: if nature does nothing in vain, and if we are naturally inclined towards truth, it follows that we do, for the most part, attain the truth. . . . We cannot infer that whatever a man believes, is true: nor even that whatever all men believe, is true." I take it that Smith could agree with most of this, shy of any "metaphysical" interpretation of "nature" or "truth." Nevertheless, Smith is not subject to what Barnes calls "the conservative parochialism of Common Sense" (p. 510). See also J. Cooper, "Review of M. Nussbaum's *The Fragility of Goodness: Luck and Ethics in Greek Tragedy and Philosophy*," *Philosophical Review* **97** (1988): 533: "Beliefs arrived at in that way [through the *endoxa*] will, to a considerable extent, anyhow after interpretation, contain the truth, because, as Aristotle thinks, human nature, being the nature of an intelligent animal, orients us toward finding out how things objectively are." Again, Smith is sceptical about saying much about "how things objectively are."

[33] I owe the phrase "external perspective" with reference to the ethical context to J. Lear, *Aristotle: the Desire to Understand* (Cambridge: Cambridge University Press, 1988), p. 158.

[34] See the *Wealth of Nations*, vol. 2 bk. V. p. 771. Smith argues in those pages of the *WN* that philosophy was taught, in the great medieval universities of Europe, as subservient to theology, as therefore concerned primarily with the essence of soul and of God, i.e.,

"metaphysicks" and "pneumaticks," the study encompassing the two being "the cobweb science of Ontology." As Smith also writes here: "In the modern philosophy it [the perfection of virtue] was frequently represented as generally, or rather as almost always inconsistent with any degree of happiness in this life; and heaven was to be earned only by penance and mortification, by the austerities and abasement of a monk; not by the liberal, generous, and spirited conduct of a man. Casuistry and an ascetic morality made up, in most cases, the greater part of the moral philosophy of the schools" (p. 771).

[35] In the *LRBL* (pp. 25–27), Smith speaks critically of a complex system of rhetorical tropes built on distinctions of the "grammarians," yielding a 'grammar' of rhetoric full of rules dictating how one should write and speak well. Smith has very little patience with that approach, and instead orients himself by the demands of the given effort at communication via "sympathy" (in the *TMS*' sense of the term "sympathy" – below).

[36] "We are not at present examining upon what principles a perfect being would approve of the punishment of bad actions; but upon what principles so weak and imperfect a creature as man actually and in fact approves of it" (p. 77). The distinction between the two standards recurs on p. 216 with respect to prudence. Smith seems content to lavish praise on the less heroic form of prudence (p. 215).

[37] The metaphor of the theater is not so much Aristotelian as it is Stoic. E. g., see M. Aurelius' *Meditations* x.27, xxi.36; and Epictetus' *Encheiridion* 17.

[38] For some discussion, see David Marshall, "Adam Smith and the Theatricality of Moral Sentiments." *Critical Inquiry* 10 (1984): 592–613: and, for some general reflections, J.A. Barish's *The Antitheatrical Prejudice* (Berkeley CA: University of California Press, 1981), pp. 243–255. J.J. Spengler notes that the famous "invisible hand" was compared by Fontenelle (in a work Smith knew) to "that of the Engineer who, hidden in the pit of a French Theatre, operated 'the Machines of the Theatre' in motion on the stage." Spengler, "Smith versus Hobbes: Economy Versus Polity," in *Adam Smith and the Wealth of Nations: 1776–1976 Bicentennial Essays*, F.R. Glahe (ed.) (Boulder CO: Colorado Associated University Press, 1978), p. 43.

[39] "Impartiality" is a controverted term in ethics, of course, and I cannot here attempt fully to define Smith's understanding of the term. I note only that for Smith the meaning of the term is illustrated by the examples of drama and the activity of the critic as well as by law and the activity of the judge.

[40] See the "History of Astronomy," in *EPS*, pp. 31–53. Smith writes "Philosophy, therefore, may be regarded as one of those arts which address themselves to the imagination; and whose theory and history, upon that account, fall properly within the circumference of our subject." Smith finds it fruitful to examine philosophies to see "how far each of them was fitted to sooth the imagination, and to render the theatre of nature a more coherent, and therefore a more magnificent spectacle, than otherwise it would have appeared to be. According as they have failed or succeeded in this, they have constantly failed or succeeded in gaining reputation and renown to their authors; and this will be found to be the clew that is most capable of conducting us through all the labyrinths of philosophical history . . ." (p. 46); and ". . . the repose and tranquillity of the imagination is the ultimate end of philosophy . . ." (p. 61). While Smith uses the term "philosophy" to include what we would call "science," it is broad enough to include also the sort of theorizing that produces the "theory" of moral sentiments. For the analogy between moral and natural philosophy, see also *WN*, vol . 2, bk. V, pp. 768–769.

⁴¹ The term "theoria" can mean the being a spectator at the theater, as at *Crito* 52b4. See the Liddell, Scott, and Jones lexicon, *ad loc.*

⁴² Hence in the *TMS* Smith studiously avoids creating a technical vocabulary, though he gives prominence to particular terms (such as "spectator") and occasionally restricts the meaning of an ordinary term to suit his ends (such as "sympathy").

⁴³ Not that Smith wished to suppress altogether any traces of the academic origins of the book. He identifies himself on the title page of the first edition as "Professor of Moral Philosophy in the University of Glasgow." Perhaps Smith succeeded too well: the *TMS* has been almost entirely removed from the academy's canon of works in moral philosophy for over a century.

⁴⁴ It is crucial to Smith's account of moral judgment that spectators understand the situation of the person being judged (p. 18 *et passim*). In various ways, that 'understanding' combines imagination and reason. Hence Smith's position can be described as anti-rationalist only "with qualifications." If Smith thought that reason had no function in moral understanding, his various prescriptions concerning liberal education would make no sense. For the prescriptions concerning philosophy and science, see *WN* vol. 2, bk. V, p. 796.

⁴⁵ Smith's frequent remarks about the benevolent intentions of nature, and his doctrine that even irrational behavior may exhibit nature's benevolent design (e.g., p. 53), support this last point. Why will philosophical theorizing about moral psychology not change our ethical behavior? To make a long story painfully short, the answer is that Smith's moral psychology makes no room for a passion for knowledge (for something like Platonic eros, that is).

⁴⁶ See Hume's *An Enquiry Concerning Human Understanding.* Sec. XII. 128; and *A Treasure of Human Nature*, L.A. Selby-Bigge (ed.) (Oxford: Oxford University Press, 1973), pp. 183, 264–274.

⁴⁷ I refer to Hume's "The Sceptic," in *David Hume: Essays Moral, Political, and Literary*, p. 162: "If we can depend upon any principle, which we learn from philosophy, this, I think, may be considered as certain and undoubted, that there is nothing, in itself, valuable or despicable, desirable or hateful, beautiful or deformed; but that these attributes arise from the particular constitution and fabric of human sentiment and affection" (cf. pp. 169, 180).

⁴⁸ I borrow the term "intellectual sentiments" from J. Cropsey's *Polity and Economy: an Interpretation of the Principles of Adam Smith* (Westport, CT: Greenwood Press, 1977), p. 43, n. 3.

⁴⁹ See the "History of Astronomy" essay, *EPS* p. 105: "And even we, while we have been endeavouring to represent all philosophical systems as mere inventions of the imagination, to connect together the otherwise disjointed and discordant phaenomena of nature, have insensibly been drawn in, to make use of language expressing the connecting principles of this one [Newton's system], as if they were the real chains which Nature makes use of to bind together her several operations."

⁵⁰ The term "insulation" is taken from M. Burnyeat's "The Sceptic in his Place and Time," in *Philosophy in History*, R. Rorty. J.B. Schneewind, and Q. Skinner (eds.) (New York: Cambridge University Press, 1984), p. 225.

⁵¹ See Sextus Empiricus *Outlines of Pyrrhonism* **II**, 10.

⁵² See Burnyeat's "Can the Skeptic Live his Skepticism?" in *The Skeptical Tradition*, p. 127, for the references.

⁵³ See M. Frede's "The Sceptic's two Kinds of Assent" in *Philosophy in History*, p. 258 *et passim*, for a description of this point and more generally of the two kinds of Scepticism I am adumbrating.
⁵⁴ The various contrasts between modern and ancient, localized and comprehensive, dogmatic and non-dogmatic Scepticism, are explored in J. Annas' "Doing without Objective Values: Ancient and Modern Strategies," in *The Norms of Nature*, M. Schofield and G. Striker (eds.) (Cambridge: Cambridge University Press, 1986, pp. 3–29).
⁵⁵ Hume, *Treatise* p. 186. R.J. Fogelin glosses Hume's argument as follows: "That is, skeptical arguments are self-refuting, but this only puts us on a treadmill, since setting aside our skepticism and returning to the canons of reason inevitably puts us on the road to yet another skeptical impasse." "The Tendency of Hume's Skepticism," in *The Skeptical Tradition*, p. 402.
⁵⁶ The point about following out Hume's program consistently would depend on interpreting Hume as a dogmatic Sceptic. For some support of that interpretation, see M. Frede's "The Sceptic's two Kinds of Assent," in *Philosophy in History*, p. 277. For some general support of my interpretation of Smith as a Sceptic, see J.B. Schneewind's "Natural Law, Skepticism, and Methods of Ethics," *Journal of the History of Ideas* 52 (1991): 289–308.
⁵⁷ Of course, all this is far from a demonstration that the position I am attributing to Smith is ultimately sound. In our post-Wittgensteinian age, the general problem of returning to 'ordinary life' while philosophizing is now a well-established one. For a penetrating study of the problem, see J. Lear's "Transcendental Anthropology," in *Subject, Object, and Thought*, P. Petit and J. McDowell (eds.) (Oxford: Clarendon Press, 1986), pp. 267–298.
⁵⁸ Hume, *Treatise*, p. 253: "The mind is a kind of theatre, where several perceptions successively make their appearance: pass, re-pass, glide away, and mingle in an infinite variety of postures and situations."
⁵⁹ See Burnyeat, "Can the Skeptic Live his Skepticism?", in *The Skeptical Tradition*, p. 132.
⁶⁰ See n. 40 above.
⁶¹ Smith's Scepticism would have to be reconciled with his Stoicism, of course, given the traditional antipathy between the two schools. While that is a topic too complex to enter into here, I note that there are some obvious places where Smith might have taken Stoicism and Scepticism to overlap. The notion that happiness consists in tranquillity provides an example; and some Stoics, such as Marcus Aurelius (of whose thought Smith clearly is fond), evince a strongly anti-metaphysical prejudice (e.g., *Meditations*, i.6, ii.13). Smith ignores Stoic metaphysics and epistemology, and concentrates instead on three themes in Stoic ethics and theology, viz. that tranquillity is the *summum bonum*, that self-command is the capstone of the virtues, and that the universe is guided by Providence. I believe that Smith could say more about both themes without violating his Scepticism. If my reconstruction of Smith is well directed, in any event, Smith is an unusually eclectic and complex thinker who attempts to unite elements of a variety of ancient schools (those of Aristotle, the Stoics, the Sceptics) along with elements of modern philosophy (I have mentioned Hume).
⁶² *LRBL* pp. 145–146. Smith there tells us that both Aristotelian and Newtonian methods are "didacticall," and that as such neither is likely to elicit much interest among non-

theorists. Smith's interweaving of "rhetorical" and "didactical" characteristics in the *TMS* seems intended to draw theorists as well as non-theorists into the discussion.

[63] MacIntyre. *After Virtue* (Notre Dame, IN: University of Notre Dame, 1984), p. 231: "What Hume identifies as the standpoint of universal human nature turns out in fact to be that of the prejudices of the Hanoverian ruling elite." For similar comments about Hume, see MacIntyre's *Whose Justice? Which Rationality?* (Notre Dame, IN: University of Notre Dame Press, 1988), p. 295: and for comments about the tradition-bound character of Aristotle's thought, see pp. 391–392 *et passim*.

[64] This paper was presented to a meeting of the Eighteenth Century Scottish Studies Society on "Glasgow and the Enlightenment." University of Strathclyde, Glasgow, July 30, 1990. I am grateful to the participants at that meeting for their helpful comments, as well as to Doug Den Uyl, Knud Haakonssen, David Roochnik and Ian Ross for their suggestions. I am pleased to acknowledge support form the National Endowment for the Humanities, the Woodrow Wilson International Center for Scholars, and the Earhart Foundation, that made work leading to this essay possible. This essay is drawn from a book tentatively titled *Liberalism, Virtue Ethics, and Moral Psychology: Adam Smith's Stoic Modernity*, forthcoming.

PATSY HALLEN

# ECOFEMINISM AS RECONSTRUCTION: MAKING PEACE WITH NATURE*

### INTRODUCTION

This paper begins by articulating why a new metaphysical and cultural reconstruction, based on the unity of nature, is vital. It then strives to show how feminism can contribute to ecological sanity. Feminism attempts to combat sexism which is one expression of a psychology of domination and repression. Sexism spells pollution of the mind and the body and as such it is fundamentally linked to environmental pollution or ecological destructiveness. I suggest that ecology needs feminism to overcome the illusion that we can successfully dominate nature and to embrace the holistic and integrative dimensions of knowledge and the universe. Ecology as a science needs feminism to challenge the assumptions of a mechanical world-view which has fundamentally influenced scientific development. Ecology as life science needs feminism to illuminate the 'masculine' nature of modern science and to demonstrate how this has contributed to environmental destruction. Ecology as a practice needs feminism to insure that reform environmentalism, which sees nature in instrumental terms, is transformed into deep ecology whereby one sees and feels onself as intimately inter-related to nature. Unless we incorporate the personal, emotional and sexual dimensions of experience into our explanations, unless our self-understanding is mediated by the scientifically revolutionary perspectives of feminism, we will not make peace with nature.

### THE PHILOSOPHICAL FRAMEWORK: WHY A FUNDAMENTAL RE-ORIENTATION IS NECESSARY

In my paper, I would like to show how ecophilosophy and ecofeminism,[1] integrated and drawing upon the richness of the perennial philosophy[2] can overcome the life-threatening separation of science from care and provide a metaphysical basis for making peace with nature.

For if the earth is to survive, we need a "fundamental re-orientation of perspective", as Skolimowski urges,[3] from a mechanistic, materialistic world-view with its frontier ethics,[4] to a holistic, ecological world-view complimented by environmental ethics, and contributed to by ecofeminism.

OTHERWISE JUST REARRANGING THE TITANIC'S DECK CHAIRS

Unless we undertake this metaphysical and cultural reconstruction, we will be picking up litter rather than attacking the production of junk in the first place, or to use a popular analogy, we will be simply rearranging the deck chairs on the Titanic. To give but one illustration, the "cunning of unreason"[5] is such that pollution control companies are subsidiaries of the polluters (Dow Chemical, Alco, Dupont).[6] So the more waste the better, as far as they are concerned. We need government regulation to offset this conflict between interest and obligation. But – and this is the important point – such regulation will not eventuate unless we understand what we are doing and why. For the environmental crisis is not just a result of maladjusted economic power, military insanity, population pressure or social injustice whereby one Australian uses fifty times the resources of one Kenyan. It is also a crisis of the human spirit. Hence to solve the grave problems posed by the environmental crisis we need not only a just and sustainable economic system, a peace force, equitable population controls, eco-community science and appropriate technology. We need new ways of seeing.

So our quest for a future must begin at start. We must begin by coming to terms with the foundations of our thought and by recognizing the powerful, foundational metaphors (such as the world is like a machine) which allow pollution to 'pay', Such a metaphysical starting point is necessary not only to arouse us from our "dogmatic slumber"[7] but to securely anchor our praxis. To alter our political economy, to transform the military-industrial complex, a new world view is essential. As Skolimowski succinctly expresses it: "Once we know what we are doing and *why*, other forms of reconstruction, including the economic one, will follow more swiftly".[8]

THE POWER OF IDEAS

Such an approach might seem idealistic compared to the real hegemony of the ruling class, but to test our hypothesis of the non-exclusive but vital importance of ideas, all we need do is step back from our culture and engage in the phenomenological task of bracketing[9] to let the meaning of our mental set stand out. If we then contrast the prevailing orthodoxy with another orthodoxy, we can clearly see how world-views prevent or encourage exploitation of the earth. Consider a culture which regarded agriculture and mining as rape.[10] Consider a culture that was without wants.[11] Such thought experiments illuminate the dominant thought habits of our age. These thought-forms are so deeply engrained they go unnoticed but they nonetheless fix the boundaries of our mental set and structure the universe of discourse in which our decisions are taken. As Wittgenstein quips, "one thinks that one is tracing the outline of a thing's nature ... and one is merely tracing round the frame through which we look at it".[12] And again,

> The aspects of things that are most important for us are hidden because of their simplicity and familiarity. (One is unable to notice something because it is always before one's eyes). The real foundations of his enquiry do not strike a man at all. Unless *that* fact has at some time struck him. And this means we fail to be struck by what, once seen, is most striking and most powerful.[13]

By focusing on foundational thought forms, I do not mean to undervalue the tremendous struggles in the economic and political spheres. But I also believe that we cannot underestimate the power struggles that exist between competing world-views. As John Maynard Keynes expresses it:

> The ideas of economists and ... philosophers, both when they are right and when they are wrong, are more powerful than is commonly understood. Indeed the world is ruled by little else. Practical men, who believe themselves to be quite exempt from any intellectual influences, are usually the slaves of some defunct economist. Madmen in authority, who hear voices in the air, are distilling their frenzy from some academic scribbler of a few years back. I am sure that the power of vested interests is vastly exaggerated compared with the gradual encroachment of ideas.[14]

## WHY WESTERN MORAL PHILOSOPHY IS LIMITED

We need *new* ways of seeing. Our traditional Western ethic is not adequate for this ecocentric metaphysical reconstruction for several reasons. Firstly, because we have a radically new situation. Modern technology has given birth to awesome and novel powers which have no precedent. A traditional moral philosopher such as Kant thought that in matters of morality, human reason can easily be brought to a high degree of accuracy and completeness.[15] But with a collective agent (the corporation), fearsomely powerful deeds (behaviour control to exterminism) and an enormous time-scale (an incomprehensible quarter of a billion years for example), accuracy and completeness are a fool's errand. Item: Two million whales killed in the last fifty years. Item: Two hundred million animals used in one year alone for experimentation. Item: The destruction of eleven million hectares of rainforest annually, resulting in the extinction of forty-eight species a day.[16] Regan comments:

The numbers are like distances in astronomy. We can write them down, compare them, add and subtract them but like light years stacked on light years, we lack the intellectual or imaginative wherewithal to hold them in steady focus.[17]

Traditional moral philosophy is also not adequate because as Lynn White[18] and many others have pointed out, our traditional ethics have contributed to the environmental crisis. Western moral philosophy is riddled with an imperalistic stance towards nature, singling out humans as the only beings who are morally considerable. Immanuel Kant, one of the giants of moral philosophy, argued that we should not torment animals because it might lead to tormenting humans; it might develop habits of callousness which would affect humans. For Kant, ill treating animals is wrong because it degrades humans, not because it causes animals to suffer. "Our duties towards animals" to quote Kant, "are merely indirect duties towards humanity".[19] Such assumptions of human moral superiority help to promote a non-ecological view of the world which regards humans as separate from and superior to the rest of nature and which regards environmental values as secondary and derivative.

## WHAT MORAL INTEGRITY MEANS

One of the underlying themes of this new reconstruction is the unity of nature. This is what moral integrity means, recognizing that we are a "complex of relations"[20] and that, more intimately "the world is my body".[21] Defending a wilderness area then becomes self-defense, for we are the rocks dancing.[22] This profound comic hook-up means that humans and nature cannot be separated. Their truth is their relation – *co-naissance* – and, as a result, to locate value only in the human realm is to devalue humans.

## SELF-REALIZATION

A pressing question posed by the thesis of the unity of nature is: how can we, as humans, transcend human chauvinism? How can we genuinely not put ourselves first? I would argue that ultimately this is neither possible nor necessary. Our self interest may be considerably enlightened but it can never be completely extinguished.

Nor should it be! Ecocentricism is not misanthropic. As Fox points out, it is opposed to human-centredness, a legitimating ideology, but not to humans.[23] Moreover, to see self-interest only in terms of self is to extinguish it. Let me try to explain.

At a fundamental level it is impossible to escape some form of human chauvinism. Although one cannot reduce ecological concern to human interests (since something like a Karri forest is an end-in-itself and not just a means to enlarge human sensibilities, capabilities or productivity), it is to human interests that we must appeal to ensure the well-being of such ecological richness. But this limitation of having to appeal to human interest, an in-built human chauvinism, is a paradoxical limitation since it turns out to be a source of strength. For in typical Hegelian fashion it is only as humans grant an intrinsic integrity to nature that we discover our true nature. It is only when we accept our dependency on nature and see ourselves as part of nature that we can be in touch with our sources, our spring, and realize our potential. To care[24] for the environment is to realize ourselves.[25] Humans and ecosystematic interests coincide. In the words of Loren Eiseley:

> I saw, had many times seen, both mentally and in the seams of exposed strata, the long backward stretch of time whose recovery is one of the great feats of modern science. I saw the drifting cells of the early seas from which all life, including our own, has arisen. The salt of the ancient seas is in our blood, its lime is in our bones. Every time we walk along a beach some ancient urge disturbs us so that we find ourselves shedding shoes and garments, or scavenging among seaweed and whitened timbers like the homesick refugees of a long war... The human brain, so frail, so perishable, so full of inexhaustible dreams and hungers burns by the power of the leaf.[26]

## THE WISDOM OF LOVING

It is this relational view of the world that is elucidated by feminist scholarship. A feminist epistemology attempts to "transcend dichotomies ... [and] emphasises holism, harmony and complexity"[27] As such it has an important contribution to make to ecophilosophy.

Perhaps what feminism can do is to help complement philosophia, the love of wisdom, with sophophilia, the wisdom of loving.[28] For as Aldo Leopold notes, "we can be ethical only in relation to something we can see, feel, understand, love or otherwise have faith in".[29] The real work of ecofeminism is becoming lovers. Einstein puts it well:

> A human being is a part of the whole, called by us the 'universe,' a part limited in time and space. He experiences himself, his thoughts and feelings, as something separated from the rest – a kind of optical delusion of his consciousness. This delusion is a kind of prison for us, restricting us to our personal desires and to affection for a few persons nearest to us. Our task must be to free ourselves from this prison by widening our circle of compassion to embrace all living creatures and the whole of nature in its beauty.[30]

Having established a *prima facie* case for the necessity and viability of a metaphysical and cultural reconstruction, it is to these feminist sources and the wisdom of loving that I now wish to turn for weaving our ecocentric philosophy. But first some remarks about the language I will be using.

## PROBLEMATIC TALK

In this paper I talk about women being more in touch with nurturing than men, more in touch with caring than men. This talk is problematic because it might reinforce traditional sterotypes of women and patriarchal fantasies about the 'caring', soft, compliant female!

## EXISTENCE PRECEEDS ESSENCE, BUT DOES NOT PRECLUDE EXPERIENCE

But in speaking about women's experience and their caring labour I am not (I hope) playing into the hands of the essentialists. For I do not believe that "anatomy is destiny".

We must not fall prey to sex-based stereotyping that is evident in some ecofeminist writing. Women are not "gurus in compassion"[31] nor do they "flow with the system of nature" by virtue of their essential nature.[32] It is these very attitudes that have maintained patriarchy and prejudice.[33]

With Simone de Beauvoir,[34] I believe that one is not born, one becomes a woman. I reject essentialism which purports all women to have a specific essential nature that differentiates them from men. In its place, I posit the ethics of existentialism: women should be free to choose, to create their nature through their actions. With J.P. Sartre I declare that "existence preceeds essence".[35] This dictum is borne out when one considers that despite cultural pressures to conform to a sexual stereotype, variability *among* men and women in a wide range of diverse traits (from resourcefulness to creativity) is much greater than the difference *between* men and women.[36]

Women might be conditioned to be the carers, but that does not necessarily make them good at caring. As well, men are not necessarily ill-equipped to help solve the environmental crisis. We must treat each other as partners in the challenging venture of reconstructing metaphysics.

But this existential, co-operative posture does not preclude women's past. And historically we have been the carers. I maintain that women can claim the past without being chained to it; we can recognise our past without being defined by it. Women's experience (their labour and their life) has been more involved with nurturing than men's, and it is this experience that is so important for us to draw upon if we are to recover a right relationship with nature.

## GENDER REPLACES SEX BUT IS NOT UNAMBIVALENT

In this paper I also draw upon the distinction between sex and gender. This is a very useful distinction separating our biology from our socialisation, our sex from our sexuality, natural processes from the social construction of meaning.

But it is also a problematic distinction, highlighted by the age-old argument about the role of nature versus nurture. Women are biologically different from men. But what difference does this make? No one knows. Despite the claims of the most conservative sociobiologist, despite the claims of the most radical proponent of total freedom, no one can know.

While sociobiologists remind us of our genetic preconditions, existentialists alert us to our essential freedom. Both perspectives are necessary, yet each is incomplete.

An analysis in terms of gender insures something very important. For it is by means of gender (among other things) that our society assigns value and meaning. Gender is a way of organising human relations. As such it is a fundamental category. As Sandra Harding points out, whether it applies at the individual level (assigning dualistic characteristics to various perceived dichotomies that rarely have anything to do with sex differences), at the structural level (division of labour by gender) or at the symbolic level (the level of meaning) gender is always asymmetrical, with the feminine defined as inferior.[37] Unless we re-formulate the agenda to insure gender a primary place on all three levels of analysis, it will get submerged, swallowed up, lost. Feminists cogently argue that a partriarchal society *needs* inferiors (and so finds gender analysis threatening). Thus a persistent tactic has been just this – repression of gender as a category, with the result that 'joining with' (for example, merging feminism and socialism as Simone de Beauvoir attempted in the 1960s, to her regret in the 1970s[38]) equates to 'being submerged within' and forgotten.

But while useful, gender is (like all other concepts) fictional. While at one level there is a woman's voice and it must be heard, at another level there are only individual women and at this individual level gender has no fixed referents. 'Woman's voice' gets deconstructed into 'women's voices' by factors such as class, race and country. What do I as a middle-class professional Australian woman of Irish-Canadian origins have in common with a Kimberley Aboriginal woman living at Milyoonga Yumpup? Race, class, country and environment often more deeply restrict the life opportunities of individuals than does sex. We can easily see this if we compare the different life opportunities available to women of the same race but different classes or women of the same class but different races. Hence gender is a racial category and race a gender category, as Harding's analysis shows.[39] For example, sexist public policies are different for people of the same gender but different race (a white female and a black female) and racist policies are different for people of the same race but different genders (white men and women).

Gender is a complex category, full of tensions, contradictions and ambivalences. But nonetheless, it is a valuable and irreplaceable tool of analysis. So just as an ecocentric outlook must dismantle anthropocentrism, so too androgyny must transform androcentrism: the two projects while not similar, are complementary. While sisterhood is necessary, it is not sufficient. It must be related back to life itself and evolutionary processes: hence the term: *eco*-feminism.

> Who is in my network,
> What links us, to be exact?
> Better ask to understand the force
> that cuts through rock the water's course,
> and binding like to like
> makes also opposites attract.
> Susan Saxe, "Ask A Stupid Question"[40]

## THE MAIN THESIS: MULTI-DIMENSIONAL

> The 'control of nature' is a phrase conceived in arrogance, born of the Neanderthal age of biology and philosophy, when it was supposed that nature exists for the convenience of man. The concepts and practices of applied entomology for the most part date from that Stone Age of science. It is our alarming misfortune that so primitive a science has armed itself with the most modern and terrible weapons, and that in turning them against the insects, it has also turned them against the earth.
> – Rachel Carson, Silent Spring

This paper aspires to illuminate pathways of making peace with nature. This is not an easy task since everywhere we are so much at war with the earth, from DDT to deforestation, from acid rain to radioactive fallout. We practice a "metaphysic of biological apartheid".[41] To help topple this metaphysic, we need the stunning clarity of people like Rachel Carson.

My central thesis is that if we are to make peace with the environment ecology needs to transformed by the knowledges[42] of feminism. This is a complex and multi-dimensional thesis involving our individual and collective psyche (the psycho-sexual realm), our ways of knowledge (science), and our ways of being (society and the environment). To support this thesis I will draw from philosophical, psychological, sociological and historical sources not in a linear, combative way (statement of main thesis, arguments for main thesis, defeat of objections to main thesis, and so on) but in a spiral, processive way. I do not propose to argue. "Any kind of polemics" writes Heidegger "fails from the outset to assume the attitude of thinking".[43] I do not propose to be polemical. What I hope to do is to evoke and share a vision.[44] This invitation-to-look approach is a deliberate strategy designed to stand as a testimony to the complexity and the vitality of the deceptively simple claim: ecology needs feminism.

## SCIENCE MUST BE MEDIATED BY FEMINIST KNOWLEDGES: CARING LABOUR VERSUS THE NEED TO DOMINATE

Feminist scholarship has shown how modern science is basically a masculine endeavour. As a result, I will suggest, science is one-sided, and potentially life-threatening. We will not overcome this one-sidedness, I will maintain, until our scientific understanding of the living world is mediated by the content of feminist epistemologies – the sensual, the relational, the intimate. Unless our science is grounded in the "caring labour" articulated by feminism,[45] science will continue to be, as Rachel Carson suggest, "against the earth".

Furthermore, I will maintain that these feminist perspectives (which have their roots in traditional thinkers, from Plato through Kierkegaard to Polanyi[46])have been systematically ignored because of our individual and collective psychotic need to dominate. Ecology's wholesomeness, I will posit, is linked in a profound way to the absence of a need to dominate. This claim, then, necessitates an investigation of the psycho-sexual dimension to uproot the causes of domination.

Thus a related thesis which supports my main thesis is this: that the current ecological crisis has psycho-sexual roots. The ecological crisis has many other causes as well (economic and political), but we have underplayed, even ignored the psychosexual causes until feminist theory invited us to investigate this dimension.[47]

## THE DOMINATION OF NATURE SPELLS DESTRUCTION

Following feminist writers I will maintain that sexism is the expression of a basic psychology of domination and repression. Ecological inbalance is, in part, due to our mistaken[48] belief that we can successfully dominate nature. So sexism (mind and body pollution) is fundamentally linked to ecological destructiveness (environmental pollution).

I will suggest that ecology as a *science* needs feminism to balance a myopic, mechanical world-view which has fundamentally influenced scientific development; ecology as a *life science* needs feminism to reveal how patriarchal thinking contributes to environmental destruction; ecology as a *practice* needs feminism to ensure that shallow ecology is transformed into a deep ecological perspective. The shallow/deep distinction was coined by the Norwegian philosopher, Arne Naess, to contrast reform environmentalism with deep ecology. Reform environmentalism aims to manage the environment based on its use-value while deep ecology seeks to help us see and feel ourselves as intimately interrelated to an intrinsically valuable nature, so that when we harm nature we diminish ourselves.[49]

Hence I will posit that one of the pathways to making peace with the environment is through feminism. Feminism is helpful to ecology as a science, as a life science and to its practitioners because the multiplicity of differences within feminism offers new ways of experiencing and understanding the world.

## THE HIGHEST COMMON FACTOR OF FEMINISM AND HOW IT ENLARGES OUR SCIENTIFIC UNDERSTANDING

Despite the many theoretical differences between feminists (liberal, Marxist, socialist, post-structuralist, radical), feminists adhere to a few basic tenets, as outlined by Marilyn French:[50] that the two sexes are at least equal in all significant ways and that this equality must be publicly recognised; that the qualities traditionally associated with women (nurturing, receptivity) are at least equal in value to those traditionally associated with men (self-assertion, power-seeking) and that this equality must be publicly recognised, not ignored or viewed as irrelevant; finally that the personal is the political, the bedroom is as relevant as the boardroom.

What I hope to show is that each of these tenets of feminism can help to enlarge our scientific understanding of the environment. The first tenet will ensure that more women will participate in the scientific community.

## MORE WOMEN IN SCIENCE

This is crucial. For as Harding points out, science is *the* model in our culture of a masculine activity. To quote Harding, "women have been more systematically excluded from doing serious science than from performing any other social activity, except perhaps frontline warfare".[51] Hence it is vital to redress the scarcity of women in the peculiarly masculine occupation of scientist.

But this, while necessary, is not sufficient. We do not just need more women in science; the whole nature of the system, the dominant ideology of science, needs to be challenged.

## MORE EQUALLY RECOGNISED WOMEN IN SCIENCE

As the researchers of women's struggles to enter science show,[52] even if horizontal segregation (science as an exclusively male domain) is to some degree overcome, vertical segregation (women scientists confined to low status positions) has yet to be eliminated. We not only need more women scientists, we need women to be equally recognised practitioners of science. As Hilary Rose points out,[53] the majority of people actually practicing science – technicians – *are* women, but their work is marginalised, trivialised, made invisible and this is so even when the content of the scientific work done by women is objectively indistinguishable from men's work. This indicates that other, deeper factors are engaged. Are masculine gender identities so fragile that they cannot afford to have women as equals to men in science, Harding asks?[54] As Harding points out until the 'emotional labour' of childcare and housework is seen as desirable for men, the 'intellectual labor' of science and public life will not be perceived as desirable for women. To transform science will require revolutionary changes in the social relations between the sexes.

## A SCIENCE WHERE SUCCESS MEANS NOT SEIZING POWER BUT TRANSFORMING SENSE

Even if there are more equally recognised female practioners of science, science's relations to society need to be altered. For women do not want to become 'just like men' in an enterprise that, in the USA for example, devotes 72 per cent of its federal funding for scientific research and development to defence.[55] Furthermore women should not want to become puzzle-solvers in vast industrialised empires devoted to material accumulation, species extermination, social control and exploitation. It is not only bad science (hastily generalised science with inappropriate data bases) but 'normal' science that needs to be criticised since, as the critics of modern science show,[56] so much of it facilitates militarism, ecological disasters and desperate poverty.

By the very same token, we do not just need more ecologists; for ecology can be exclusively reductionist, 'understanding' life in terms of non-life, and so failing to see living things in process, in relation and in context, in terms of dynamic energy patterns. In addition, all-too-often environmental impact statements are fronts, hoaxes. As Neil Evernden points out,[57] Universities willingly disgorge troops of environmental scientists and managerial environmentalists and while they appear to be tools of environmental defense, they turn out – lo and behold! – to serve the interests of the developer. So for truly emancipatory knowledge-seeking, the first tenet of feminism, more women ecologists, is not sufficient.

### CHERISHING THE 'FEMININE'

We need the second tenet of feminism whereby the so-called 'feminine' qualities will be cherished and female experience can be incorporated into explanations. This will help to offset what Evelyn Fox Keller calls the "hegemony in science",[58] the 'masculine', virile, domineering nature of science, and to reclaim science as a human, not a masculine project, and so to, in Keller's words, "transform the very possibility of creative vision".[59]

This key argument will hopefully be further substantiated in the course of this paper. For now, suffice it to say that, as Keller points out,[60] despite the wide diversity in the practice of science, there is a monolithic ideology of detachment and domination which crystalised in the 17th Century, and this ideology has deeply influenced the selection of goals, the methods, values and explanations that operate in contemporary science.

### BUT WHY HAS ITS IMPORTANCE BEEN SURPRESSED?

While this view draws our attention to how so-called 'feminine' thinking – initiuitive and relational thinking – has advanced scientific comprehension, the perspective of valuing feminine qualities is also not sufficient. It is vital to point out the importance of pluralism in the intellectual pursuit of knowledge, but until we understand what motivates the exclusion of these non-macho elements in our scientific story-telling, we will not understand why some ideas gain legitimacy and others do not. So it is necessary to investigate the psychological profile of masculinity: a third level of analysis – the psycho-sexual.

### COSMOGENY RECAPITULATES EROGENY

This third tenet of feminism – the personal is the political or the sexual is the scientific – announces a three-storied method.

Firstly, as feminists point out, the public sphere depends upon the private resource base. But this private base is often not acknowledged or if it is, it is trivialised. Feminism insists that we ask why it is trivialised and that we come to terms with the importance of this private dimension. For as Ben McNaughton argues,[61] the businessmen's lack of house-training is a major cause of world pollution. The tycoon makes multi-million dollar deals but if he has never cleaned a toilet or washed a shirt, how can he be expected to realize how his factories are making our planet a ghastly cesspit?

Secondly, feminist analysis shows us that who we are (in the bedroom) shapes our politics (in the boardroom). So feminism insists that our sexuality, our sense of self, be articulated in any intellectual endeavour. We must 'begin at start'. We must start from the personal since it is so formative.

Thirdly, feminism seeks to enlarge our understanding of nature and the world not just by including female experience in explanations, but also by insisting on including those domains of *human* experience that have been relegated to women: namely the personal, the emotional, the sexual. An example of this method is Brian Easlea's book, *Fathering the Unthinkable*,[62] which analyses a scientific achievement – the building of the bomb – in psychosexual terms. I will go into this example as illuminating the third tenet of feminism in more detail later; for now the announcements must suffice: namely that including the personal, emotional, sexual dimensions of experience in our explanations will and does make us better scientists because they alert us to our signatures and in the process help to enlarge our sense of what is possible.

But such theories calling for the integration of the personal and the scientific will be 'fated to remain mere intellectual curiosities – like the ancient Greek ideas about atoms'[63] until we understand the unparalleled importance of our sexuality and the relation between sexual repression and domination.[64] Just as ontogeny (the development of an individual) recapitulates phylogeny (the evolution of the species) – for example a human embryo mirrors our reptilean and amphibean past in its life history – so too cosmogeny (world-building) recapitulates erogeny (sexual building).[65] Until we understand these principles, we will not achieve psychic health or scientific wholeness.

## A RELATIONAL VIEW OF NATURE; WE ARE INTIMATELY CONNECTED TO NATURE

Feminism acknowledges the central importance of the erotic, the private, the personal and as a result it speaks "in a different voice", to borrow Carol Gilligan's phrase.[66] It offers us a mode of thought at variance with the dominant ideology. Its central ontological category is not substance, but relation. This category transcends mere relativism, goes beyond and beneath divisive dualisms and dichotomies and insists on the scientific importance of the personal. To quote Hilary Rose,

> [a feminist epistemology] transcends dichotomies, insists on the scientific validity of the subjective, on the need to unite cognitive and affective domains; it emphasises holism, harmony and complexity rather than reductionism, domination and linearity.[67]

Rose argues that this relational view is easier for women to attain because of their labour. Women's labour is more "in touch with necessity".[68] It is caring labour reflecting the unification of mental, manual and emotional activities ("hand, brain and heart") characteristic of women's work. As such it is more complete and therefore "truer knowledge".[69]

In addition, I surmise this relational view might be easier for women to attain because of their gender identity. Psychologists and social theorists who have studied gender identity (as distinct from sexual identity) tell us that femininity is defined through attachment and connection, while masculinity is defined by independence and separation.[70] The male gender is threatened by intimacy; the female gender is threatened by separation. So gender sensitive theories show us how to better relate to our environment, show us how to more fruitfully connect with nature.

It is no accident that ecology, which etymologically means 'a study of the household', needs the experience of women. To see the earth as a life-sustaining home is a vision of ecology which, I believe, is accessible to women.

## FEMINISM: SCIENTIFICALLY REVOLUTIONARY CONSCIOUSNESS

As a result of women's caring labour and women's gender identities, I propose to you that feminist consciousness is a scientifically revolutionary consciousness. I see feminism and ecology as sharing the same perspective, which represents a new (yet very old) way of seeing, a way of making peace with (rather than war on) the environment. Such a perspective is holistic: everything is connected to everything else and each aspect is defined by and dependent upon the whole, the total context. Life is interconnected and interdependent: we are not above nature, we are an intimate part of it.

We have been blinded to holism by the gender ideology of science, by our arrogance, by our false sense of superiority, by our fragile identities rendered vulnerable by projecting our inner tensions outward onto 'the second sex' rather than working them through, by our incomplete notions of survival of the fittest with its resultant stress on competitiveness. Even David Attenborough's superb television series, 'Life on Earth', was biased in this way. Its scenes of intense competition far outweighed its scenes of co-operation in the evolutionary story. As John Livingston points out, most wildlife biologists carry around an extraordinary baggage of assumptions: concepts like dominance, aggression and competition – a "long roster of marketplace concepts applied to nature".[71] In addition to this excess baggage, we have been blinded to holism by the spatial metaphor with which we grew up, God a ruling class male above, then man with woman and nature (mythed as female) below. This hierarchial picture has dominated Western thought for hundreds of years and it is still very much alive despite the 20th Century revolutions in scientific thinking. I propose that we unconsciously imbibe this world view, we smuggle it into our conscious mind and let it structure our thoughts.

## NO FREE LUNCH

Both ecology and feminism challenge this hierarchial picture. Both reject the divisive dualism that this world view implies: heaven – earth, male – female, mind – matter, reason – emotion, objective – subjective. To develop the masculine and the feminine in each of us, to become whole people, to be able to see objects as subjects, to see nature as a thou, to experience the earth as a live presence, to feel nature as part of ourselves – the visions of ecology and feminism dovetail. Co-operation is stressed (not competition), understanding is vital (not power), appreciation is important (not domination). The watchwords are solidarity and sharing, not rivalry and ruling. As Mary O'Brien puts it, what is important is "reciprocal intimacy and not conquest".[72] Both ecology and feminism share a non-hierarchical, egalitarian perspective. Both participate in a common philosophy whereby process and participation are primary. Process philosophy is a deep and complex topic with its sources in the philosophies of Alfred North Whitehead and Henri Bergson.[73] For our purposes it is sufficient to say that feminism and ecology both stress creative activity over inert matter, dynamic order over static laws, partial autonomy over determinism, relation over substance, objects as subjects over subjects as objects. Finally for both ecology and feminism, as Carolyn Merchant points out,[74] there is no free lunch. This is one of the four laws of ecology articulated by Barry Commoner.[75] We may think that we are getting a cheap digital watch at $4.99 but the real cost must include the Formosan women who go blind producing the liquid crystals. There *is* no free lunch. To produce organised matter, energy in the form of work is needed. Reciprocity and co-operation oiling the feedback loops and closing the energy circuits is what is needed, not free lunches.

## THE CATEGORY OF THE OTHER TRANSCENDED

Whereas patriarchical society needs an inferior 'other' in terms of which to define itself – an insight so clearly articulated by Simone de Beauvoir – (whether that other be 'woman' or 'African' or 'black'),[76] a society transformed by feminist consciousness would oppose such hegemony. Feminism argues for a plurality of discourses and for a vision of wholeness based not on divisive dichotomies or defensive competitiveness but on a genuine appreciation of difference.

The field of primatology has attracted a disproportionate number (compared to other related fields) of woman researchers and it is here that we can witness feminist consciousness in action: the explosion of the old models of male-dominated power structure, the realisation of the importance of co-operation and the undermining of this category of 'other'.[77] Likewise as Livingstone points out, study of short grass prairie songbirds in New Mexico came to the (surprised) conclusion that "competition is not the ubiquitous force that many ecologists have believed".[78]

Both the female primatologists and John Wiens' papers on the short grass songbirds challenge the notions of competition, dominance and the concept of 'the other'. They show us how in nature the self/other distinction is often inappropriate. Success for a species depends upon individual bonding with the environment – an extended consciousness of self that transcends an isolated self. This "participating consciousness" is crucial for the survival of many species and it is crucial for human survival. And the perception in sensitive science that the environment or the non-self is not 'other', is music to feminist ears.

## A STRIKING EXAMPLE OF RECIPROCAL INTIMACY IN SCIENCE

Let me share with you a striking example of how science might be different, of how science might look when it is based on an attitude of reciprocity and co-operation rather than a competitive 'other'. This attitude of "reciprocal intimacy" in science is illustrated by Barbara McClintock, who in 1983, won a Nobel Prize for her work in biology in the 1940s. Evelyn Fox Keller has written a moving biography of McClintock called *The Feeling for the Organism*[79] and in it she asks:

> What is it in an individual scientists's relation to nature that facilitates the kind of seeing that eventually leads to productive discourse? What enabled McClintock to see further and deeper into the mysteries of genetics than her colleagues?

Her answer, Keller tells us, is simple:

> Over and over again McClintock tells us one must have the time to look, the patience to hear what the material has to say to you, the openness to let it come to you. Above all one must have a feeling for the organism. And McClintock goes on: 'No two plants are exactly alike. They are all different. And as a consequence you have to know that difference.' She explains: 'I start with a seedling and I don't want to leave it. I don't feel I really know the story if I don't watch the plant all the way along. So I know every plant in the field. I know them intimately and I find it a great pleasure to know them'.

## TO EMBRACE THE WORLD, NOT TO CONQUER IT

McClintock calls herself "a mystic in science", this women whose work has been recognised with a Nobel Prize, and she says that her aim is "to embrace the world", according to Evelyn Keller, "in its very being, through reason, and beyond". Now to embrace the world is something very different, of course, than the desire to conquer it.

Barbara McClintock's articulated desire to respect and embrace the world stands in stark contrast to Francis Bacon's expressed desire to "put nature on the rack and torture her"[80] so that nature – the ultimate 'other' – will reveal her secrets. Hence there is within science a very different tradition than that of domination, manipulation and control exemplified by one of the founding fathers of modern science, Francis Bacon. And it is this alternative tradition of science, represented by Barbara McClintock's love of plants, that we need to understand and emulate if we are going to have an ecologically sound and sustainable society.

McClintock shows us how to open our other eye, how to overcome our long history of one-sidedness. To McClintock science is not premised on detachment, on domination, on a division between subject and object, between self and 'other'. In a more recent book, *Reflections on Gender and Science*, Keller states that McClintock's love of nature allowed for "intimacy without the annihilation of difference".[81] As Keller says, "division is relinquished without generating chaos".[82] A vivid illustration of this love, this form of attention to things, comes from McClintock's own account of how a deeper understanding in one particularly incomprehensible piece of analysis was realized. McClintock describes the state of mind accompanying the crucial shift in orientation that enabled her to identify chromosomes she had earlier not been able to distinguish.

> I found the more I worked with them, the bigger and bigger the chromosomes got and when I was really working with them, I wasn't outside. I was part of the [system] . . . it surprised me because I actually felt as if I was right down there and these were my friends . . . As you look at these things they become part of you . . .[83]

## A REAL FEELING FOR THE ORGANISM

This account of matter is a far cry from the dead inert matter of the mechanical model. As Keller points out, McClintock allows us to see the profound *kinship* between us and nature. She encourages us to witness the astonishing diversity and unimaginable resourcefulness of the natural order. She inspires a real feeling for the organism. McClintock sees nature not as blind, simple and obedient, but as self-generating, complex and resourceful. Nature, for McClintock, is not more complex than we know, but more complex than we *can* know. Nature is ingenious. 'Anything you can think of, you will find', declares McClintock.[84] Hence for McClintock the goal of science is not the power to manipulate but *empowerment*, the power to understand, the power to appreciate, the power to be humble.

## DIFFERENT VOICES SHOWING OUR IDEOLOGICAL BIASES

There are real limitations in contemporary science and feminist writers like Keller, and scientists such as McClintock, are helping us to see the ideological biases of science, and so to transcend them. Keller notes that McClintock is not a feminist scientist[85] since McClintock's vision was of a science not based on sex or gender. But here Keller misses the point: feminism wishes just that. Feminism seeks to transcend the dividing dichtomies between masculine and feminine and to found a new science based on McClintock's vision. What is encouraging is that there is a shift in emphasis in numerous fields which gives weight to the argument developed here.

Nel Noddings revolutionised ethics through her book, *Caring*,[86] by showing how morality was dominated by certain values and concerns, by talk of principles and justification rather than by talk of caring and receptivity, relatedness and responsiveness. In her Introduction she says:

> One might say that ethics has been discussed largely in the language of the father: in principles and propositions, in terms such as justification, fairness and justice. The mother's voice has been largely silent. Human caring and the memory of caring and being cared for, which I shall argue form the foundation of ethical behaviour, have not received attention, except as outcomes of ethical behaviour.[87]

If care is the foundation of our understanding we will include both categories 'mother' and 'father' as Noddings does, to embrace a new way of relating to nature.

Likewise, Erazim Kohak in his book *The Embers and the Stars*,[88] states that the aim of the book is to shift the burden of philosophising from the making of arguments to the aiding of vision. Not cunningly devised theories but encountering the wonder of being, this is the objective.

We need these different voices. We need them urgently. Theories can be very influential but to help them be so, as Herbert Marcuse argued in *Eros and Civilization*,[89] they must move from the surface and delve into the deepest biological layers of human energies, the well-springs of human action. Otherwise our theories of political, economic, social or ecological change are rootless; we are picking up litter rather than attacking the production of unusable waste in the first place. And this delving into the biological layers is precisely what feminist theory is doing.

## BIRTH WITHOUT WOMEN

Is it an accident that the telegram relating the first successful test of the hydrogen bomb read "it's a boy". The release of a monstrous force of destruction is represented by the birth of a boy, birth without a woman. Is it an accident that Edward Teller was called 'father' of the H-bomb? Is it an accident that the first atomic bombs were called, 'Little Boy', 'Fat Man', 'George and Mike'?[90] Is it an accident that the names that the military use are laden with psychosexual overtones: missile erector, deep penetration, soft lay downs? The real reason Trident was purchased by the British said historian E.P. Thompson, with a smile, was because we are in a "post imperial phallic system" and there is a need to show we British can "still get it up".[91] Dr John Gilbert, the British Labour Backbench Defense Minister, referred critically to the "Trident Virility Symbol".[92] The Chairman of the Committee on Public Information during WWI compared military training to national virility.[93]

Helen Caldicott's book, *Missile Envy*, explores this theme[94] and Brian Easlea, for better or worse, takes these images literally in his book *Fathering the Unthinkable: Masculinity, Science and the Nuclear Arms Race*.[95] Why? Because if you look at the nuclear arms race, the intensity could not be explained just by economic factors. It *is* intense. According to the Swedish Peace Research Foundation, it is an enterprise that engages over 45 per cent of all scientists and technologists alive. To Easlea something else is going on: male-female factors are engaged. Men play a relatively small part in the creation of life in our culture, so do they try and overcompensate elsewhere? Not in creating life, but in risking it – *this* is the key", says Simone De Beauvoir, and she continues, "it is why superiority has been given not to those who bring forth life, but to those who kill . . .".[96]

Brian Easlea begins his book *Fathering the Unthinkable*, simply but powerfully:

The single most important issue facing human kind today is surely the nuclear arms race. It prevents other problems being addressed, like the eradication of desperate poverty, and it will sooner or later lead to unmeasurable catastrophe both for humanity and for much of the wild life with whom we ought to share this planet.[97]

The aim of Easlea's book is to investigate *one* determinant, not the only determinant, but one insufficiently examined determinant of this insane race, namely the overall 'masculine' nature of modern science and weapons science.

Easlea was a nuclear physicist. While working as an English expatriate in Brazil in the 1960s he became aware of the futility of making nuclear weapons. So he switched fields, from science to the history and philosophy of science. This new discipline led him to study the persecution of witches in the 17th Century. "I came to believe", Easlea remarks, "that economic causes were not sufficient to explain the intensity and brutality of many of these persecutions".[98]

As E.W. Monter has pointed out,[99] witchcraft was by far the most important capital crime for women in early modern Europe. More women were put to death between 1500 and 1700 for the crime of witchcraft than for all other crimes put together. Why? Of course, economic causes were operative – witches competed with the newly formed class of physicians and medical doctors. But the intensity and duration of these persecutions compels us to search for additional motivations. Easlea argues that: "... Non-economic factors such as gender identity and sexual attitudes were important to understand the ferocity of the persecutions and the underlying causes".[100] By the same token if one investigates the nuclear arms race, one must look at the male-female factors involved in order to satisfactorily understand it.

Modern science is basically a masculine endeavour and as such it is lopsided. The whole nature of the scientific system and its assumptions needs to be challenged otherwise we will not make peace with the environment. One way to do this is to contrast 'normal' science[101] with alternative traditions within science, such as McClintock's in which science serves the interests of preservative love. Another way is to pinpoint alternative traditions within philosophy such as Noddings and Kohak, which take account of those areas of human experience deemed 'feminine'. A third way is to attempt to show how a psycho-sexual analysis of science such as Easlea's illuminates the patriarchal construction of knowledge and how this blocks our realisation of a sound ecology.

## SCIENCE SERVING THE INTERESTS OF POWER, NOT PRESERVATIVE LOVE

All modern, scientific thinking is at bottom power thinking, that is to say, the fundamental human impulse to which it appeals is the love of power, or, to express the matter in other terms, the desire to be the cause of as many and as large effects as possible.[102]

Ever since Bacon announced that knowledge is power, science has been dominated by a will to power and so has been domineering, as Bertrand Russell argues. Bacon talks about how the new science will be a rebirth. He jeers at the Greeks, whom he calls "mere boys" and calls for a masculine birth of time, a rebirth without indebtedness to women.[103] Through this rebirth, the scientific imagery goes, one can conquer the universe. And what does conquering the universe involve? The "death of nature", as Carolyn Merchant points out.[104] At the same time as Bacon, Descartes declared Mother Nature dead – mere matter in motion without sentience, without consciousness. Animals are mere machines, in the Cartesian framework.

Hence, in 17th Century we have a cognitive attack on Mother Nature through the writings of such influential thinkers as Bacon and Descartes.

Coupled with this assault we also have a physical attack on women. In 16th and 17th Century scores of thousands of women were hunted and killed for witchcraft. This link is, Easlea argues,[105] no accident. It spells a deep antagonism to women which men cope with by being violent towards nature. So anti-feminist sentiment feeds ecological disaster, from the testing of atomic bombs to using animals as tools of trivial research projects.

## THE CAUSES AND CONSEQUENCES OF AN ELUSIVE MASCULINITY

Brian Easlea in his book, *Science and Sexual Oppression*,[106] argues that sexuality defines the kind of existence one lives and vice-versa. The person who wants to live sensuously and joyously must also love senuously and joyously. The person committed to frenetic thrusting, for mastery in sexual intercourse, whether male or female, is often also committed to domination and agression in social relations and to relations of brutality with the natural environment. Easlea's principle argument is that hyperaggressive people seek power over other men and other women and over nature, "not *solely* because of harsh material conditions . . . but *also* because they seek through that power to underwrite and demonstrate an elusive masculinity".[107]

Why do men in our culture face an elusive masculinity? Because of the way our society defines male sexuality exclusively in terms of a partial and limited (and marketable) sexuality – genital sexuality; because in our culture only mothers parent and so men are not equally integrated into the reproductive process, hence there is a cultural alienation regarding paternity; finally, because of the natural alienation men experience from the living products of their own sexuality – men can never be sure if the baby is their child.

Herbert Marcuse argues in *Eros and Civilization*,[108] that there is a direct link between the repression of sexuality and the eruption of agression. Marcuse argues that our 'permissive' society is in fact sexually repressive. It is dominated by the tyranny of genital sexuality where permissiveness is linked to marketable commodities. Thus a partial and limited sexuality has been put to work in service of the established order. This myopic view of sexuality feeds an elusive masculinity which in turn exacerbates aggressive behaviour.

The second cause of elusive masculinity has been developed by Nancy Chodorow in her book, *Mothering*.[109] If only women parent, she argues, it is far more difficult for a male to establish his identity than for a female to establish hers. Why? Because in our culture the male model is relatively inaccessible. In many middle class families, Dad is out of the house nine to five, five days a week. In a boy's search for a male identity, he over-rejects the receptivity, softness and connectedness present in the female model and identifies manhood as an achievement in which he must appear to be hard, separate, relatively unaccessible and superindependent. As a result of this social conditioning, men grow up with blocked access to their receptive, empathetic and compliant aspects and so they continually try to dominate. Unless men parent, Chodorow argues, men will be less likely to recognise their dependence, specifically their dependence on nature which is ultimately the source of our being. If men parent, boys will be more able to be freed from the demands of being aloof and will be more likely to be receptive, empathetic, compliant, playful and tender. Likewise, girls will be liberated from the demands of being exclusively emotional and intuitive and will be able to exhuberant, active and creative.

This theory is limited, Elizabeth Fee notes,[110] because it assigns a primary role to the structure of family relationships as the cause and eventually the solution of other social problems. But while shared parenting between men and women will not be the magic wand to transform the social organisation of gender and to eliminate sexual inequality it would surely help. Though not a panacea, it provides a useful perspective and a partial solution.

The shared parenting theory has also been criticised[111] because it is based on modern, Western, middle-class nuclear families in which an isolated, full-time mother takes direct responsibility for childcare and housework while the absent father is occupied in the labour market. But even though the theory is ethnocentric and classist, it must be remembered that it is predominately white, middle-class men reared in exactly this family situation who hold positions of some power in our society and who, as a result, define the meaning of gender. So the theory alerts us to an important aspect of one culturally bound, class bound but influential world.

The third cause of an elusive masculinity has been well articulated by Mary O'Brien.[112] The moment of ejaculation creates a natural alienation from the male creative process. And this natural alienation which men experience heightens their culturally induced paternal alienation.

Hence our thesis is that the insecure identity of sexist, hyperaggressive men is a significant (but not the only) force underlying the irrationality of our society intent on plunging the world into ecological or nuclear holocaust. The evidence for this thesis comes from three sources: 1) psychological theories which make intuitive sense about the exaggerated role of genital sexuality and the links to aggressive behaviour; 2) sociological theories about the importance of our primary identification with our parents, which tend to tie men to a role of aloof independence; and 3) philosophical theories about a man's natural alienation regarding fatherhood which fans the flames of their insecurity.

### SCIENCE: A SURROGATE SEXUAL ACTIVITY

Evidence also is found in the history of science. Brian Easlea's book, *Science and Sexual Oppression*, amasses considerable evidence to show how "so often scientific investigation into the properties of the natural world has been viewed metaphonically as active male penetration into the innermost recesses of a passive female nature".[113]

Scientists often describe their quest of understanding nature in sexual terms: scientists have to be 'rigorous' in their thinking, 'hard' in their questioning to unfold nature's secrets. According to Francis Bacon, nature is very much a woman whose secrets "need to penetrated".[114] Again, says Bacon, we need "to storm and occupy her castles and strongholds".[115]

The imagery of male, sexual wooing, conquest and penetration of female nature is very explicit in the 17th Century, as Easlea's scholarship shows. To quote Henry More, a Fellow of the Royal Society: "We must break open her private closet and pierce into her very centre".[116] Henry Vaughan, another notable 17th Century scientist, succinctly identified the sexual imprint of the new scientific spirit of inquiry:

> I summoned nature; pierced through all her store:
> broke up some seals, which none had touched before;
> her womb, her bosom, and her head,
> where all her secrets lay a-bed.[117]

As a result of this evidence, Easlea sees

science as a kind of surrogate sexual activity in which scientists can penetrate to the hidden secrets of an essentially female nature, thereby proving their manhood and virility without necessarily running the risk of attempting the same with real, live and perhaps far from passive women.[118]

The sexual language of science has repeatedly been demonstrated by other scholars.[119] Since the 16th Century the message of science has been clear: mind is male, nature is female and scientific knowledge must be an aggressive act to penetrate nature's mysteries, an act for which women are ill-suited. And lest we think we have outgrown the idea that to be a true scientist is to be non-feminine, contemporary examples abound.[120]

### OBJECTION: SURELY THE KNOWLEDGE-CLAIMS OF SCIENCE ARE NOT SEXIST

At this point the reader might object and declare: the language used by scientists contains a surprising number of sexual metaphors, the practice of science was sexist, as it excluded women, but surely there is nothing remotely sexist about the knowledge-claims of science! At first sight this objection might seem correct but consider an apparently a-sexual neutral scientific law, Newton's law of universal gravitation. Newton viewed the existence of gravitational attraction as a manifestation of God's direct agency in nature.[121] It was through God's action that the earth was attracted to the sun. In Newton's mechanical model of the universe, no action-at-a-distance was possible. Only direct, contact action was conceivable. So, God became the all-important medium for gravitational pull. And Newton's God was God the Father. So inevitably male-female relationships entered into the interpretation of gravitational theory.

The reader might continue to object and say these debates about the nature of God were extraneous to science. But such debates were central to the scientific community at the time. Physics and mathematics might be the most removed of all the sciences from the sphere of masculine bias but why have they become paradigmatic.[122] Furthermore even these so-called 'hard' sciences (the very label is telling) are not just sets of symbols; they are actual practices whose research purposes, activities and knowledges are socially produced and as such are determined by values. Hence even in physics the *context* and *meaning* of gravitation related to gender.

### MALE DOMINANCE PROGRAMMED INTO THE CONTEXT OF SCIENCE

As Easlea points out,[123] how much more so must male-female relationships enter into scientific pronouncements on the nature of life, how it originates and how it reproduces. Surely all kinds of hopes and fears about sexual reproduction will be projected onto these theories, despite claims to objectivity. Stop to consider the 'master' molecule theory in biology (DNA – the material of which genes are composed), or the exaggerated importance given to male dominance and male initiative in ethology. The concept of the 'master' molecule allowed us to overlook the importance of the cell and to see it as a subordinate, passive recipient of directions and orders from the master (the organism's genes or the genome).[124] And the exaggerated role given to male hierarchies in controlling primate troop behaviour lead to excessive claims like this:

> the [primate] females were incapable of 'governing' the group . . . the introduction of but one adult male . . . corrected the situation immediately . . . primate females seem biologically unprogrammed to dominate political systems . . .[125]

The feminist perspective helps us to see how scientific theories support masculine dominance. As Fee writes,

> There is not . . . a single feminist science that represents *only* the interests of women as a unified group. There *is* however, a feminist perspective on science that shows the ways in which gender-based dominance relations have been programmed into the production, scope and structure of natural knowledge, distorting the content, meaning and uses of knowledge.[126]

## MASCULINE FEARS

Men tend to fear women's affinity with nature, as the witch-hunt craze shows. Male insecurity and the resultant need to dominate nature led to the persecution of scores of thousands of women who had some understanding of nature.

Men tend to fear dependence, as the history of science and its language and metaphors show. Therefore, it is not easy for them to think clearly and feel positively about our human dependence upon the ecosystems of the biosphere.

Finally, men tend to fear women's ability to reproduce life and hence they declared that nature, mythed as female, was really barren. Nature does not have life-giving powers; she is just a machine to be exploited, a "standing reserve" to use Heidegger's phrase.[127]

## FROM MOTHER EARTH (NATURE ALIVE) TO AN INERT MACHINE (NATURE DEAD)

The patriarchy's envy and fear of women and the earth was transformed symbolically into a mechanistic world view. As Merchant illustrates,[128] the earth was no longer regarded as a nurturing mother to be cherished but as a machine to be manipulated and exploited. The transformation from earth mother to barren, inert matter had been accomplished through the scientific revolution of the 17th Century and men hoped to become, through the medium of science, "masters and possessors of nature".[129]

This mechanical world-view accelerated the exploitation of humans, both men and women, and the exploitation of natural resources. Mechanism saw nature not as a live presence but as a system of dead, inert particles moved by external forces. As such, it bears no moral self-examination. Nature is merely material for man's appropriation, a view which suited commercial capitalism.

## THE FEMINIST CHALLENGE TO MECHANISM

Both feminism and ecology challenge mechanism's assumptions which make the manipulation and control of nature seem possible and acceptable. The ontological assumption of mechanism is that matter is not living, interdependent bundles of energy, but that matter is composed of dead, discrete particles. The epistemological assumption is that knowledge and information can be abstracted without distortion from the natural world because they are independent of context and value-free. The methodological assumption is that real-life problems can be successfully analysed into parts that can be manipulated by mathematics. The moral assumption is that humans (especially white, upper class males) are more valuable than the rest of nature.

The mechanical paradigm, though outdated, persists. States Francis Crick: "The ultimate aim of the modern movement in biology is, in fact, to explain *all* biology in terms of physics and chemistry".[130] The project is to explain life (humans, brains) in terms of non-life (machines, computers). The commitment is that nature is lifeless. And this allows us to perform gruesome experiments on living animals and to write them up "as if the experiments had been done on inert matter".[131]

Contrast two descriptions: Woodridge, a modern psychological researcher, convinces himself that when a monkey with an electric current passing through its 'pain centre' bites objects so hard as to wrench teeth from its jaw – then the monkey really is experiencing pain and not exhibiting 'meaningless or automatic physical symptoms'.[132] Compare this to a description of a chimpanzee being studied by Jane Goodall in the Tanzanian National Park of Gombe Stream. Once a chimpanzee grasped her hand "firmly and gently with his own before scurrying off into the forest". Goodall writes that "at that moment there was no need of any scientific knowledge to understand this communication of reassurance".[133]

### NATURE: A LIVE PRESENCE AND 'OUR BODY'

Goodall's attitude to non-human life forms is, while superior, not a female perrogative. Martin Buber, for example, has an I-Thou relationship while looking into the eyes of his cat.[134] And people who live close to nature see nature not as an 'it', but as a 'thou'. Nature is for so-called 'primitive' people a live presence, each part of which is unique, and not satisfactorily describable by a universal law.[135] Hence their relationship with a tree, for example, is emotional, direct, not fully able to be articulated. For many traditional peoples, "nature is their body".[136]

> Communication, at its best, is called love; when it breaks down completely, we call it war. And it is a sort of war that is going on now between human beings and the earth. It's not that nature refuses to communicate with us, but that we no longer have a way to communicate with it. For millennia, primitives communicated with the earth and all its beings by means of rituals and festivals where all levels of the human were open to all levels of nature.[137]

### CARNAL KNOWLEDGE

We need to recapture this reality. We need, as A.N. Whitehead urges, to develop "a science based on an erotic sense of reality rather than an aggressive, dominating attitude towards reality".[138]

We need to relate the crisis of production (and overproduction) to the crisis of reproduction.[139] We will not solve our environmental crisis unless we allow the earth to reproduce.

We need to appreciate reproductive values, such as a recognition of our own death. Freud did not consider aggression a basic psychological fact. Aggression for Freud was rather secondary manifestation of a more fundamental instructural force, the death instinct.[140] Norman O.

Brown, analysing Freud in *Life Against Death*,[141] felt that we could learn to channel our aggressive behaviour positively if only we could learn to contain death within life, that is, if we were not "fugitives from our own death",[142] in short, if we learnt how to die with dignity.[143]

As Barbara McClintock reminds us, we need a new scientific basis whereby we recognise that each organism is unique and that it is an integral part of an ecosystem. We need a new psychological basis whereby we see, as Carl Jung reminds us,[144] that to heal is to make whole, as in wholesome, to reunify our split selves, to integrate the masculine and the feminine in each of us. We need a new epistemological basis whereby we realise, as Norman O. Brown reminds us,[145] that we must have carnal knowledge, a copulation of subject and object. We need a new ethical basis whereby we recognise, as the Aboriginal peoples show us, the intrinsic value of and our dependence upon the non-human aspects of Nature. Finally, we need a new ontological basis whereby we experience, as Hegel details,[146] that reality is a process and that the truth is just as much subject as substance – that the truth is the whole.

We need to recognise that while reductionism is a powerful and an important tool, it is limited. For the whole is more than the sum of the parts and the parts themselves take meaning from the whole. Each part is defined by and dependent upon the total context. Isolation (as in a laboratory) distorts the truth because it distorts the whole. As it is valid to interpret the higher (life) in terms of the lower (non-life) so it is also valid to interpret the lower in terms of the higher. 'Aim' can be applied to cells; 'enjoyment' can be applied to gorillas. Reality is a complex and dynamic web of energy. Nature is alive and active and its parts are fundamentally interconnected by cyclical processes. We need to realize, as gatherer-hunter peoples do, that if we spit on the earth, we spit on ourselves. Defending the earth is self-defense.

## WHY WE NEED EACH OTHER

We need a reversal of mainstream, malestream values and the triumph of feminism, a revolution in economic priorities and a steady-state economy,[147] a peace force for a just and sustainable society,[148] a social force for voluntary simplicity,[149] and collective action for the ecological reconstruction of society.[150] We need a new metaphysical and cultural reconstruction to empower us to help create this new reality.

We need to overcome our dichotomies and to discover our deep sources, our springs, as Rachel Carson did.

We need each other.

*School of Social Sciences*
*Murdoch University*
*Western Australia*

## NOTES

\* Published in a shorter version by *Canadian Woman Studies*, Vol. 9, No. 1, Spring '88, pp. 9–18, Under The Title 'Making Peace With The Environment: Why Ecology Needs Feminism'.

[1] Eco-feminism is a term coined by Francoise D'Eabonne, in her book *La Femme Avant Le Patriarchat*, in 1975.

[2] Aldous Huxley, 1963, Introduction to *The Song of God: Bha'gavad-gita*, New York: Mentor, pp. 11–22.

[3] Henryk Skolimowski, 1981, *Eco-philosophy: Designing New Tactics for Living*, Boston: Marion Boyars.

[4] K.S. Shrader-Frechette, 1981, *Environmental Ethics*, Pacific Grove, CA: The Boxwood Press, pp. 28–56.

[5] G.W.F. Hegel, 1964, *The Phenomenology of Mind*, translated by J.B. Baillie, London: George, Allen and Unwin, p. 17.

[6] See such analysis of the political economy as Barry Commoner, *The Closing Circle*, London: Jonathan Cape, 1973.

[7] David Hume, 1955, *An Inquiry Concerning Human Understanding*, New York: The Bobbs-Merrill Co., Inc., p. 9.

[8] Henryk Skolimowski, *Eco-philosophy*, op. cit., p. 51.

[9] Edmund Husserl, 1962, *Ideas: General Introduction to Pure Phenomenology*, translated by W.R. Boyce Gibson, New York: Collier Books, pp. 38–39.

[10] "You ask me to plough the ground. Shall I take a knife and tear my mother's breast?" Smohalla of the Sokulk Tribe, Priest Rapids, Columbia River, Washington, *Touch the Earth: A Self-Portrait of Indian Experience*, compiled by T.C. McLuhan, London: Abacus, 1976, p. 56.

[11] "In short they seem'd to set no value upon any thing we gave them . . . this in my opinion argues that they think themselves provided with all the necessarys of life . . .". Captain James Cook, August, 1770 commenting upon his contact with "The Natives of New-Holland" (or Australian Aboriginal people), *The Journals of Captain James Cook*, edited by J.C. Beaglehole, The Voyage of the Endeavour, 1768–1771, Cambridge University Press for the Hakluyt Society, 1968, p. 399.

[12] Ludwig Wittgenstein, 1958, *Philosophical Investigations*, second edition, New York: The Macmillan Company, 48e, paragraph 114.

[13] *Ibid.*, 50e, paragraph 129.

[14] John Maynard Keynes, 1936, *The General Theory of Employment, Interest and Money*, London: Macmillan, p. 17.
[15] Immanuel Kant, 1909, *The Critique of Practical Reason*, translated by J.K. Abbott, New York: Longmans, Green and Co., p. 321.
[16] John Seed, 1987, 'Alternatives to the Rainforest Timbers', *Australian Wellbeing*, No. 23, pp. 91–92.
[17] Tom Regan, 1982, *All That Dwell Therein: Essays on Animal Rights and Environmental Ethics*, Berkeley: University of Califronia Press, p. 72.
[18] Lynn White, 1967, 'The Historical Roots of our Ecologic Crisis', *Science*, 10 March, 1967, Vol. 155, No. 3767.
[19] Immanuel Kant, *The Critique of Practical Reason*, op. cit., p. 347.
[20] G.W.F. Hegel, 1964, *The Science of Logic*, translated by A.V. Miller, London: George Allen and Unwin, Ltd., p. 22 and p. 245ff.

For a fuller development of this thesis please see, Patsy Hallen, 'How the Hegelian Notion of Relation Answers the Question: "What's Wrong With Plastic Trees?"', S206 Environmental Ethics Study Guide, Murdoch University, 1988, pp. 52–70.

[21] Karl Marx, 1971, *1844 Manuscripts*, in *The Early Texts*, edited by D. McLellan, Oxford: Oxford University Press, p. 175.
[22] John Seed, 1985, 'Anthropocentrism', Appendix E in Bill Devall and George Sessions, *Deep Ecology: Living as if Nature Mattered*, Salt Lake City: Peregrine Smith Books, pp. 243–246.
[23] Warwick Fox, 1989, 'The Deep Ecology – Ecofeminism Debate and its Parallels: A Defense of Deep Ecology's Concern with Anthropocentrism', *Environmental Ethics* **11**: 5–25.
[24] For a fully developed notion of care as an ontologically prior aspect of human beings see Martin Heidegger, *Being and Time*, New York: Harper and Ros, 1962, Part One, Division One, VI, 'Care as the being of desire', H 180ff.
[25] Arne Naess has articulated and developed two ultimate norms or institutions of ecological consciousness, self-realization and biocentric equality. Mature people know that their own realization lies in embracing the non-human world. See Arne Naess, 'Self-realization in mixed communities of humans, bears, sheep and wolves', *Inquiry* **22** (1979): 231–241.
[26] Loren Eiseley, 1970, *The Unexpected Universe*, London: Victor Golcancz Ltd., p. 51.
[27] Hilary Rose, 1986, 'Beyond Masculinist Realities: A Feminist Epistemology for the Sciences' in Ruth Bleier (ed.), *Feminist Approaches to Science*, New York: Pergamon Press, p. 60.
[28] Mary O'Brien, 1987, 'Loving Wisdom', in *Resources for Feminist Research* **16**(3): 7.
[29] Aldo Leopold, 1949, *A Sand County Almanac*, Oxford: Oxford University Press, p. 268.
[30] Albert Einstein, 1982, quoted in Michael Nagler, *America Without Violence*, Covelo, California: Island Press, p. 11.
[31] J. Elgin, 1984, 'Women and Non-Violence', *Social Alternatives* **4**(1): 23.
[32] Ariel Kay Salleh, 1984, 'Deeper Than Deep Ecology: The Eco-Feminist Connection', *Environmental Ethics* **6**: 340.
[33] For some informed criticisms of this essentialist view, see Marti Kheel, 'Ecofeminism

and Deep Ecology: Reflections on Identity and Difference', paper presented to 'Echofeminist Perspectives: Cuture, Nature and Theory' Conference, University of California, 27–29 March, 1987; Michael Zimmerman, 'Feminism, Deep Ecology, and Environmental Ethics', *Environmental Ethics* **9** (1987): 21–44; Jim Cheney, 'Eco-Feminism and Deep Ecology', *Environmental Ethics* **9** (1987): 115–145; Karen J. Warren, 'Feminism and Ecology', *Environmental Ethics* **9** (1987): 3–20; Alan E. Wittbecker, 'Deep Anthropology: Ecology and the Human Order', *Environmental Ethics* **8** (1986): 261–270; and Jean Grimshaw, *Philosophy and Feminist Thinking*, Minneapolis: University of Minnesota Press, 1986.

[34] Simone de Beauvoir, 1975, *The Second Sex*, translated and edited by H.M. Parchley, Harmondsworth, Middlesex, England: Penguin Books.

[35] Jean Paul Sartre, 1956, 'Existentialism is a Humanism' in *Existentialism* from *Dostoevsky to Sartre*, edited by Walter Kaufmann, New York: Meridian Books, pp. 287–311.

[36] Marion Namenwirth, 'Science Seen Through a Feminist Prism' in Ruth Bleier (ed.), *Feminist Approaches to Science*, op. cit.

[37] Sandra Harding, 1986, *The Science Question in Feminism*, Ithaca: Cornell University Press, pp. 52–57.

[38] 'Simone de Beauvoir: 'The Second Sex 25 years Later', and interview with John Gerassi, *Society* **13** (January–February 1976): 84.

[39] Harding, *The Science Question in Feminism*, op cit., p. 17.

[40] Susan Saxe, 1983, 'Ask a Stupid Question' in *Reclaim the Earth: Women Speak Out for Life on Earth*, London: The Women's Press, p. 66.

[41] Barbara Porter, 1987, 'Management Strategies for Endangered Australian Fauna', M.Sc. Thesis, Faculty of Agriculture, University of Western Australia.

[42] Michel Foucault, 1980, *A History of Sexuality* I, New York: Random House.

[43] Martin Heidegger, 1977, 'What calls for thinking?' in D.F. Krell *et al.* (eds), *Basic Writings*, New York: Harper and Row, p. 354.

[44] Erazim Kohak's goal in his eloquent book *The Embers and the Stars: A Philosophical Inquiry into the Moral Sense of Nature*, Chicago: University of Chicago Press, 1984, p. xiii.

[45] See, for example, Hilary Rose, 'Beyond Masculinist Realities: A Feminist Epistemology for the Sciences' in Ruth Bleier (ed.), *Feminist Approaches to Science*, op cit., pp. 57–76; or Hilary Rose, 'Hand, Brain and Heart: a feminist epistemology for the natural sciences', *Signs: Journal of Women in Culture and Society* **9**, No. 1, 1983.

[46] Plato, *The Collected Dialogues including the Letters*, edited by Edith Hamilton and Huntington Cairns, New York: Pantheon Books, Bollingen Series LXXI, 1961; Soren Kierkegaard, *Concluding Unscientific Postscript*, translated by Walter Lowrie, Princeton: Princeton Univesity Press, 1956; Michael Polanyi, *Personal Knowedge*, New York: Harper Torchbooks, 1958.

[47] For example Simone de Beauvoir's fruitful framework of analysing women's status as 'the other'. See *The Second Sex*, op cit.

[48] For instance, the treadmill situation of pesticide use in California. See Robert van den Bosch, *The Pesticide Conspiracy*, New York: Doubleday, 1978. Bosch documents how pesticides poison humans, other animals and birds, not pests. We have not success-

fully eliminated one species of pests, only their predators. Hence we are using vast quantities of pesticides and still losing more crops to the resiliant and genetically variable pests than ever before: a treadmill situation.

[49] Arne Naess, 1973, 'The shallow and the deep, long-range ecology movement', *Inquiry* **16**: 95–100.
[50] Marilyn French, 1985, *Beyond Power*, New York: Harper and Row.
[51] Harding, op cit., p. 31.
[52] Michele Aldrich, 'Women in Sciences', *Signs: Journal of Women in Culture and Society* **4**, No. 1, 1978; Rita Arditti, 'Women drink water: Men drink wine' in R. Arditti, P. Brennan and S. Cavrak (eds.), *Science and Liberation*, Boston: Solithend Press, 1980; Alice Rossi, 'Why so few?', *Science* **148** (1965), 1196; Margaret Rossiter, *Women Scientists in America: Struggles and Strategies to 1940*, Baltimore, MD: John Hopkins University Press, 1982.
[53] Rose, 'Beyond Mascilline Realities', op cit., pp. 60–62.
[54] Harding, op cit., p. 64.
[55] *Science*, 9 April 1985.
[56] Hilary and Steven Rose (eds.), 1976, *Ideology of/in the Natural Sciences*, Cambridge, Mass.: Schenkman.
[57] Neil Evernden, 1984, *The Paradox of Environmentalism*, symposium proceedings, the Centre for Environmental Studies, Toronto: York University, p. 9.
[58] Evelyn Fox Keller, 1985, *Reflections on Gender and Science*, New Haven: Yale University Press, p. 4.
[59] *Ibid.*, p. 178.
[60] *Ibid.*, p. 134.
[61] Ben McNaughton, 'Is He House-Trained?', *New Internationalist*, March 1988, p. 23.
[62] Brian Easlea, 1983, *Fathering the Unthinkable: Masculinity, Scientists and the Nuclear Arms Race*, London: Pluto Press.
[63] Harding, op cit., p. 146.
[64] Norman O. Brown, 1959, *Life Against Death*, New York: Random House, Vintage Books; Herbert Marcuse, 1964, *Eros and Civilization*, Boston: Beacon Press.
[65] These ideas were born for me in listening to the lectures of Professor John Raser, a Foundation Professor of Murdoch University, Perth, Western Australia.
[66] Carol Gilligan, 1982, *In a Different Voice*, Cambridge: Harvard University Press.
[67] Hilary Rose, 'Beyond Masculinist Realities', op cit., p. 72.
[68] Marilyn French, *The Women's Room*, quoted by Rose, *ibid.*, p. 68.
[69] Rose, *ibid.*, p. 72.
[70] Nancy Chodorow, Dorothy Dinnerstein, Jessica Benjamin, Jane Flax and Ann Oakley, among others.
[71] John Livingston, 'The Dilemma of the Deep Ecologist' in Neil Evernden, *The Paradox of Environmentalism*, op cit., p. 63.
[72] Mary O'Brien, 1983, *The Politics of Reproduction*, Boston: Routledge and Kegan Paul.
[73] Alfred North Whitehead, *Process and Reality*, corrected edition edited by David Ray Griffin and Donald W. Sherburne, New York: Collier, Macmillan, 1978; Henri Bergson, *Creative Evolution* translated by Arthur Mitchell, New York: Random House, The Modern Library, 1944.

[74] Carolyn Merchant, 1985, 'Feminism and Ecology', Appendix B, pp. 229–231 in Bill Devall and George Sessions, *Deep Ecology: Living as if Nature Mattered*, Salt Lake City, Peregrine Smith Books.
[75] Barry Commoner, 1972, *The Closing Circle: Nature, Man and Technology*, New York: Alfred A. Knopf, Inc., pp. 41–42.
[76] Harding, *The Science Question in Feminism*, chapter 7, pp. 163–196.
[77] Donna Haraway, 'Primatology is Politics by Other Means' in Ruth Bleier, *Feminist Approaches to Science*, op cit., pp. 77–118.
[78] Livingston, 'The Dilemna of the Deep Ecologist', op cit., p. 67.
[79] Evelyn Fox Keller, 1983, *A Feeling for the Organism: The Life and Work of Barbara McClintock*, New York: W.H. Freeman. All quotes in this paragraph are from pp. 99–102.
[80] Francis Bacon, 1970, 'Thoughts and conclusions' in Benjamin Farrington, *The Philosophy of Francis Bacon*, Liverpool: Liverpool University Press, p. 92; see also F. Bacon, *Essays*, Ward, Lock and Bourden, 1894.
[81] Keller, *Reflections on Gender and Science*, op. cit., p. 164.
[82] *Ibid.*, p. 165.
[83] *Ibid.*
[84] *Ibid.*, p. 162.
[85] Evelyn Fox Keller, 'Women, Science and Popular Mythology' in *Machina ex Dea: Feminist Perspectives on Technology* (cited in a personal correspondence).
[86] Nel Noddings, 1984, *Caring: A Feminine Approach to Ethics and Moral Education*, Berkeley: University of California Press.
[87] *Ibid.*, Introduction, p. 1.
[88] Kohak, *The Embers and the Stars*, op cit.
[89] Marcuse, *Eros and Civilization*, op cit.
[90] Easlea, *Fathering the Unthinkable*, op. cit., pp. 93–96.
[91] *Ibid.*, Introduction, p. 8.
[92] *Ibid.*
[93] *Ibid.*, p. 146.
[94] Helen Caldicott, 1985, *Missile Envy*, New York: Doubleday.
[95] Easlea, *Fathering the Unthinkable*, op. cit., p. 8.
[96] De Beauvoir, *The Second Sex*, op. cit., p. 95.
[97] Easlea, *Fathering the Unthinkable*, op. cit., p. 3.
[98] *Ibid.*, p. 5; see also, Brian Easlea, 1980, *Witch-Hunting, Magic and the New Philosophy: An Introduction to Debates of the Scientific Revolution 1450–1750*, Brighton, Sussex: Harvester Press.
[99] E.W. Monter, 1976, *Witchcraft in France and Switzerland*, Ithaca, N. Y.: Cornell University Press.
[100] Easlea, *Witch-hunting, Magic and the New Philosophy*, op. cit., p. 7.
[101] Thomas S. Kuhn, 1962, *The Structure of Scientific Revolutions*, Chicago: University of Chicago Press, p. 10., ff.
[102] Bertrand Russell, 1949, *The Scientific Outlook*, London, Allen and Unwin.
[103] Francis Bacon, 'The Masculine Birth of Time and the Great Insaturation of the dominion of man over the universe' in B. Farrington, *The Philosophy of Francis Bacon*, op. cit.

[104] Carolyn Merchant, 1980, *The Death of Nature: Women, Ecology and the Scientific Revolution*, San Francisco: Harper and Row.
[105] Easlea, *Witch-hunting, Magic and the New Philosophy*, op. cit., p. 10.
[106] Brian Easlea, 1981, *Science and Sexual Oppression: Patriarchy's Confrontation with Woman and Nature*, London: Weidenfeld, Nicolson. Introduction, p. x.
[107] *Ibid.*, pp. 28–29.
[108] Marcuse, *Eros and Civilization*, op. cit.
[109] Nancy Chodorow, 1978, *The Reproduction of Mothering: Psychoanalysis and the Sociology of Gender*, Berkeley, California: University of California Press.
[110] Elizabeth Fee, 'Critiques of Modern Science: the Relationship of Feminism to Other Radical Epistemologies' in Bleier, *Feminist Approaches to Science*, op. cit., pp. 48–49.
[111] *Ibid.*, p. 49.
[112] O'Brien, *The Politics of Reproduction*, op. cit., p. 12.
[113] Easlea, *Science and Sexual Oppression*, op. cit., p. 88.
[114] Francis Bacon, 1960, *The New Organon*, Book I, aphorism 109, New York: Bobbs-Merrill, p. 101.
[115] *Ibid.*, Book I, Aphorism 130, p. 119.
[116] Easlea, *Science and Sexual Oppression*, op cit., p. 84.
[117] *Ibid.*, p. 89.
[118] *Ibid.*, p. 83.
[119] Carolyn Merchant, Donna Haraway, Ludi Jordanova and Evelyn Fox Keller among others.
[120] Fee, 'Critiques of Modern Science', op cit., p. 45.
[121] Issac Newton, 1964, 'Mathematical Principles of Natural Philosophy' in *Problems of Space and Time*, edited by J.J.C. Smart, New York: The Macmillan Company, pp. 81–88; see also M. Jammer, 1954, *Concepts of Space*, Cambridge, Mass.: Harvard University Press.
[122] Sandra Harding questions the validity of assuming that physics is the paradigm of science in *The Science Question in Feminism*, op cit., p. 43ff.
[123] Easlea, *Science and Sexual Oppression*, op. cit., p. 90.
[124] Namenwirth, 'Science Seen through a Feminist Prism', op cit., p. 26.
[125] *Ibid.*
[126] Fee, 'Critiques of Modern Science', op cit., p. 54.
[127] Martin Heidegger, 1977, *The Question Concerning Technology*, translated by William Lovitt, New York: Garland Publishing Co., p. 22.
[128] Merchant, *The Death of Nature*, op. cit., p. 67.
[129] Bacon, 'The Masculine Birth of Time', op. cit., p. 28.
[130] Francis Crick, 1968, *Of Molecules and Men*, New York: Pan Books.
[131] Brian Easlea, 1973, *Liberation and the Aims of Science*, London: Chatto and Windus Ltd., p. 263.
[132] D.E. Wooldridge, 1963, *The Machinery of the Brain*, New York, McGraw Hill, p. 130.
[133] Jane van Lawick-Goodall, 1974, *In the Shadow of Man*, Glasgow, William Collins, Fontana Books, p. 184.
[134] Martin Buber, 1970, *I and Thou*, New York: Clark.
[135] Henry Frankfort *et al.*, 1959, *Before Philosophy*, Hammodsworth, Middlesex, Penguin Books, pp. 12–13.

[136] Karl Marx, '1844 Manuscripts' in *The Early Texts*, op. cit., p. 117.
[137] Dolores Lachappelle, 1978, *Earth Wisdom*, Silverton Co., Guild of Tutor Press, p. 96.
[138] Alfred North Whitehead, 1958, *Modes of Thought*, New York, Capricorn Books, p. 202.
[139] O'Brien, *The Politics of Reproduction*, op. cit.
[140] Paul Robinson, 1972, *The Sexual Radicals*, London: Paladin, p. 169.
[141] Norman O. Brown, *Life Against Death*, op. cit., p. 110 ff.
[142] Martin Heidegger, 1962, *Being and Time* translated by John Macquarrie and Edward Robinson, New York: Harper and Row, Part One, Division Two, H235 ff, pp. 279–311.
[143] Elisabeth Kubler Ross, 1969, *On Death and Dying*, London: Tavistock Publications.
[144] Carl G. Jung, 1959, *The Undiscovered Self*, translated by R.F.C. Hull, New York: The New American Library, Mentor Books, p. 119 ff.
[145] Norman O. Brown, 1966, *Love's Body*, New York: Random House, p. 17.
[146] G.W.F. Hegel, 1964, *The Phenomenology of Mind*, translated by J.B. Baillie, London: George Allen and Unwin Ltd., Preface.
[147] Herman Daly, 1977, *Steady State Economics*, San Francisco: W.H. Freeman.
[148] A phrase coined by the World Council of Churches; see P. Abrecht, *Faith, Science and the Future*, Geneva: World Council of Churches, 1978.
[149] K.S. Shrader-Frechette, 1981, 'Voluntary Simplicity and the Duty to Limit Consumption', in K.S. Shrader-Frechette ·(ed.), *Environmental Ethics, Pacific Grove*, California: The Boxwood Press, pp. 169–191.
[150] Commoner, *The Closing Circle*, op. cit., p. 299.

DAVID B. ZILBERMAN

# ON CULTURAL RELATIVISM AND "RADICAL DOUBT"

Alfred Kroeber and Clyde Kluckhohn, in their study of culture,[1] mention in passing that Descartes, the man who commenced the new Western philosophy by introducing his method of "radical doubt", was also the first one who detected the problem of cultural relativism. And indeed, as we read in Chapter 1 of his "Discourse on Method":

> It is true that while I only considered that manners of other men I found in them nothing to give me settled convictions; and I remarked in them almost as much diversity as I had formerly seen in the opinions of philosophers. So much was this the case that the greatest profit I derived from their study was that, in seeing many things which, although seem to us very extravagant and ridiculous, were yet commonly received and approved by other great nations, I learned to believe nothing to certainty of which I had only been convinced by example and custom.[2]

This observation is important, in two different aspects. It reaches us as the voice of the epoch when impressions of far travels and overseas discoveries were fresh and agitated curiosity, the epoch particularly rich with talks about striking, sometimes fantastically exaggerated, variety of manners. What is significant is that Descartes drew his conclusions by no means in favor of the uncompromising empiricism of his elder co-founder of the new philosophy, Francis Bacon, who described this diversity of manners as no more than the variously decorated "idols of the crowd", overshadowing the reality of experience which unveils itself only to the unvarnished judgment of the investigator, after scientifically controlled experiment.

On the contrary, for Descartes, his observation of the variety of manners was only the last step before falling into doubt about the legitimacy of his "metacultural" claim to see reality as it is. He was far from scoffing at strange customs as deceptive fantasies; he acknowledged their reality and ranked his own mental habits alongside, and possibly below theirs – as his mind was open enough to learn from other cultures. What a striking contrast indeed, with Bacon, the forefather of the modern "objectifying empiricism" and with epistemological attitudes so paradoxically predominant among professional anthropologists of today. But Kroeber and Kluckhohn are quite right: Descartes was not just a "curious onlooker of cultural variety." He was the first who recognized the kind of philosophical problem that is posed in cultural relativism, and thus proved anthropology *possible* as a *science*: Insofar as "strange manners" are *objectively* different, they can be investigated.

Now to the second aspect: the subject of scientific consciousness. The protagonists of modern phenomenology considered two sides of the problem. Understanding of a culture can differ not only because cultures themselves are different but also owing to the variability of the explorer's comprehension of the same cultural content when different methods are used. Properly speaking, Descartes was more preoccupied with the second possibility than with the first. So he can be considered a founder of the philosophy and methodology of anthropological science which, quite symptomatically, emerged *before* the science itself did.

He tirelessly pondered over many things which could be called in question until he finally stopped before the only thing which looked indubitable, that is, doubting itself. Then Descartes tried to scrutinize its existence, in the two basic modes of "extension" and "thinking", the second being acceptable as certain, (because it was certainly a "doubting thinking") while the first as only contingent (because the "extension" of doubting is always uncertain). Thus, for Descartes, the variety of *cultures*, though objective, is just a contingency, while the explorer's subjective doubting about each of them is evidential.

But here Descartes happened to meet the opponent who nearly forced him to take the just introduced division back. Thomas Hobbes translated and published in London, in 1680, "The Six Metaphysical Meditations" of Descartes, with sixteen letters appended containing his critical remarks and accompanied by Descartes' answers. In his criticisms, Hobbes repeatedly stated that an intently directed thought is immediately "materialized" by its extension – and that makes its existence doubtful, no less radically than existence of "non-thought" material things (which incidentally can be someone else's previously materialized "extensions" of intentional thinking, or even previously expressed thoughts of the same subject.) Hobbes the founder of political science and stout supporter of "nominalistic voluntarism" in philosophical interpretation, pointed out that:

> ... we collect by reasoning nothing of or concerning the nature of things, but of the names of things, that is to say, we only discover whether or not we join the names of things according to the agreements which at pleasure we have made concerning their significations; if it be so (and so it may be) ratiocination will depend on words, words on

imagination, and perhaps imagination, as also sense, on the motion of corporeal parts of an organic body.[3]

In response to this objection, Descartes seemed almost ready to take his suggestion on cultural relativism back, or more properly and meaningfully, he confined it to the sphere of *language* only:

... There is in ratiocination, a conjunction not only of words, but of things signified by words; and I much admire that the contrary could possibly enter any man's thoughts; for whoever doubted but that a Frenchman and a German may argue about the same things, though they use very different words and does not the Philosopher disprove himself when he speaks of the agreements which at pleasure we have made about the signification of words? For if he grants that something is signified by words, why will he not admit that our ratiocinations are rather about this something, than about words only?[4]

But it was not doubting *things* that made it possible for Descartes to recognize some of them (the manners) as "cultural" facts? It was not the suggestive activity of his thinking that created them as such, but it was his *qualified* readiness to comprehend that necessitated his capacity to *judge* about them as such.

Let us take a simple example. The proposition, $2 \times 2 = 4$, is not a statement of cultural fact. But the judgment, "I know (or: "It is necessary", ..., probably, ..., etc.), $2 \times 2 = 4$" is comprehensible as a statement of cultural fact: because of the qualified type of judging attitude, i.e. owing to the *modality* of judgment. Some would probably argue that statements like "I know that ...", as well as notions of knowledge, are merely redundant, a kind of "information noise" rather than inculcations of "objective facts". But that is not true. These statement are *signifiers* of the particular fact that the subject of investigation *had* assimilated and transcended his subject-matter in a ready-to-report manner.

But what then, if his very "transcendability" is doubted, as in the case of cultural relativism? Let us assume that we succeeded in calculating all of the possible modalities of that kind. Does it mean calculation of probabilities of the cultural relativism of *all* possible ways of realization of certain facts as "cultural"? Probably not. But perhaps, the accomplishment of this operation would reproduce the structure of the anthropologist's consciousness, not the individual of course, but as the notion socialized into science: a double of the explored variety of cultures and explorative operations that anticipates and imitates each of them in various ways, thus making the investigation of the real variety somewhat useless.

This second aspect of cultural relativism has hitherto escaped the attention of cultural anthropologists. In even the most sophisticated version of cognitive (ethics/emics)[5] and structural anthropology,[6] the mentality of the investigating subject is stiffened dogmatically in a monomodal representation of its relation to a self-evident empiricity. Cartesian doubt never touches it, and is never turned against its integrity.

There certainly must be objective reasons for this neglect, and they can be easily found. As a matter of fact, the modalization of judgment which, as it was noticed above, is the only way of distinguishing some facts as "cultural", can never bring forth knowledge by itself. Kant was the first who epitomized this circumstance in his "Critique of Judgment". Only loosely associated observations of taste and purpose can be drawn instead.[7] Many anthropologists have since certified that Kant was right in his conclusions.

Thus the representatives of the school of "cultural relativism" (Ruth Benedict, Melville Herskovitz) tried quite outspokenly to approach other cultures with the criteria of taste. What is even more notable, the same criteria were applied to estimation both of their field findings and of their theoretical generalizations.[8] On the other side of the spectrum, we meet with the rigoristic functionalist anthropology whose basic postulates are just slightly reformulated teleological judgments.[9] But neither the first nor the second extreme can provide *knowledge* of another culture. What they report is just a set of "either/or" judgments of taste and purpose, quite non-discriminating as to their own modalization. Nobody can tell where the uncertainty of fact belongs, whether to the cultural thing or to the culturing mentality. So far, modern anthropology is still unable to provide scientific knowledge in the strict sense, either of other cultures or of its own method.

It seems, however, that the cognitive approach and structural anthropology must be exempted, because they rest on, and are concerned with the linguistic and language-like aspects of culture. And even Descartes acknowledged the legitimacy of language relativism. But if someone ever succeeds in correlating *all* cultural facts with the linguistic calculus of modal judgments about these facts, he will thereby turn cultural facts back into the extra-cultural existence of explicit things, as soon as the last "I know, that . . ." is removed. Thus every attempt of modal logic is concluded by reducing it to one of non-modal kind.[10] What is the real subtext of this necessity? from where did doubt and modalized judgments appear?

Kant, in the same "Critique", pointed out that, simply, judgments of taste and purpose spring up where the *difference* in tastes and purposes is unavoidable, i.e. due to the *principle of multipositionality* at work. This state of affairs is again, two-sided. For example, in the case of judgment about cultures: There *are* many of them (at least two, a culture of the investigator and a culture of the investigated), as well as of judgments about *one* culture and its possible realizations. In so far as this relativism is internally duplicated, it cannot be eliminated from our knowledge.

Anthropologists have learned with difficulty how to cope with the plurality of cultures. They come armed with the plainly positivistic, comparativist methods of detecting diffusions and borrowings, as well as with the quasimetaphysical, universalistic postulates of functionalism and "strict" structuralist models. But they are still unable to overcome the inner inconsistency of their investigatory position. From this side, the discipline of anthropology grows sporadically. Until some new "facts" are found which disprove the previously held theory, or until some new conception dawns upon some lucky anthropologist, which entails a change in the treatment of the old "facts" to the degree that the previous theory, having once been satisfactory, becomes on the contrary, unapproachable, i.e. "non-falsifiable" by the same facts.[11] Of course, things can be settled one way or another by splitting the anthropological discipline into two or many different sciences, or by giving up its epistemological commitments and their replacement by aesthetic, socially functional, and other incentives. But all these moves cannot overshadow the basic fact of non-realization of the subjective, modal relativism of anthropology itself.

Attempts at a new approach to the process of doubting and the phenomenon of modalized consciousness were undertaken by Edmund Husserl.[12] In his "Experience and Judgment" we find the following description of the "ego which is at odds with itself";

> 'Poly-ego-ism' underlying the nature of the doubting modalized consciousness is a disagreement of the ego with itself. It is here no longer a question of the mere phenomenon of the cleavage of perception, but of a disagreement of the ego with itself, though a disagreement founded and motivated by these passive occurrences. The ego is now at odds with itself, is in dissension with itself, inasmuch as it is inclined to believe now this, now that. This being-inclined, then does not merely signify the affective pull of the attracting possibilities; rather they attract me in their being, and I go along now with one, now with the other, in an active position-taking which to be sure, is obstructed again and again. This going-along-with of the ego is motivated by the weight of the

possibilities themselves. These possibilities, as attractions, issue in a tendency toward judgment, which I actively followed for a certain time, and which entails that I bring about something like a momentary personal decision in its favor. But then, in consequence of the effective claims of the opposing possibilities, I remain stuck fast . . . This being-inclined-to as an impulse to act; an inclination to act as a feeling-oneself-drawn into an act of judging thus or so, belongs to the phenomena of reaching out, tending, of striving in the broader sense and must be distinguished from the position-taking of the ego, from the act of judging, which (as in active doubt) may be accomplished only momentarily, but by which I espouse one of the two sides . . . The ego, as it were, goes along part way in the accomplishment, but it does not go all the way to firm resoluteness of belief.[13]

Here, as at a focal point, two different faces of the problem are superimposed, with this description (which is, indeed, nothing more than description so far) projected against the problem of anthropological understanding. Two alternative moves can be foreseen: either into the dull infinitude of psychoanalysis (fortunately there seem to be no more volunteers) or into the new fashionable games of the anthropologist with his one and indivisible self. (That is probably why Wittgenstein, the genius of truisms, is getting so popular among anthropologists nowadays. No wonder. His advice is "not to think").[14] In any case, it is clear that no accumulation of scientific knowledge is possible with the investigatory position devoid of its inner cultural reflection. just another endless version of Hamlet's situation . . .

In order to realize *this* situation as a "cultural" fact, here it is proposed to consider the "planned misfit understanding" within the anthropological approach. Anthropology is a modern science and, like all modern scientific disciplines, it rests on the "presumption of understanding". According to this presumption, there can be no *facts* which ever escape comprehension: at least by *someone*, in *some* specific position, *some* time and in *some* part.

But, a considerable part of modern anthropology consists in studies of traditional societies. Besides being a new science, it is useful from time to time to consider itself related to a certain scientific *tradition*. So it would not do harm to find out what makes the traditionalist mentality so different from the modernist outlook. Both theoretical and pragmatic activity of traditionalist thinking rests, on the "presumption of misfit understanding". According to this second presumption, there must be some things which always escape understanding: at least by *anybody*, from *any* position, *any* time, in *any* part.

The anthropologist is not invited thereby to imitate his respondents and forerunners. But the basic congeniality of the "inner" anthropological problem and the "presumption of misfit understanding" is noticeable. This congeniality is objective and being such, it necessitates introduction of the "planned misfit understanding" into the anthropological method.

If anthropologists refuse to introduce it, Hobbes will supersede Descartes in their three-centuries-long duel, and political science will engulf anthropology just as it has almost devoured sociology. For sociologists were more reckless in their bare methodological issues and bare concern with "social facts". To get to this revision, a special domain of anthropological studies should be demarcated. It must be revised for models of realization of the above-described situation. This would be a subject-matter of the "misfit understanding" anthropology.

The operation of its models can be compared to the principle of uncertainly in physics, and their incentive, to the productive force of "avidya" ('ne-science') in Advaita Vedānta. But these comparisons are good only as initial landmarks. Consequently, all the cultural facts are to be composed as correlative to a few axial working principles. There is no more need to doubt *these* principles in their turn, as initial facts, for they are to be deliberately introduced as 'doubtful', controversial ones, and this makes their double-doubting redundant. For that very reason, their study must constitute an inherent and unalienable subject of a reformed anthropology.

To prove this, let us place anthropology among some related disciplines. Thus, theology is the actual discipline concerned with *definite* cultural statements. Linguistics is interested in those phenomena and processes of communication which are indefinite in principle, i.e. usually fail to become facts of realization. Hence this discipline cannot be complementary to theology, but can easily complement anthropology. Political science and sociology ignore both definite cultural statements and indefinite facts. History, of course, should not be missing from this list of comparables.

Its *disciplinary* function (which is certainly different from the feeling of "lived history") consists in studying *possibilities* (i.e. social *dynamics*, in the etymological meaning of "dynamics" as possibility) rather than *actualities* (which are bound to social activity and thus belong to anthropology) of arrangement of the totality of cultural facts grouped around some particular axial principle. In this case, the power and order of every grouping would completely determine the concreteness of its individuals: which is the main ambition of the science of history as the "idiographic" discipline.

Now philosophy alone seems missing from our list. But its subject matter, though being critical and reflective and in these aspects somewhat resembling the principle of "misfit understanding", is not specifically associated with the studies of culture. Philosophers must be preoccupied with philosophy itself, rather than with the "philosophical consciousness" naturalizing itself in various critical realizations of other disciplines. With this latter consideration in mind, it should be stated with all possible clarity that our anthropology of the "planned misfit understanding" *must* be different in principle from what is called "dialectical" anthropology, mainly because the anthropological *method*, rather than the matter of *understanding* is the basic issue of its concern; and for this reason, anthropology cannot be subordinate either to pure philosophy or to ideological considerations.

Of course, anthropological thinking, once doubt is accepted into its inner constitution and transformed therein into a method, cannot help but reflect its own controversies by way of splitting "facts" before approaching them, in this sense, regardless of whether the nature of facts themselves is controversial or not. This thinking can be easily branded as "conflicting", "antithetical", etc., and "planned misfit understanding" can be mistaken for a paronymic of "contradiction" in thinking about nature itself.

But in this case, this reflective thinking, though being pure negativity in itself, is only able to activize itself after our insistence, as a negativity doubting even this very negativity rather than denying cultural facts. Only after that doubting being is taken into itself, may it return to the facts and thus take them positively, critically or dialectically. To prove the specificity of anthropology as a scientific discipline *appropriating* the self-critical and in-advance-of itself-critical *understanding* as crucial for its demarcation from an "anthropologizing" philosophy, especially the dialectical one, let us take an excursion into Hegel's treatment of the same problem of *Understanding*, which, as we shall see, is axial for anthropology as a science. *sui generis.*

In his *Phenomenology of Mind*, Hegel places Understanding as a mediatory link between Perception and Consciousness, with its own and their contradictions being negated and sublated into the idea of Self-Consciousness. Now how does *this* Understanding work?

Analysis of an idea, as it is used to be carried out, did in fact consist in nothing else than doing away with its *character of familiarity*. To break up *an idea into its ultimate elements means returning upon its moments*, which at least do not have the form of the given idea when found, but are the immediate property of the self. Doubtless this analysis only arrives at thoughts which are themselves familiar elements, fixed inert determinations. But what is thus separated, and in a sense is unreal, is itself an essential moment, for just because the concrete fact is self-divided, and turns into unreality, it is something self-moving, self-active. *The action of separating the elements is the exercise of the force of Understanding*, the most astonishing and greatest of all powers or rather the *absolute power*. The circle which is self-enclosed and at rest, and, *qua* substance, holds its own moments, is an immediate relation, the immediate, continuous relation of elements with their unity, and hence arouses no sense of wonderment. But that an accident as such, when cut loose from its continuous circumference – that which is bound and held by something else and actual only by being connected with it – should obtain an existence all its own, gain freedom and independence on its own account – this is the portentous *power of the negative*; it is the energy of thought, of pure ego.[15]

This interpretation looks in some aspects similar to the phenomenological description of the radically doubting "ego at odds with itself", but actually, its conceptual treatment is different. Here, Understanding is *opposed* to the nature of fact, while in Husserl's phenomenology, understanding, even though impaired by doubt, is treated as a *natural fact about consciousness itself*. In terms of the philosophy of the Notion, the subjective activity of Understanding meets its objective counterpart in *Force* of Nature, or, in the case of anthropology, of Culture.

Thus, the action of *forces of culture*, i.e. cultural norms, is objectively the same in its relation to the unity of differences which is being subjectively realized when the mind understands about the circle and its parts or elements. Force is a self-conditioned principle of unity; the differences are the "expressions of force", the unity evolves the differences out of itself. Understanding similarly is a self-conditioned process; it consists in reducing differences to "explaining" these differences from itself. Owing to this parallelism of performance, Understanding insists on the establishment of a "kingdom of laws" in nature or culture, which in its entirety is the meaning of world so far as Understanding goes.

But, if there is a failure in understanding, this automatism would induce the understanding mind to *intrude* and "correct" the natural state of affairs in some other culture, i.e. Understanding is being transformed into Force, and the commitment to scientific objectivity is abandoned. The secret of this too easy transformation is, that, from the "scientific" point of view (i.e. philosophically speaking, from the point of view of Kantianism, positivistically oriented and replete with logicism), "laws" *per se* are looked on as an inner realm, which merely "appears" in the detailed particulars which those laws control, and in which those laws are made manifest. The differences, in fact, are "phenomena", the laws themselves are behind the scenes. The world as a whole thus becomes distinguished into a realm of phenomena and the realm of noumena.

These two realms set a new problem to the mind, and must again be brought together in a more complete way than understanding can do. This new state of consciousness is self-consciousness. But if self-consciousness unites two positions, the empirical and the theoretical, it is also free to judge what is "regularity", and even to impose its own preconditions upon external reality. This for example is, exactly what happened, as some assure us, when theoreticians of the British school of social anthropology extended their theoretical presuppositions (functionalism as a method of understanding) into the field of force – and thus identified themselves as functionaries and defenders of the British colonial administration.

In this role it is only natural for them to abandon their previous, "theoretical" position of passive understanding of other cultures, in order to "regulate" the corresponding societies in the functionalist' preconditions. Perhaps many functionalists themselves would deny this transformation. But we have a much more convincing example of the unmasked transfusion of Understanding into a Cultural Force, exactly in accordance with the Hegelian scheme, in the case of Soviet "cultural engineering". That was the most radical attempt of "understanding" other cultures by forcing them to change without any functionalist commitments. The attempt was not only controversial in its external, objective consequences, but also very dramatic for anthropology itself as a science, which was arrested for a few decades at the primitive "ethnographic" stage.[16]

Now it must be clear that the chain of "regularity" and convertible correspondence between Understanding and Force *can* be broken without damaging the scientific status of anthropology, only if this is performed within the science itself, by the device of "planned misfit understanding" as a quite legitimate and fruitful way for anthropology to extricate itself from the immanent temptation of "cultural imperialism". This means turning the anthropological mind towards its own culture. The more critically and forcefully the anthropologists doubt their own society and, in the microcosm-macrocosm way, their own anthropological community within that society, the better it can be both for anthropologists and for society. Anthropology would be disciplined in its method and society would be taught to respect other cultures.

As for the first aspect, it must be clear that the "planned misfit understanding" constitutes an inherent distinction and structuring feature of the anthropologist's mentality rather than of the discipline itself. Figuratively speaking, it is an inner spring which repeatedly impels the anthropologist to look into other cultures not only for "facts" but for motives for doing his science. For Descartes, the fact of the variety of manners was the final stroke that turned his thoughts toward "radical doubt". In a similar way, the anthropologist, who departs more or less deliberately for other cultures in search of what seems important to his doubting mind, can penetrate into them only to the degree of his *doubt* being respondent.

Thus for example, Clifford Geertz drew the concept of religion as a cultural system from a series of facts – irreconcilable cognitive, physical, and moral contradictions – which aroused doubts; and concluded his analysis with a most explicit statement that the only way to "think" towards an understanding of other cultures is a symbol-free readiness of the anthropologists to grasp the cultural facts as they are displayed by the informants and thus to enter the culture as deeply as he is invited to (cf. in particular his "think" description of *"kapar"*, the gesture of fear among Balinese).[17]

This, of course, is only one possible resolution for the doubting mind, in the plain *apodictic* sense. Victor Turner resolved it differently by placing the socio-psychological conflict in the sources of the cultural organization of ritual. From this, he inferred the notion of "anti-structure", as not implying "structure" as something demonstrable at all. The locus and source of the "antithetical" cultural conflicts are always *hypothetical* both for the subjects of culture and the anthropologist, but the deontic force of conflict is beyond doubt.[18]

Milton Singer found the most natural approach to the understanding of culture in those situations where the participants themselves are aware of the artificial and deliberate nature of their actions and explain them to the anthropologist correspondingly, i.e. as "cultural performances". Hence, the anthropologist never touches the "reality": he always has to hypothesize what seems to be performed and explained as the "cultural meaning" taken as such by the performers.[19]

These three theoretical approaches, different as they are, still have something in common. This "something" is a striving towards a "positional relativism", towards taking a displaced position for understanding the strange culture by *learning beforehand* how to understand. That is why we find them corresponding to the first, and the most indispensable, step in "radical doubt" turned upon itself. The specific importance of this step, as liberating the mind from dogmatism and customary ways of thinking, was also stressed by Descartes. He recollected that, ". . . while travelling, having realized that all those who have attitudes very different from out own are not for that reason barbarians or savages but are as rational or more so than ourselves, and having considered how greatly the self-same person with the self-same mind who had grown up from infancy among the French or Germans would become different from what he would have been if he had always lived among the Chinese or the cannibals . . . I found myself forced to try myself to see things from their point of view".[20]

But the spring can propel only if it is wound in the opposite direction. There is another possibility hidden in the anthropological mentality which still remains unnoticed and unappreciated in terms of its own worth. This is a peculiar ability to be repelled by a certain treatment of material, to avoid unilateral judgment on the problems under consideration. Take for example, the immediate reactive movement towards "impressionism" which arose from A.R. Radcliffe-Brown's attempts to mathematize the concept of "social structure"[21] or from C. Levi-Strauss' logistic exercises in "dual oppositions".[22] These reactions imparted an adverse moment in the development of anthropological thought.

But the same "counterpoint" in this development becomes by far more meaningful and important for anthropology when associated with the specific critical reactions against the "Western parochialism" of anthropological theory itself. Shortly after appearance of Louis Dumont's "Homo Hierarchicus", based on the structural analysis of Indian caste society, it was followed by a host of critical rejoinders and doubts about the applicability to Indian civilizations of such typically Western contra positions as "authority" and "power", "hierarchy" and "equality"; "individualism" and "wholism"; "democracy" and "totalitarianism", etc.[23,24] All these objections are quite legitimate and make sense of a greater appeal to the indigenous ways of understanding other cultures. But they also have another, motive, usually overshadowed: to preserve the state of the "ego at odds with itself" by overruling the universalistic solutions which would eradicate doubt. Quite naturally, the crucial reflective step should be taken right here. It must consist in casting doubt on such and similar contrapositions not only as applicable to Indian civilization but to the Western one as well. And this would mean exactly the "planned misfit understanding" of one's own cultural position.

Learning one's culture by its critical sublation: this peculiar anthropological asceticism must be the second step of "radical doubt". Otherwise the anthropological discipline will be never established in its individuality, and will be reduced to imitating the non-congenial creation of a phony "polycultural mess", and to borrowing "analogs" and methods from other sciences. All these are links of the same chain. It should be remembered that when Roberto de Nobili appeared before the Madurai Brahmins, with his head shaved and clad in the Brahmic robe but preaching the Gospels, he was rejected not because he poorly imitated a brahmin or was not conversant both in the Brahmanic philosophy and his Gospel (he succeeded very much in both), but because he was branded a poor Christian.

The creation of a self-sufficient and self-respecting anthropology is apparently on the agenda. It must include a study of infrastructures of the discipline, i.e. of the modal alternatives of a "planned misfit understanding" of the previously studied facts, and must also give due consideration to the uncertainty of the anthropologist's investigatory position, both in the sense of "role complementarity" and division of the epistemic functions within the body of unified science, as well as in the sense of a qualification of anthropology's "cultural individuality", which is as it were, incomprehensible by itself, until it arrives at the moment of "knowing the Other". The essential attainment would be a realization of the methodological project first drafted by Descartes. "Radical doubt", if practiced by anthropologists in the situation of both "outer" and "inner" cultural relativism, would result in the maturation of their science, relieved from the fetters of empiricism and positivism, and transformed by a "planned misfit understanding".

## NOTES

[1] Alfred L. Kroeber and Clyde Kluckhohn, 1952, 'Culture: A Critical Review of Concepts and Definitions', *Harvard University Peabody Museum of American Archeology and Ethnology Papers* **47**(1), Cambridge, Mass.

[2] Rene Descartes, 1904, *Discourse de la methode, Oeuvres de Descartes*, publieés par Ch. Adam et P. Tannéry, t. IV, Paris.

[3] Thomas Hobbes, 1680, *Six Metaphysical Meditations by Renatus Descartes with Objections of Sir Thomas Hobbes*, London, pp. 126–127.

[4] *Ibid.*, pp. 127–128.
[5] F.G. Lounsbury, 1965, 'Another view of the Trobriand Kinship categories', in: *American Anthropologist* **67**(5), Part 2 (special publication), *Formal Semantic Analysis*, E.A. Hammel (ed.).
[6] Claude Lévi-Strauss, 1963, *Structural Anthropology*, Basic Books, New York.
[7] Immanuel Kant, 1914, *Critique of Judgement*, translated by Bernard, London.
[8] Alfred R. Lindesmith and Anselm L. Strauss, 1950, 'A Critique of Culture-Personality Writings", in: *American Sociological Review* **15**, October, 1950.
[9] Karl Hempel, 1965, *Aspects of Scientific Explanation and Other Essays*, in: *The Philosophy of Science*, New York.
[10] B. Donchenko, 1964, 'Modal Logic', in: *The Philosophical Encyclopedia* **3**, Moscow.
[11] Karl Popper, *The Logic of Scientific Discovery*, London, 1959.
[12] Edmund Husserl, 1972, *Ideas: General Introduction to Pure Phenomenology*, translated by W.R. Boyce Gibson, New York, London: Collier Books, pp. 104–139.
[13] Edmund Husserl, 1973, *Experience and Judgment: Investigations in a Genealogy of Logic*, Evanston: Northwestern University Press, p. 303.
[14] Ludwig Wittgenstein, 1968, *Philosophical Investigations*, translated by G.E.M. Anscombe, New York: The Macmillan Co.
[15] G.W.F. Hegel, 1967, *The Phenomenology of Mind*, translated by J.B. Baillie, New York: Harper Colophon Books, pp. 92–93.
[16] David B. Zilberman, 1976, 'Ethnography in the Communist Society', in: *Dialectical Anthropology* **1**(2), Winter 1976, pp. 135–160.
[17] Clifford Geertz, 1966, 'Religion as a Cultural System', in Michael Banton (ed.), *Anthropological Approaches to the Study of Religion*, London: Tavistock; New York Praeger.
[18] Victor W. Turner, 1969, *The Ritual Process*, London, New York.
[19] Milton Singer, 1968, 'Culture', in: *International Encyclopedia of the Social Sciences*, New York: The Macmillan Co. and The Free Press.
[20] René Descartes, op. cit.
[21] David B. Zilberman, 1971, 'Social Anthropology: The Dynamics of Development and Perspectives', in: *Voprosy filosofii* **2**, Moscow.
[22] Bob Scholte, 1971, 'Penelopean Efforts of Claude Levi-Strauss', in: *Ethnology* **4**.
[23] Louis Dumont, 1970, *Homo Hierarchicus*, The University of Chicago Press, Chicago.
[24] McKim Marriott and Ronald B. Inden, (in press), *An Ethnosociology of South Asian Caste Systems*.

PAOLA ZAMBELLI

# FROM THE *QUAESTIONES* TO THE *ESSAIS*: ON THE AUTONOMY AND METHODS OF THE HISTORY OF PHILOSOPHY*

## 1. HISTORIOGRAPHY OF RENAISSANCE THOUGHT AND NON-CLASSICAL SOURCES.

History of philosophy and history of science have methods quite different from those of economic and social history and also – the distinction means much to me – of *histoire des mentalités*. Where do we find history of philosophy in the three volumes of essays published in 1974 by Jacques Le Goff and Pierre Nora with the aim of defining "the new problems, objects, and approaches" of the *nouvelle histoire*? Do not misunderstand me: if history of philosophy is absent from these pages, the fault may lie with historians of philosophy, and I have no intention here of lodging a complaint.

It is good to remember how historians of the Renaissance worked in the nineteenth Century. While Burckhardt, in any case, had advanced along the path leading to a *histoire des mentalités*, his contribution was slight – here and everywhere else – to the history of philosophical thought. The creator of *Kulturgeschichte*, unlike his contemporary Hegel, did not see philosophical thought as a moving force in history.[1] But also the few pages which Hegel devoted to "Renaissance and Reformation" are sketchy. So they appear in the way they were arranged by Karl-Ludwig Michelet in 1833–36, and they are even more so in the very recent critical text.[2] Giordano Bruno (whom before Hegel the erudite historians of atheism, as well as Schelling and Jacobi, had interpreted as a precursor of Spinoza) was the only Renaissance thinker to interest and intrigue Hegel, but he honestly confessed that he had difficulties in mastering Bruno's "Nolan" (personal and vernacular) language.

In my opinion two other names should be brought into any discussion of nineteenth-century interpretations of the history of philosophy in the Renaissance. First, along the interpretative line which might be termed Hegelian, Bertrando Spaventa. I imagine that he is now forgotten, but I think he is much more worthy of note than Moritz Carrière or, for the Italian area, than Fiorentino.[3] Spaventa unearths in Nicholas of

Cusa, Pico, and Ficino moments of Hegelian "philosophy of the spirit", which Hegel himself had not traced back so far in time as the fifteenth century, and he was probably the first to pinpoint "dignitas hominis" in Pico and the latter's dynamic idea of the microcosm. He continued with monographs on Bruno and Campanella and distilled the dialectical moments of an extremely abstract speculative philosophy which could be defined "professional philosophy".[4] And yet his hero was Giordano Bruno, "accademico di nessuna accademia" (a "member of no academy"). Without Spaventa we would not have had Fiorentino's documentary contributions, nor would we have had Gentile's *Il pensiero italiano del Rinascimento*.

Something analogous must also be said about Wilhelm Dilthey. He not only inspired such great historians as Hans Baron and a number of specialists in the history of the Reformation and of ideas of tolerance, but he was also rediscovered and used as a model by an anthropologist like Victor Turner.[5] Therefore, he must be remembered, and not only in connection with the origins of the philosophy of hermeneutics.

However, there is an "object" in our field: philosophical thought, which can not be changed and which must be studied in compliance with demanding and formal rules. This does not mean that no need is felt today for new "problems and approaches" – anthropological, sociological, and sociolinguistic ones – nor that various attempts are not being made in those directions, although unfortunately such departures will never be sufficient in themselves to allow one to complete a piece of research in our field. For the Ancient Age, anthropological techniques and the new decoding of myth are acquired procedures at this point. On the other hand, the work, most of it on a Jungian line, which has been done on the Renaissance seems neither conclusive nor convincing.

However, it has not proved impossible to find some "new objects" of study for historians of philosophy: besides topical method developed by Valla and Ramus, critical and philological conscience before and after Erasmus, let me use as example the theories of astrology and various magical practices. While it has been observed that several decades ago witchcraft and sorcery had got thematically in the foreground of social and religious history itself,[6] it is also true that the theme of their theories and foundations is strongly present today as well in the debate on the history of Renaissance philosophy. Which schools of thought differed concerning the two kinds of magic, spiritual and natural? Which schools of thought defined themselves in the debate on the basic tenets of astrology? What philosophical sources were particularly dear to the Renaissance theoreticians of magic? Some scholars go so far as to ask which currents of Hellenistic thought concerning magic have a right to be considered philosophical. Can the theory of celestial influence, recently studied by John North and particularly well by Edward Grant, be held to be a theoretical premise of Natural magic because of its astral-elementary classifications of every single being in the sublunary world? Or is divination, like alchemy and theurgy or spiritual magic, merely a game or a somatisation, the end-product of an evocative word or of the imagination? What did philosophers think about witches, and why did philosophical discussions concerning these intensify during the period of their persecution and within the historical context of the crisis brought about by the Reformation? The last point has been studied by Keith Thomas, who makes excellent use of Evans-Pritchard and other social anthropologists.

I have been studying for a certain number of years the problems of magic and witchcraft,[7] but precisely for that reason I prefer not to deal exclusively with them now. Moreover, it is not in the area of magic that the 1990s student of history of philosophy or history of ideas differs most significantly from Michelet or Burckhardt. Let us consider *La Sorcière* and the second to the last chapter of *The Civilization of the Renaissance in Italy*: both discuss a number of what they called "superstitions" without asking what theories Renaissance thinkers formulated about them. Burckhardt, for example, uses Cornelius Agrippa, a German rather than an Italian, as his yardstick. Although he cites *De occulta philosophia*, he does not discuss the theories contained in its first chapters, and he uses instead the shorter and different review contained in *De vanitate*, a text which he did not mention, but also possessed in his private library. This is because he was only interested in describing social customs, and the existence of a "philosophy of magic and astrology" never occurred to him. We can read about this now in a chapter of the recent *Cambridge History of Renaissance Philosophy*; however, it seems a shame that it opens emphasizing both Agrippa and Burckhardt without mentioning such striking characteristics.[8]

How much light has been newly thrown on a number of issues by studies concerning astrology and magic? If we take another look at the sources of certain fifteenth – and sixteenth – century philosophical discussions – on free will and astral determinism, on the absolute freedom of the single contingent event, which upon its realization becomes necessary, on constellations and especially on the conjunctions which determine "great changes" and cycles in the universal history of societies, states, and religions – we realize that almost all the Renaissance authors use al Kindi, Avicenna, and Albumasar and that these three are quoted with as much respect as Ptolemy and Aristotle. The same is true for magic. In their discussions of imagination (or *phantasia*) in terms of its psychosomatic and transitive action based on what Fazlur Rahman called a "psychological law of symbolization", Renaissance philosophers commented on a number of texts of Aristotle's *De anima* using not only Avicenna, but al Kindi and al Farabi, in addition to the *Picatrix* and books of recipes or prayers transmitted in Greek from the Hellenized East or in oriental languages from lands under Islamic rule. The Middle Ages and the Toledo school of translators had not passed in vain. However, I am listing these exotic names not in order to seem erudite, but because there is no agreement in recognizing both their presence and their importance in the Renaissance, the "period of the rebirth of classical culture". Burckhardt had discarded this definition by showing the organic nature of the civilization of the Renaissance in Italy and by placing its roots in the life of that society. Today we are ready to recognize its more composite nature, not merely classical, "Western", Euro-centric. Of course no one can deny that the recovery of classical texts, the mastering of the Greek language, and the creation of textual criticism were fundamental elements of Renaissance culture. But one of the most subtle philologians and grammarians, Johannes Reuchlin, together with Pico, brought a heretofore unknown patrimony into the Latin world: the Kabbalah. Reuchlin not only compiled the first Hebrew grammar but, in the name of cultural pluralism, he defended the Kabbalah and the Talmud, when they were threatened by one of the Dominican-sponsored bookburnings. Many humanists supported him, and the polemic which ensued produced the famous *Epistolae obscurorum virorum* which was one of the first philosophical pamphlets, or at least one of the first intellectual arguments using satire and maccheronic language as weapons to reach a large public. Why did the humanists take action? To save not a Greek but a Jewish library.

Even Byzantine thought contained elements which were not exclusively Greek. Ficino's translation of a treatise by Psellus (or more likely by a pseudo-Psellus), for example, transmitted the current popular demonology, as did Gemisthus Pletho the religion of Mithras. But I want to end this part of my discussion by bringing in an author who was entirely philosophical and had little to do with the occult sciences: Averroes. His treatise against al Ghazali, the *Destructio destructionum*, saw theology as a non-cognitive, but pedagogical faculty and gave priests, ceremonies, and sermons the political role of ensuring the consensus of believers. This text, already heterodox in the Muslim context and translated into Latin in 1328, was not very well known until Agostino Nifo put it back into circulation with his 1497 printed commentary. Whereupon it had considerable influence on theories concerning religion and the state, from Machiavelli and Pomponazzi to the period of *libertinage érudit* and the discussions on the so-called "frauds of religions".

## 2. THE PRINTING PRESS AS AN AGENT OF CHANGE IN PHILOSOPHICAL LANGUAGE.

How can the Renaissance, deprived of these non-Western elements, still be seen as a whole? Let us leave the heritage of the East and the Middle Ages for now. In any case, that discussion would not be complete without bringing in the awareness of the New World and philosophical speculation concerning its inhabitants. It must also be added that my research interests do not coincide with the period of much-debated civic humanism, but cover, instead, the historical epoch about which discussion continues as to whether it can be called "the age of the printers". Marshall Mc Luhan's and Elizabeth Eisenstein's thesis of "the printing press as an agent of change" seems to me acquired knowledge at this point, even though I subscribe to many of Anthony Grafton's criticisms of the latter. I would like to discuss a point raised by Grafton: the question of the "process of publication" of printed or manuscript writings during the age when the two techniques coexisted – actually at its beginning, since manuscript publication "reserved for the few" continued until at least the eighteenth century. Grafton probably has good reasons for asking whether it was true that "the process of publication itself changed so radically as Eisenstein holds, especially from the author's point of view". He thereby admits implicitly that from the readers' point of view more might have changed, but does not take the matter further, probably because such a problem does not enter into his methodology. One of the many and great contributions unanimously recognized as coming from P.O. Kristeller and his school is the attention paid to the material support or framework of a philosophical writing. In this connection Grafton reminds us that Kristeller "showed long ago that publication followed the same course for a fifteenth-century author whether the book in question was to be copied or printed".[9] Kristeller is also invoked by both Eisenstein and the authors of the *Cambridge History of Medieval Philosophy* in their chapters on "Manuscripts", "Printing and Censorship", "The Availability of Ancient Texts", and "The Rise of Philosophical Textbooks".[10] However, only Eisenstein poses the problem of the readers or users of these messages.

Of course it is unacceptable to see in printing an "agent of change" for what concerns the content of philosophical ideas. I, too, obviously find such vulgar historical materialism repugnant. But I must confess that, in studying the history of philosophy, I discover that my interest extends beyond its doctrinal content. While this does not seem to me to be resorting to *histoire des mentalités* – I like, instead, to keep anchored to a specific (but not exactly "professional") definition of philosophy – I nevertheless still find its impact on society important: the way in which the forms of circulation vary and the answers it sometimes gives to the problems of the community and the common man. Such a conviction has a number of reasons. I will not mention every reading, which influenced me, but just two of them. In any case, the first scholar I want to mention in this connection is a non Marxist philosopher, Chaïm Perelman, who studied argumentation in order to analyse "the adaptation of the speaker to the audience" and demonstrate how the nature of the audience has a reaction-effect on the author and the nature of the discourse.[11] More pertinent, perhaps, is research based on Marshall McLuhan's findings, and, better still, Jack Goody's.[12] Why not try and study the consequences of extended literacy in the history of philosophy as well? What one might expect to find after adapting Eisenstein's hypothesis involves the circulation and interaction of philosophical ideas rather than that mysterious process consisting in the original conception of a new philosophical doctrine. But who can truly isolate and fix that instant? If the audience also has a reaction-effect on the author philosopher, on his possibility of communicating, and on the different messages which he wishes or is able to convey, then new questions are also forcing themselves upon the attention of the historian of philosophy.

Some examples. Why did Scholastic philosophy and its *quaestio* begin to fade from the age of Gutenberg on?[13] Why, when both religious and, precisely, philosophical theses could finally be printed and used for propaganda purposes, do we find the most rigid and systematic forms of censorship, and finish up with the Index of prohibited books? Why is it that the specific possibility of attaining different levels (from publication in print to clandestine circulation), is a process which led to the birth of the essay?

Kristeller himself has noted that this literary genre was born in the Renaissance. In my opinion the moment of crossover from the Middle Ages to the Renaissance, or "from the age of scribes to the age of printers", lies in the difference between a *quaestio* (be it by Aquinas or Buridan, Petrus Abanus or Pomponazzi) and an *essai* (especially if it is by Michel de Montaigne).[14] This is not meant to reduce an entire philosophical periodization to the perspective of one format, or even stylistic aspect. It is obvious that there are other fundamental differences beyond those of literary genre. But even a genre can be revealing. There are, of course, particularly in the pre-Scholastic age, texts which could be loosely termed essays: John of Salisbury's *Policraticus*, for example. Then we have Petrarch's *Secretum* and his *De sui ipsius et aliorum ignorantia*. But Petrarch's name reminds us that he and Dante and Boccaccio would never have used vernacular prose to write about lofty subjects.

The same was true in the Germanic area for Nicholas of Cusa, Gutenberg's contemporary. Here is something else to watch for then: the slow and geographically varied use of the vernacular in treating such a high subject as philosophy. (In this tendency Italy proved to be precocious). Leonardo Olschki did the groundwork in 1919–1922 for a history of scientific literature in the modern languages, but, seventy years later, his outline is still waiting to be filled in and brought up to date according to new working methods.[15]

The following could be some additions to Olschki. One concerns Simone Porzio, a mid-sixteenth-century professor who, after cautiously keeping his Pomponazzi-inspired writings in the bottom of a drawer for years and years, decided to publish them all together at the same time, both in his elegant Latin and in the still more elegant Tuscan versions of G.B. Gelli. However, the anonymous translation of the treatise *De mente humana*, concerning the ticklist question of the immortality of the individual soul, was prepared but stayed in manuscript form. It is clear that at some point in the publication process, considerations intervened concerning censorship and being reported to the Inquisition. A calculatedly bold treatise could be published in Latin because it was thought to be intended for a more restricted and cultivated, almost specialist audience, but the same was not true of vernacular writings. They belonged to the field open to ladies and shoemakers (such was the profession of G.B. Gelli, Porzio's translator).[16]

However, it is Pomponazzi himself from the ranks of the philosophy teachers who provides an opportunity for a classification and typology of philosophical writings which is particularly interesting and full of developments. I will not treat here the rôle which he tried to assume, or which Speroni attributed to him, language he used for his lectures. A discussion of maccheronic language would be valuable since it was used so often in philosophical polemics:[17] Hutten and the *Epistolae obscurorum virorum* I mentioned earlier, a few of Erasmus's *Colloquia*, and Agrippa's pseudonymous *Dialogus de vanitate scientiarum et ruina christiane religionis*, up to Rabelais and far beyond. However, let us stay with Pomponazzi. He did not have the stylistic and linguistic competence of a humanist and confessed this openly insofar as the Greek classics were concerned; it was not necessary to do so for his style. However, he was an up-to-date thinker and his philosophical views had evolved considerably, to the point where from 1513 on he stood on advanced radical positions. That is, he expounded a theory of the intellect influenced by Alexander of Aphrodisias and his politico-religious ideas were derived from Averroes' *Destructio*. Harmony with the Inquisitors was difficult under these circumstances. And yet he dared **A** to publish and publicly defend in printed debates his famous *Tractatus de immortalitate animae*, and it is the only one of his works to be written in a readable Latin. Pomponazzi does not abandon the procedures typical of *quaestiones*, but he exploits their best aspect in distinguishing all of the different hypotheses on the nature and parts of the soul. The work stands apart from many of his other works for its accurate syntax, and he probably had some help in writing and revising it. In this case, we can assume that he and his friends wanted the *Tractatus* to reach a wider than university audience and to be read by fastidious Latinists. He obtained this result as we know, to the point of involving the refined Gaspare Contarini as an anonymous "Contradictor"; on the opposite side of the baricades, he unleashed furious sermons against himself in the *campielli* of Venice.

Nevertheless, for many years Pomponazzi had permitted himself blasphemous maccheronic jokes in his university lectures **B**, which were much heavier and more theologically compromising. In that setting, however, he enjoyed the so called *patavina libertas*. The courses which were "reported" by his students and copied out in manuscript form, probably with his consent, were a model for many teachers of Aristotle for decades. Here is a second type of circulation intended for a specialized public and probably circulating clandestinely.

**C** Pomponazzi's published works, or at least the *Quaestiones* prior to the *Tractatus* consitute a third case: there the usual problems of the Aristotelian schools are examined without maccheronic language and anti-clerical jokes: they are for the eyes of colleagues only.

**D** However, two books constitute an exception: the *De incantationibus* and the *De fato* written in 1520 were purposely left unpublished because of their excessively daring doctrinal content. (Did an analogous caution inspire Machiavelli keeping the *Prince* unprinted for almost twenty years until his own death?). However Pomponazzi referred in these works to his former ones, thus avowing his authorship and letting them be circulated among friends. When they were published much later in Reformation territory, the editor Guglielmo Gratarol had to have them completely rewritten in a more correct Latin. A comparison between this case and the first one mentioned shows, on the one hand, the difficulties of a Scholastic teacher in coming to terms with a new literary style, and, on the other, the changing resources and levels of publicity to which he recurred in order to make his ideas known. **C** The *Quaestiones* were certainly the least effective, both at the time and today. The other three cases worked notably well: **B** manuscript copies of the *reportationes* (notes from courses) circulated throughout the sixteenth century and afterwards; **D** clandestine copies of the two 1520 treatises made it possible for them to circulate widely, and also before their printing (1556; 1567) a number of readers also knew them in manuscript form; and finally **A** the cleverly orchestrated polemic about the *Tractatus de immortalitate* is an almost unique case of the use of printing in order to concentrate attention on a philosophical thesis which had just been condemned by the Fifth Lateran Council. This polemic will fade from memory, together with the attacks on the author and his apologies, but the Alexandrist thesis, for the modern reader, still remains linked with the name of Pomponazzi.[18]

## 3. LATIN AND VERNACULAR, ELITIST AND POPULAR, CLANDESTINE OR PUBLISHED PHILOSOPHY IN THE RENAISSANCE

He was not the only good navigator. Setting a course between readers and Inquisitors was initially easier for Humanists and Platonists who used a less technical language in order to introduce inexperienced readers to philosophy. Several chose the vernacular: Ficino wrote several of his works in Italian or else had them translated by friends; L.B. Alberti wrote in Italian *De la Famiglia*; Pico his *Comento alla canzona d'amore*; Leone Ebreo the *Dialoghi d'amore*; Campanella *La Citta del sole*. However, it must be recognized that the vernacular writing was not the only one to reach a large audience. It has been shown again recently that Erasmus, the humanist *par excellence*, by means of a "selective" and "radicalizing" reading by school teachers, who were often heretics, reached a large popular audience not only in Spain but also in Italy. Here we are talking of indirect circulation and word of mouth, whereby "Erasmus the grammarian guilty of placing the Holy Spirit under the rod of Donatus" in his edition of the New Testament, and Humanism in general actually seem to some historians to be "the communicative dimension of the Reformation".[19] However Erasmus also had a direct impact, and his splendid, unpedantic Latin served as a model for the written presentation of an argument. Thomas More, Vives and Cornelius Agrippa followed his example. Without sacrificing discipline and coherence of philosophical reasoning, they chose to write in a low register, cultivating a style which would be comprehensible to the common reader and also creating a type of polemical writing which was both allusive and ironic. Think of Erasmus's *Enchiridion militis christiani*, of his *Ratio seu methodus* and last but not least of his *Colloquia* (designed as an exercise book for students of spoken Latin), but so felicitous both in terms of plot and religious or philosophical polemic. They are said to be among the most important models for Rabelais, for the *Viaje de Turquia* by Doctor Laguna, a high-level scientist and intellectual,[20] and for the *Apologie de Raymond Sebond* and all of Montaigne's *Essais*. By the time the latter were written, a type of philosophy had been created which was both colloquial and militant – insofar as one could be militant during the Counter Reformation.

Why were the Index and the Inquisitors always so hard on Erasmus and his heirs? As a writer his conduct had been responsible, and he had shown good sense in judging what and how to print. Another case confirms and perhaps explains this censure: that is the fact that Giordano Bruno, a decidedly superior speculative philosopher with a popular style, began his "heretical" meditations reading, precisely, writings and editions by Erasmus. For fear of being discovered, he hid the compromising material in the latrines when escaping from his Naples convent.[21]

In conclusion I would like to mention two very great figures in both science and philosophy: Copernicus and Galileo. Galileo's magisterial prose is a milestone in the history of the Italian language. Hoping to deceive the censors, he had purposely chosen the dialogue form for the *Massimi Sistemi*, but to no avail. Is it not possible that the very clarity of his vernacular exposition contributed to worsening his position? Copernicus was writing in less severe times, and he is a case of even more extreme caution. Having decided not to publish the *De revolutionibus orbium caelestium*, he did not let it circulate at all in manuscript form, even to such a qualified reader as Cardinal Nicholas Schönberg who had offered to pay for having it copied. He only released a very few copies of the *Commentariolus*, an outline written in 1514. If it had not been for the persuasive efforts of Bishop Tiedemann Giese and of a Wittenberg professor Georg Joachim Rheticus, the Copernican system would still lie within the pages of a manuscript in the small town of Frombork and might even have been lost altogether. Copernicus finally decided to authorize Rheticus's *Narratio prima*, a sort of preprint, and it was submitted for safety's sake to Melanchthon beforehand and only allowed to circulate after it had been approved. In the end the author consented to the publication of his masterpiece. The place of publication is typical of the times: While Rheticus's edition of Copernicus' technical work *De lateribus et angulis triangulorum* was published without problems at Wittenberg, Nuremberg, a less official center of Lutheran culture, and a friend as printer (Petreius) were chosen for the *De revolutionibus*. In any case, it was the "conventionalist" preface written by another protestant, the theologian Osiander, which, by acting as an antidote, ensured the treatise's publication.[22] As Bruno later deplored, the Copernican system was only published because it was passed off as a fable.

Philosophy and science are commonly supposed to be élitist forms of knowledge. However, they are not élitist by nature, and it may occasionally happen that they become involuntarily involved at levels which have nothing élitist about them. In some instances, as in the case of the esoteric "Pythagoric" penchant sometimes exhibited by Copernicus, censorship and the Inquisition encouraged a circulation limited to a few elect souls as well as obscure or secret expressions of thought. The Sixteenth Century (mainly the Counter-Reformation) was, after all, the period which gave birth to the emblem. Even if in emblems the image seems to prevail over the text, it is really the latter which gives the key of its (nicodemitic?) meaning. Word gives the message, image helps to remember it.

But even in the face of adverse circumstances, philosophy obeys an ideal of communication which is clear and distinct and therefore accessible to any healthy mind. Descartes said this although he too was certainly cautious. The clarity and distinctness[23] being sought after in the age of printing is no longer that of the syllogisms and of *quaestiones*. An author's choice of forms which are more and less accessible to his readers depends on a number of extremely complicated factors in this period: and the different cultural levels at which his messages are received and bear fruit certainly do not depend only upon the author. Could Giovanni Pico ever have predicted that his *Conclusiones*, which he wanted on display in all the universities and which he would had wanted discussed in Rome, would have had a relatively late circulation (in the same way as his clandestinely printed *Apologia*)? Or that the rather ponderous *Disputationes adversus astrologiam divinatricem* would have been so widely discussed, not only in the universities, but in the numerous popular prognostications unleashed by the fear of the end of the world in 1524?

The sociological study of reading – increased literacy in the age of printing, cost, demand, and organizational structures for reading, etc. – can help fill in the picture of the history of philosophy. It also gives new meaning to data about manuscripts, printing and censorship, the availability of the classics, and manual-writing. But such an approach will be particularly useful if, as we examine a thinker's text, we ask ourselves: was he engaged in his time or did he prefer not to get involved in its battles? How strongly did he react to contemporary religious, political, or social problems or were they extraneous to him? How hard did he try to reach his public or to what extent was he prevented from doing so? History of philosophy is now a part of history, no longer a part of philosophy, nor its completion. Of course the importance of the historical context and the clash with fundamental philosophical ideas vary from case to case. Perhaps this is why today one sees difficulties everywhere in writing general histories of philosophy, even for a limited period such as the Renaissance.

We see that historians have acquired a critical distance which enables them no longer to identify themselves with the Renaissance, as Burckhardt's followers used to do, and this turn of mind developed gradually after World War I.[24]

Are historians of philosophy today about to acquire another type of critical distance? I mean that they are probably becoming aware that history of philosophy need no longer be identified with philosophy, nor philosophy with its history, as Hegel, Spaventa, Gentile and their followers used to maintain. If this way of doing the history of philosophy is finished, as many have written before now, why not try to give some modest, documentary and chronologically limited contributions to an *histoire à part entière*?

*University of Florence*
*Department of Philosophy*

### NOTES

\* I read a first draft this paper in a panel-discussion with Nathalie Zemon Davis, Anthony Molho and other scholars opening the 50th Meeting of the Renaissance Society of America in Boston 1989; Guido Oldrini, Alessandro Pizzorno and Ronald G. Witt very kindly read and commented on it; Ann Vivarelli translated my paper: I am very warmly grateful to all of them.

[1] Cfr. W.K. Ferguson, 1948, chapters VII and VIII. M. Ghelardi, 1991; this author kindly informed me about the copy of Agrippa's *De vanitate* owned by Burckhardt.
[2] G.W.F. Hegel, 1986, pp. 49–51, 232, 238–239.
[3] Cfr. P.O. Kristeller, 1979, pp. 154–155.
[4] *Ibid.*, 154.
[5] Cfr. V. Turner, 1982 (see Introduction).
[6] E.W. Monter, 1969, p. 205.
[7] P. Zambelli, 1991.
[8] B. Copenhaver, 1988, *Astrology and Magic*, in Ch. B. Schmitt and Q. Skinner (eds.), 1988, pp. 264–267; cf. P. Zambelli, 1992b.
[9] A. Grafton, 1980, p. 280–281. Grafton refers to P. O. Kristeller, 1937. See also Kristeller, 1956, pp. 473–493.
[10] Ch. B. Schmitt and Q. Skinner (eds.), 1988, see the chapters by J.F. D'Amico, Grendler, Grafton, Ch. S. Burnett.
[11] Ch. Perelman et L. Obrechts-Tyteca, 1958; cfr. also Toulmin.
[12] I am mainly referring to Goody (1962–63): the new developments of Goody (1977),

(1986: see preface) and (1987) do not modify the issues which interest me in the first, seminal essay.

[13] I am not persuaded by the thesis enounced by Paul Grendler, *Printing and Censorship*, in: Ch. B. Schmitt and Q. Skinner (eds.), 1988, 38 that "the massive printing of mediaeval philosophy during the first fifty years of printing ensured its survival and continuity": scholastic books were still needed for university teaching and this is a sufficient explanation for being printed, not a reason for cultural survival.

[14] T. Cave, 1979; Beaujour, 1980; A. Tournon, 1983; N.Z. Davis, 1985.

[15] Olschki, 1919–1922; Ferguson, 1962, p. 306 f.; Rice, 1970, pp. 8–10; Pörksen, 1983 and 1986.

[16] A. de Gaetano, 1976; P. Zambelli, 1980, pp. 6, 30–31.

[17] L. Lazzerini, 1988, pp. 83–88.

[18] B. Nardi, 1965; P. Zambelli, 1992.

[19] S. Seidel Menchi, 1987, pp. 124 and 55.

[20] M. Bataillon, 1950, p. 682; M. Bataillon, 1958.

[21] Ciliberto, 1990, pp. 10–11.

[22] E. Rosen, 1971, pp. 281–288; N. Copernicus, 1975, pp. 28–31; R.S. Westman, 1975; J.J. Rheticus, 1982.

[23] R. Descartes, 1966, p. 41 (Discours de la Méthode, Cinquième Partie): "ne recevoir aucune chose pour vraie, qui ne me semblait plus claire et plus certaine que n'avaient fait auparavant les démonstrations des géometres".

[24] G. Ritter, 1923, pp. 396–397; W.K. Ferguson, 1948; P.O. Kristeller, 1979, p. 153.

## BIBLIOGRAPHY

H. C. Agrippa, 1970, *Opera omnia*, with an Introd. by R.H. Popkin, Hildesheim-New York: Olms.

M.L. Altieri Biagi, 1965, *Galileo e la terminologia tecnico-scientifica*, Firenze: Olschki.

S. Anglo (ed.), 1977, *The Damned Art*, London: Routledge and Kegan Paul.

A. Asor Rosa (ed.), 1983, *Letteratura italiana*, II: Produzione e consumo, Torino: Einaudi (see papers by G.R. Cardona, I. Paccagnella, A. Petrucci, A. Quondam).

M. Bataillon, 1950 [1937] *Erasmo y España*, Mexico, Fondo de cultura economica.

M. Bataillon, 1958, *Le docteur Laguna auteur du Voyage en Turquie*, Paris: Libraire des Editions espagnoles.

M. Beaujour, 1980, *Miroir d'encre. Rhétorique de l'autoportait*, Paris: Seuil.

G. Bossong, 'Science in the vernacular Language: the case of Alfonso X el Sabio', in M. Comes., Puig R. y Samsó J. (eds.), *De astronomia Alphonsi Regis*, Barcelona: Universidad de Barcelona, Inst. Millás Vallicrosa.

J. Burckhardt, 1860, *Die Cultur der Renaissance in Italien*, Basel.

P. Burke, 1987, 'The uses of literacy in early modern Italy', in P. Burke and R. Porter (eds.), *The Social history of Language*, Cambridge: Cambridge U. P.

P. Burke, 1990, *Linguaggio, società e storia*, Bari: Laterza.

P. Burke, 1990, *The French historical Revolution: the Annales school, 1929–1989*, Cambridge: Polity Press.

P. Burke, 1991, *Küchenlatein, Sprache und Umgangssprache in den frühen Neuzeit*, Berlin: Wagenbach.

T. Cave, 1979, *The cornucopian Text. Problems of Writing in the French Renaissance*, Oxford: Clarendon Press.
T. Cave and G. Castor (eds.), 1984, *Neo-latin and the vernacular in Renaissance France*, Oxford: Oxford U.P.
R. Chartier, 1982, 'Intellectual History or Sociocultural History?', in D. La Capra and S.L. Kaplan (eds.), *Modern European Intellectual History. Reappraisals and new Perspectives*, Ithaca: Cornell U.P., pp.13–46; ital. transl. together with 'Filosofia e Storia' (inedito) and other papers in his *La rappresentazione del sociale*: Torino, Bollati Boringhieri, 1989.
R. Chartier, 1987, *Lectures et lecteurs dans la France d'ancien Régime*, Paris: Seuil.
M. Ciliberto, 1990, *Giordano Bruno*, Roma-Bari: Laterza.
N. Copernic, 1970, *Des révolutions des orbes célestes*, Introduction, traduction et notes par A. Koyré, deuxième ed. par E. Rosen, Paris.
N. Copernicus, 1975, *Commentariolus*; J.J. Rheticus, *Narratio prima*. Introduction, traduction francaise et commentaire par H. Hugonnard-Roche, E. Rosen et J.-P. Verdet. Préface de R. Taton, Paris: Blanchard.
N.Z. Davis, 1985, 'A Renaissance text to historians's eye: the gifts of Montaigne', *The Journal of Medieval and Renaissance Studies* **XV**(1): 47–56.
A. de Gaetano, 1976, *Giambattista Gelli and the Florentine Academy*, Firenze: Nuova Italia.
R. Descartes, 1966, *Discours de la méthode*, Texte et commentaire par E. Gilson, Paris: Vrin.
A. C. Dionisotti, A. Grafton and J. Kraye (eds.), 1988, *The Uses of Greek and Latin*, London: The Warburg Institute.
C. Dionisotti, 1968, *Gli umanisti e il volgare*, Firenze: Le Monnier.
E. Eisenstein, 1979, *The printing press as an agent of change*, Cambridge: Cambridge U.P.
E. Eisenstein, 1982, '*La culture de l'imprimé* presenté par J. Revel', in: *Le débat* **22**: 178–192.
L. Febvre et H.J. Martin, 1958, *L'apparition du livre*, Paris: Albin Michel.
W.K. Ferguson, 1962, *Europe in Transition 1300–1502*, Boston.
W.K. Ferguson, 1948, *Renaissance in Historical Thought*, Boston.
M. Gauthier, 1980, 'Le De daemonibus du Pseudo-Psellus', *Revue des études byzantines* **38**: 105–194.
C. Geertz, 1973, *Interpretations of Cultures*, New York: Basic Books.
G. Gentile, 1920, *Il pensiero italiano del Rinascimento*, Firenze: Vallecchi.
M. Ghelardi, 1991, *La scoperta del Rinascimento. L' «età di Raffaello» di Jacob Burckhardt*, Torino: Einaudi.
M.A. Gismondi, 1985, '*The Gift of Theory*: a crititque of the *histoire des mentalités*', *Social History* **X**: 211–230 (see Bibliography).
J. Goody and I. Watt, 1962–63, 'The Consequences of Literacy', *Comparative Studies in Society and History* **5**: 304–345; reprinted in J. Goody (ed.), 1968.
J. Goody (ed.), 1968, *Literacy in Traditional Society*, Cambridge: Cambridge U.P.
J. Goody, 1977, *The Domestication of the Savage Mind*, Cambridge: Cambridge U.P.
J. Goody, 1986, *The Logic of Writing and the Organization of Society*, Cambridge: Cambridge U. P.

J. Goody, 1987, *The Interface Between the Written and the Oral*, Cambridge: Cambridge U.P.
H.J. Graff, 1987, *The Legacy of Literacy. Continuities and Contradictions in Western Culture and Society*, Bloomington-Indianapolis: Indiana U.P.
E. Grant, 1987, 'Mediaeval and Renaissance Scholastic Conceptions of the Influence of the Celestial Region on the Terrestrial', *Journal of Mediaeval and Renaissance Studies* **XVII**: 1–23.
A. Grafton, 1980, 'The importance of Being Printed', *Journal of Interdisciplinary History* **XI**(2): 265–286.
A. Grafton and L. Jardine, 1986, *From Humanism to Humanities*, Cambridge Mass.: Harvard U.P.
C. Grayson, 1960, *A Renaissance Controversy. Latin or Italian?*, Oxford: Clarendon Press.
H. Grundmann, 1958, 'Literatus-illiteratus', *Archiv für Kulturgeschichte* **40**: 1–65.
E.R. Havelock, 1989, *The Muse learns to write. Reflections on Orality and Literacy from Antiquity to the Present*, New Haven and London: Yale U.P. (see Bibliography).
G.W.F. Hegel, 1986, *Vorlesungen. Ausgewählte Nachschriften und Manuskripte. IX Vorlesungen über die Geschichte der Philosophie, IV: Philosophie des Mittelalters and neueren Zeit [1825–26]*, hg. v. P. Garniron und W. Jaeschke, Hamburg.
M.C. Horowitz, 1988, 'Montaigne's Doubts on the Miraculous and the Demonic in Cases of his own Day', in: *Regnum religio and ratio*. Essays presented to R.M. Kingdon, Kirksville.
M. Idel, 1983, 'The Magical and Neoplatonic Interpretations of the Kabbalah in the Renaissance', in: *Jewish Thought in the sixteenth Century*, Cambridge, Ma.: Harvard U.P., 186–242.
M. Idel, 1987, *Kabbalah: new Perspectives*, New Haven – London: Yale U.P.
R.M. Kingdon (ed.), *Transition and Revolution. Problems and Issues of European Renaissance and Reformation History*, Minneapolis: Burgess.
H.W. Klein, 1957, *Latein und Volgare in Italien*, München: Fink.
F. Krafft, 1974, 'Der Naturwissenschaftler und das Buch in der Renaissance', in: F. Krafft und D. Wuttke D. (hg.), *Das Verhältnis der Humanisten zum Buch*, Boppard: Harald Boldt.
P.O. Kristeller, 1956 [1946], 'L' origine e lo sviluppo della prosa volgare italiana', in his *Studies in Renaissance Thought and Letters*, **I**: 473–493, Roma: Edizioni di storia e Letteratura.
P.O. Kristeller, 1979, 'Il Rinascimento nella storia del pensiero filosofico', in: *Il Rinascimento. Interpretazioni e problemi*, Roma-Bari: Laterza.
L. Lazzerini, 1988, *Il testo trasgressivo*, Milano: Franco Angeli.
J. Le Goff et P. Mora (eds.), 1974, *Faire l' histoire*, Paris: Gallimard.
D. C. Lindberg and R.S. Westman (eds.), 1990, *Reappraisals of the Scientific Revolution*, Cambridge: Cambridge U.P.
M. McLuhan, 1962, *The Gutenberg Galaxy. The Making of Typographic Man*, Toronto: University of Toronto Press.
M. McLuhan, 1966, *The Effect of printed book on Language in the 16th Century*, in E. Carpenter and H.M. McLuhan (eds.), *Exploration in Comunications*, Boston: Beacon Press, 125–135.
H.-J. Martin, 1969, *Livre, pouvoirs et société à Paris au XVIIe Siècle (1598–1701)*, Genéve: Droz.

H.-J. Martin, 1988, *Histoire et pouvoir de l'écrit*, Paris: Perrin.
H.-J. Martin, 1987, *Pour une histoire du livre (XV-XVIIIe siécle)*, Napoli: Bibliopolis.
W. Milde und Schuder (hg.), 1988, *De captu lectoris. Wirkungen des Buches im 15. und 16. Jahrhundert dargestellt an ausgewählten Handschriften und Drucken*, Berlin-New York: W. de Gruyter.
J. Michelet, 1982, *La sorcière*, Paris: Hetzel.
E.W. Monter, 1969, 'Trois historiens actuels de la sorcellerie', *Bibliothèque d'Humanisme et Renaissance* **XXXI**, 205-213.
O. Murray, 1989, 'The word is mightier than the pen', in: *TLS*, pp. 655-656.
B. Nardi, 1965, *Studi su Pietro Pomponazzi*, Firenze: Le Monnier.
J.D. North, 1986, 'Celestial Influence. The Major Premiss of Astrology', in P. Zambelli (ed.), *'Astrologi hallucinati'. Stars and the End of the World in Luther's Time*, Berlin-New York: W. de Gruyter, pp. 45-100.
J.D. North, 1987, 'Mediaeval Conceptions of celestial Influence', in P. Curry (ed.), *Astrology Science and Society*, Woodbridge/Suffolk: Boydell, pp. 5-18.
L. Olschki, 1919-1922, *Geschichte der neusprachliche wissenschaftliche Literatur*, Heidelberg.
W.J. Ong, 1958, *Ramus. Method and the Decay of Dialogue*, Cambridge, Ma.: Harvard U.P.
W.J. Ong, 1982, *Orality and Literacy*, London and New York: Methuen.
Ch. Perelman (ed.), 1963, *Raisonnement et démarches de l'historien*, Bruxelles, Ed. Inst. Sociologie. Universitè Libre.
Ch. Perelman et L. Obrechts-Tyteca, 1958, *Traité de l'argumentation*, Paris: Presses Universitaires de France.
A. Petrucci (ed.), 1977, *Libri, editori e pubblico nell'Europa moderna. Guida storico-critica*, Roma-Bari: Laterza.
U. Pörksen, 1983, 'Der Übergang vom Gelehrtenlatein zur deutschen Wissenschaftsprache', *LILI Zeitschrift für Literaturwissenschaft und Linguistik* **51/52**: 227-258.
U. Pörksen, 1986, *Deutsche Naturwissenschaftssprachen: historischen und kritischen Studien*, Tübingen: Narr.
*Le pouvoir et la plume*. (1982) Actes du colloque international organisé par le Centre Interuniversitaire de Recherche sur la Renaissance Italienne, Paris.
M. Pozzi (ed.), *Trattatisti del Cinquecento*, Milano-Napoli: Ricciardi, I, 471 f., 585-636.
P.M. Rattansi, 1972, 'The Social interpretation of Science in the 17th Century', in P. Mathias (ed.), *Science and Society*, Cambridge: Cambridge U.P.
J.J. Rheticus, 1982, *Narratio prima*. Edition critique, traduction francaise et commentaire par H. Hugonnard-Roche, et J.-P. Verdet, avec la collaboration de M.-P. Lerner et A. Segonds, Wrocław: Ossolineum.
E. Rice, 1970, *The Foundation of early modern Europe, 1460-1559*, New York and London.
G. Ritter, 1923, 'Die geschichtliche Bedeutung des deutschen Humanismus', *Historische Zeitschrift* **127** (3. F., 31, 1. H.): 396-397.
S. Rizzo, 1984, *Il lessico filologico degli umanisti*, Roma: Edizioni Storia e Letteratura.
R. Rorty, J.B. Schneewind and Q. Skinner (eds.), 1984, *Philosophy in History*, Cambridge: Cambridge U.P. (see papers by Ch. Taylor, A. MacIntyre, L. Krüger, I. Hacking, B. Kuklick, W. Lepenies, J. B. Schneewind and Q. Skinner).

E. Rosen, 1971, 'Copernicus' Attitude Toward the common People', *Journal of the History of Ideas* **XXXII**(2): 281–288.
Pietro Rossi (ed.), 1988, *La memoria del sapere*, Roma-Bari: Laterza.
J.H. Rowe, 1965, 'The Renaissance Foundations of Anthropology', *American Anthropologist* **67**: 1–20.
Ch. B. Schmitt and Q. Skinner (eds.), 1987, *Cambridge History of Renaissance Philosophy*, Cambridge: Cambridge U.P.
Ch. B. Schmitt and B. Copenhaver, 1992, *A History of Renaissance Philosophy*, Oxford: Oxford U.P.
Menchi S. Seidel, 1987, *Erasmo in Italia*, Torino: Bollati Boringhieri; cf. German transl. *Erasmus als Ketzer. Reformation und Inquisition in Italien des 16. Jahrhunderts*, Leiden: Brill, 1992.
B. Spaventa, 1908 [1860] *La filosofia italiana nelle sue relazioni con la filosofia europea*, ed. by G. Gentile, Bari: Laterza.
L. Strauss, 1952, *Persecution and the Art of Writing*, New York: The New Press.
M. Tavoni, 1984, *Latino, grammatica, volgare. Storia di una questione umanistica*, Padova: Antenore.
K. Thomas, 1971, *Religion and the Decline of Magic*, London: Weidenfeld and Nicolson.
A. Tournon, 1983, *Montaigne. La glose et l'essai*, Lyon: Presses universitaires de Lyon.
J. Tully (ed.), 1988, *Meaning and Context. Quentin Skinner and his critics*, Princeton: Princeton U.P.
V. Turner, 1982, *From Ritual to Theatre*. New York: Performing Art Journal Publ.
R.S. Westman, 1975, *The Copernican Achievement*, Berkeley: University of California Press.
R.S. Westman, 1986, *The Copernicans and the Churches*, in D.C. Lindberg and R.L. Numbers (eds.), *God and Nature*, Berkeley and Los Angeles: University of California Press, pp. 76–113.
R.G. Witt, 1988, Review of A. Grafton and L. Jardine, 1986, in: *Renaissance Quarterly* **XLI**: 479–482.
P. Zambelli, 1980, 'Scienza, filosofia, religione nella Toscana di Cosimo I', in: *Villa I Tatti. The Harvard University Center for Italian Renaissance Studies* (ed.), *Florence and Venice. Comparison and Relations*, Firenze: Nuova Italia, 1–52, 30.
P. Zambelli, 1991, *L'ambigua natura della magia*, Milano: Il Saggiatore.
P. Zambelli, 1992a, '*La metafora è conosciuta solo da chi fa la metafora*. Pomponazzi, Bessarione e Platone', *Nouvelles de la République des Lettres* **II** (1991): 75–88.
P. Zambelli, 1992b, 'Cornelio Agrippa, ein kritischer Magus', in: *Die Okkulten Wissenschaften in der Renaissance* (Arbeitskreis für Humanismusforschung, 1988), hg. v. A. Buck, Wolfenbüttel: Herzog August Bibliothek, 67–89.
E. Zilsel, 1941–42, 'The Sociological Roots of Science', *American Journal of Sociology* **47**: 544–561.

# INDEX OF NAMES

Abanus, Petrus 379
Abel-Smith 216
Ackerman, Bruce A. 249
Adler, Alfred 87
Adorno, Theodor 9
Agrippa, Cornelius 375, 380, 382
Ai, Si-qi 15
Al Farabi 376
Al Ghazali 377
Al Kindi 376
Alberti, L.B. 382
Albumasar 376
Allan, George 100, 103, 104, 108
Almond, Gabriel A. 249, 257, 258
Althusser, Louis 29
Aphrodisias, Alexander of 380
Aquinas, Thomas 379
Arato, Andrew 177
Arendt, Hannah 164, 237, 242, 259
Aristotle 235, 247, 253, 278, 279, 296, 300–303, 312, 376, 381
Attenborough, David 336
Augustine 297
Averroes 377, 380
Avicenna 376

Bacon, Francis 115, 117, 338, 342, 345, 359
Bahro, Rudolf 54, 61, 62, 64
Baltimore, David 119
Baron, Hans 374
Bayle, Pierre 301
Beauvoir, Simone de 326, 328, 337, 341
Beck, Lewis White 102
Benedict, Ruth 362
Bentham, John 103
Berdiayev 43
Bergson, Henri 336

Beria, L. 48
Beritashvili 46
Berkeley, G. 296, 301
Berlin, Isaiah 249
Bernard, Claude 130
Bikov 46
Blokhintsev 47
Blumenthal, David 125
Bobbio, Norberto 259
Boccaccio 379
Boethius 297
Bohm 200
Bohr 44, 47
Boshyan 46
Brezhnev 50
Broglie, de 47
Bronowski, Jacob 116, 120
Brown, Hanbury 115
Brown, Norman O. 11, 349, 350
Bruno, Giordano 373, 374, 383
Buber, Martin 349
Burckhardt, Jacob 373, 375, 376, 385
Buridan 379
Butlerov, A. 49

Caldicott, Helen 341
Campanella 374, 382
Camus, Albert 243
Carmon, Ira 120
Carnap, Rudolf 27–42
Carrière, Moritz 373
Carson, Rachel 329, 330, 351
Chen, Duxiu 15
Chodorow, Nancy 343, 344
Chrysippus 301
Churchill, Winston 95
Cini, Marcello 71
Cicero, Marcus Tullius 296, 297

# INDEX OF NAMES

Cohen, Robert S. 1–3, 15, 22–24, 27–42, 53, 65, 113, 118, 177, 215, 245
Colletti, Lucio 69
Commoner, Barry 117, 118, 337
Comte, A. 237
Condorcet 237
Confucius 19
Contarini, Gaspare 380
Copernicus, N. 383, 384
Crick, Francis 348
Cusa, Nicolas of 379

Dahl, Robert 249
Darwin, Charles 45
Deng, Xiaoping 19
Derrida, Jacques 155
Descartes, René 101–103, 237, 297, 342, 359–362, 365, 369–371, 384
Deutsch, K.W. 249
Dewey, John 44, 100–105, 108, 130
Dilthey, Wilhelm 144, 374
Dingle, Herbert 130
Dionysius of Halicarnassus 296
Dubinin 46
Dumont, Louis 370
Dunn, John 250
Durkheim, Emile 192, 235–245, 281
Dworkin, Ronald 249

Easlea, Brian 334, 341–343, 345, 346
Easton, David 249, 250, 255, 257, 258
Ebreo, Leone 382
Einstein, Albert 44, 48, 137, 200, 326
Eiseley, Loren 325
Eisenstein, Elizabeth 377, 378
Engels, Friedrich 22, 37, 45, 56, 64, 69, 70, 96
Epstein, Samuel 118
Erasmus, Desiderius 380, 382
Etzkowitz, Henry 124, 126
Eulau, Heinz 249
Evans-Pritchard 375

Fallot, Jean 69
Falter, Jürgen 250
Fee, Elizabeth 344, 347
Feenberg, Andrew 107

Feher, Judith 237
Feigl, Herbert 23, 29
Feng, Yulan 17
Ficino, Marsilio 376, 382
Fiorentino 373, 374
Fisichella, Domenico 260
Fock, W. 47
Foucault, Michel 107
Fox 325
Franklin, Benjamin 195
Fraser, Nancy 172, 175
Frege, Gotlob 34
French, Marilyn 331
Freud, Sigmund 11, 44, 51, 87, 237, 241, 349
Friedländer, Paul 100, 101
Fromm, Erich 7, 9, 11

Galileo, G. 383
Geertz, Clifford 369
Gelli, G.B. 379, 380
Gentile, G. 374, 385
Giese, Tiedemann 383
Gilbert, John 341
Gilligan, Carrol 334
Goethe, Johann Wolfgang von 237
Goldthorpe, John 226, 227
Goodall, Jane 349
Goodman, Nelson 135
Goody, Jack 378
Gorbachev, Michael 50, 57–59, 63
Grafton, Anthony 377
Gramsci 22
Granin, Daniil 51
Grant, Edward 375
Gratarol, Guglielmo 381
Graziano, Luigi 260

Habermas, Jürgen 163–166, 171, 176
Haeckel, Ernst 44
Haldane, J.B.S. 113, 121
Harding, Sandra 118, 327, 328, 331, 332
Harris, Errol 103
Harsanyi 262
Hartshorne, Charles 103, 104
He, Lin 17

# INDEX OF NAMES

Hegel, G.W.F. 28, 34, 36, 37, 103, 150, 163, 350, 366, 373, 374, 385
Heidegger, Martin 31, 155, 329, 347
Heims, S. 181
Heisenberg, Werner 44, 45, 47, 52
Heller, Agnes 171, 174
Herbert, A.P. 217
Herskovitz, Melville 362
Hobbes, Thomas 247, 301, 360, 365
Hodges, K.V. 205
Honecker, E. 63
Horkheimer, Max 28, 38
Hu, Feng 17
Hu, Qiaomu 21
Hu, Shi 17
Hume, David 134, 296, 300, 301, 307, 309, 310, 312
Husserl, Edmund 242, 363, 367
Hutcheson 301
Hutten 380

Ivanov-Smolensky 46

Jacobi 373
Jin, Yuelin 17
Jogiches, Leo 13
Jung, Carl Gustav 350

Kant, Immanuel 28, 102, 103, 152, 156, 237, 298, 323, 324, 362, 363
Kapitsa, P. 51, 52
Kekulé 49
Keller, Evelyn Fox 118, 331, 338, 339
Kelsen, Hans 38, 39
Keynes, John Maynard 51, 323
Khrushchev, N. 18, 50, 57, 59, 60
Khvolson, O. 44
Kierkegaard, Sören 330
Kluckhohn, Clyde 359
Kohak, Erazim 340, 342
Korneychuk, A. 50
Kroeber, Alfred 359
Krimsky, Sheldon 121, 123, 125
Kristeller, P.O. 377–379
Kroeber, A. 194
Kuhn, Thomas 250

Kurchatov, I. 48

Laguna 382
Landau 51
Lapunov, A. 51, 52
Laue, Max von 52
Le Goff, Jacques 373
Lenard 52
Lenin 23, 24, 27, 30, 34, 35, 39, 43–45, 48, 56, 87, 97
Lenz, Jean-Jacques 69
Leopold, Aldo 326
Lepeshinskaya 46
Leverrier 136
Lévi-Strauss, Claude 370
Liang, Shuming 17
Lin, Biao 19
Lin, Shu 9
Lin, Wei 9
Lindblom, Charles E. 249
Lipietz, Alain 70
Livingstone, John 336, 337
Locke, John 247, 300
Lomonosov, M. 50
Lossky 43
Lu, Dingyi 17, 18
Lucian 297
Lukacs, George 22
Lunacharsky 44
Luxemburg, Rosa 13
Lyotard, Jean-Francois 104
Lysenko, Trofim 45, 46, 50

Mach, Ernst 23, 24, 31, 39
Machiavelli, N. 247, 377, 381
MacIntyre, Alisdair 114, 249, 312
MacIver, Robert M. 238
Malinowski, B. 188
Mao, Zedong 6, 8–12, 16–19, 97
Marcuse, Herbert 10, 11, 28, 53–67, 107, 340, 343
Margenau, Henry 116
Marx, Karl 5, 15, 20–23, 27, 29, 36, 56, 64, 69–98, 104, 105, 107, 132, 247
Matteucci, Nicola 259
Maximov, A. 48
Mayr, O. 181

McAdoo 272
McClintock, Barbara 338, 339, 342, 350
McDyer, John 272, 273
McLuhan, Marshall 377, 378
McNaughton, Ben 334
Merchant, Carolyn 118, 337, 342, 348
Merton, Robert K. 113, 115, 120, 124
Mettrie, Julien Offray de la 37
Michelet, Karl-Ludwig 373, 375
Miliband, R. 232
Mill, John Stuart 103
Monod, J. 183, 187, 193
Montaigne, Michel de 379, 382
Monter, E.W. 341
Montesquieu, Ch. de 237
More, Henry 345
More, Thomas 382

Naes, Arne 330
Nagel, E. 188
Nelkin, Dorothy 126
Neurath, Otto 37–39, 261
Newton, Isaac 137, 308, 346
Nicholson, J.L. 220
Nietzsche, Friedrich 155, 237
Nifo, Agustino 377
Nobili, Roberto de 371
Noddings, Nel 340, 342
Nora, Pierre 373
North, John 375
Northrop, F.S.C. 116
Nozick, Robert 249

O'Brien, Mary 336, 344
Olschki, Leonardo 379
Oparin, Alexander 47
Orbelli, L. 46

Partridge, P.H. 249
Pasquino, Gianfranco 260, 261
Pavlov, Ivan 44, 46
Pelczynski, Zygmunt 163
Perelman, Chaïm 378
Petrarch 379
Pico, Giovanni 376, 382, 384
Plamenatz, J.P. 249
Planck, Max 52

Plato 100, 101, 103, 235, 247, 296, 297, 300, 309, 310, 330
Pletho, Gemisthus 376
Poincaré, Henri 116, 137
Polanyi, Karl 165, 330
Polzunov, I. 49
Pomponazzi, P. 377–381
Popov, Gavril 52
Popper, Karl 31, 87–98, 130, 136, 138, 254
Porzio, Simone 379, 380
Price, Don K. 116
Proctor, Robert 117
Ptashne, M. 183, 193
Ptolemy 376
Pythagoras 235, 237

Quine, Williard van Orman 28, 34–37

Rabelais 380, 382
Rahman, Fazlur 376
Ramus 374
Randall, John H. 114, 115
Rawls, John 39, 249
Regan, Tom 324
Reich, Wilhelm 9, 12
Reichenbach, Hans 132, 133
Reilly, William 197
Reuchlin, Johannes 376
Rheticus, Georg Joachim 383
Ricci, David Maria 249, 257
Rorty, Richard 155
Rose, Hilary 332, 335
Rosenblueth, A. 181
Ross, David 114
Rousseau, Jean Jacques 236, 237
Russell, Bertrand 342
Ryan, Alan 249

Saint-Simon 237
Salisbury, John of 379
Sartori, Giovanni 255–260
Sartre, Jean Paul 326
Scales, Junius 1
Scheler, M. 144
Schelling, F.W. 373
Schmitt, Carl 259, 262

# INDEX OF NAMES

Schönberg, Nicholas 383
Schrödinger, E. 44
Scriven, Michael 130
Sextus Empiricus 308, 309
Sherwin, Martin 117
Shmalhausen 46
Singer, Milton 369
Sinsheimer, Robert L. 119
Skinner, S.A. 252
Skolimowski, Henryk 321
Smith, Adam 295–320
Smith, Bruce L.R. 116
Spaventa, Betrando 373, 374, 385
Speroni 380
Spinoza, B. 297, 373
Stalin, Jozef 48–50, 75
Stamp, Josiah 222
Starck 52
Strauss, Leo 249, 259, 262
Struik, Dirk 2
Swazey, Judith 121, 125

Tamm 51
Taylor, Charles 249
Taylor, Jeremy 267, 271, 272
Taylor, Lawrence 272, 273, 288
Teilhard de Chardin, Pierre 103
Teller, Edward 340
Terletsky, Y. 47
Thatcher, Margaret 232
Thomas, Keith 375
Thompson, E.P. 340
Thurlow, Edward 189
Timofeyev-Ressovsky, Nicholai 51
Titmuss, Richard 216
Toennies, F. 188, 189
Toulmin, Stephen 115, 116, 131
Townsend 216

Toynbee, Arnold 104
Truman, David B. 249
Tscha, Hung 17
Turner, Victor 369, 374

Valla, L. 374
Vaughan, Henry 345
Vavilov, Nicholai 45, 46, 52
Vigier 47
Vives 382
Von Wright, George Henryk 133
Voegelin, Eric 249, 259, 262

Wang, Ruoshui 20, 21
Ward, Harry F. 2
Wartofsky, Marx W. 99, 105, 108, 109, 245
Watson 252
Weber, Alfred 188, 189, 238, 256
Weinberg, Alvin 197
White, Lynn 194, 324
Whitehead, A.N. 103, 104, 336, 349
Wiener, N. 49, 181
Wiens, John 337
Williams, Bernard 46, 101, 102
Windelband, Wilhelm 235
Wittgenstein, Ludwig 32, 33, 323, 364
Wolin, Sheldon S. 250
Woodridge 348
Wright Mills, C. 255
Wu, Hsun 17

Yang, Xianzhen 17, 18
Yu, Guangyuan 19
Yu, Ping-po 17

Zhang, Dongsun 15
Zhou, Yang 18, 20

# Boston Studies in the Philosophy of Science

*Editor:* Robert S. Cohen, *Boston University*

1. M.W. Wartofsky (ed.): *Proceedings of the Boston Colloquium for the Philosophy of Science, 1961/1962.* [Synthese Library 6] 1963
 ISBN 90-277-0021-4
2. R.S. Cohen and M.W. Wartofsky (eds.): *Proceedings of the Boston Colloquium for the Philosophy of Science, 1962/1964.* In Honor of P. Frank. [Synthese Library 10] 1965 ISBN 90-277-9004-0
3. R.S. Cohen and M.W. Wartofsky (eds.): *Proceedings of the Boston Colloquium for the Philosophy of Science, 1964/1966.* In Memory of Norwood Russell Hanson. [Synthese Library 14] 1967 ISBN 90-277-0013-3
4. R.S. Cohen and M.W. Wartofsky (eds.): *Proceedings of the Boston Colloquium for the Philosophy of Science, 1966/1968.* [Synthese Library 18] 1969
 ISBN 90-277-0014-1
5. R.S. Cohen and M.W. Wartofsky (eds.): *Proceedings of the Boston Colloquium for the Philosophy of Science, 1966/1968.* [Synthese Library 19] 1969
 ISBN 90-277-0015-X
6. R.S. Cohen and R.J. Seeger (eds.): *Ernst Mach, Physicist and Philosopher.* [Synthese Library 27] 1970 ISBN 90-277-0016-8
7. M. Čapek: *Bergson and Modern Physics.* A Reinterpretation and Re-evaluation. [Synthese Library 37] 1971 ISBN 90-277-0186-5
8. R.C. Buck and R.S. Cohen (eds.): *PSA 1970.* Proceedings of the 2nd Biennial Meeting of the Philosophy and Science Association (Boston, Fall 1970). In Memory of Rudolf Carnap. [Synthese Library 39] 1971
 ISBN 90-277-0187-3; Pb 90-277-0309-4
9. A.A. Zinov'ev: *Foundations of the Logical Theory of Scientific Knowledge (Complex Logic).* Translated from Russian. Revised and enlarged English Edition, with an Appendix by G.A. Smirnov, E.A. Sidorenko, A.M. Fedina and L.A. Bobrova. [Synthese Library 46] 1973
 ISBN 90-277-0193-8; Pb 90-277-0324-8
10. L. Tondl: *Scientific Procedures.* A Contribution Concerning the Methodological Problems of Scientific Concepts and Scientific Explanation. Translated from Czech. [Synthese Library 47] 1973 ISBN 90-277-0147-4; Pb 90-277-0323-X
11. R.J. Seeger and R.S. Cohen (eds.): *Philosophical Foundations of Science.* Proceedings of Section L, 1969, American Association for the Advancement of Science. [Synthese Library 58] 1974 ISBN 90-277-0390-6; Pb 90-277-0376-0
12. A. Grünbaum: *Philosophical Problems of Space and Times.* 2nd enlarged ed. [Synthese Library 55] 1973 ISBN 90-277-0357-4; Pb 90-277-0358-2
13. R.S. Cohen and M.W. Wartofsky (eds.): *Logical and Epistemological Studies in Contemporary Physics.* Proceedings of the Boston Colloquium for the Philosophy of Science, 1969/72, Part I. [Synthese Library 59] 1974
 ISBN 90-277-0391-4; Pb 90-277-0377-9

# Boston Studies in the Philosophy of Science

14. R.S. Cohen and M.W. Wartofsky (eds.): *Methodological and Historical Essays in the Natural and Social Sciences.* Proceedings of the Boston Colloquium for the Philosophy of Science, 1969/72, Part II. [Synthese Library 60] 1974
ISBN 90-277-0392-2; Pb 90-277-0378-7
15. R.S. Cohen, J.J. Stachel and M.W. Wartofsky (eds.): *For Dirk Struik.* Scientific, Historical and Political Essays in Honor of Dirk J. Struik. [Synthese Library 61] 1974  ISBN 90-277-0393-0; Pb 90-277-0379-5
16. N. Geschwind: *Selected Papers on Language and the Brains.* [Synthese Library 68] 1974  ISBN 90-277-0262-4; Pb 90-277-0263-2
17. B.G. Kuznetsov: *Reason and Being.* Translated from Russian. Edited by C.R. Fawcett and R.S. Cohen. 1987  ISBN 90-277-2181-5
18. P. Mittelstaedt: *Philosophical Problems of Modern Physics.* Translated from the revised 4th German edition by W. Riemer and edited by R.S. Cohen. [Synthese Library 95] 1976  ISBN 90-277-0285-3; Pb 90-277-0506-2
19. H. Mehlberg: *Time, Causality, and the Quantum Theory.* Studies in the Philosophy of Science. Vol. I: *Essay on the Causal Theory of Time.* Vol. II: *Time in a Quantized Universe.* Translated from French. Edited by R.S. Cohen. 1980  Vol. I: ISBN 90-277-0721-9; Pb 90-277-1074-0
Vol. II: ISBN 90-277-1075-9; Pb 90-277-1076-7
20. K.F. Schaffner and R.S. Cohen (eds.): *PSA 1972.* Proceedings of the 3rd Biennial Meeting of the Philosophy of Science Association (Lansing, Michigan, Fall 1972). [Synthese Library 64] 1974
ISBN 90-277-0408-2; Pb 90-277-0409-0
21. R.S. Cohen and J.J. Stachel (eds.): *Selected Papers of Léon Rosenfeld.* [Synthese Library 100] 1979  ISBN 90-277-0651-4; Pb 90-277-0652-2
22. M. Čapek (ed.): *The Concepts of Space and Time.* Their Structure and Their Development. [Synthese Library 74] 1976
ISBN 90-277-0355-8; Pb 90-277-0375-2
23. M. Grene: *The Understanding of Nature.* Essays in the Philosophy of Biology. [Synthese Library 66] 1974  ISBN 90-277-0462-7; Pb 90-277-0463-5
24. D. Ihde: *Technics and Praxis.* A Philosophy of Technology. [Synthese Library 130] 1979  ISBN 90-277-0953-X; Pb 90-277-0954-8
25. J. Hintikka and U. Remes: *The Method of Analysis.* Its Geometrical Origin and Its General Significance. [Synthese Library 75] 1974
ISBN 90-277-0532-1; Pb 90-277-0543-7
26. J.E. Murdoch and E.D. Sylla (eds.): *The Cultural Context of Medieval Learning.* Proceedings of the First International Colloquium on Philosophy, Science, and Theology in the Middle Ages, 1973. [Synthese Library 76] 1975
ISBN 90-277-0560-7; Pb 90-277-0587-9
27. M. Grene and E. Mendelsohn (eds.): *Topics in the Philosophy of Biology.* [Synthese Library 84] 1976  ISBN 90-277-0595-X; Pb 90-277-0596-8
28. J. Agassi: *Science in Flux.* [Synthese Library 80] 1975
ISBN 90-277-0584-4; Pb 90-277-0612-3

# Boston Studies in the Philosophy of Science

29. J.J. Wiatr (ed.): *Polish Essays in the Methodology of the Social Sciences.* [Synthese Library 131] 1979 ISBN 90-277-0723-5; Pb 90-277-0956-4
30. P. Janich: *Protophysics of Time.* Constructive Foundation and History of Time Measurement. Translated from German. 1985 ISBN 90-277-0724-3
31. R.S. Cohen and M.W. Wartofsky (eds.): *Language, Logic, and Method.* 1983 ISBN 90-277-0725-1
32. R.S. Cohen, C.A. Hooker, A.C. Michalos and J.W. van Evra (eds.): *PSA 1974.* Proceedings of the 4th Biennial Meeting of the Philosophy of Science Association. [Synthese Library 101] 1976 ISBN 90-277-0647-6; Pb 90-277-0648-4
33. G. Holton and W.A. Blanpied (eds.): *Science and Its Public.* The Changing Relationship. [Synthese Library 96] 1976 ISBN 90-277-0657-3; Pb 90-277-0658-1
34. M.D. Grmek, R.S. Cohen and G. Cimino (eds.): *On Scientific Discovery.* The 1977 Erice Lectures. 1981 ISBN 90-277-1122-4; Pb 90-277-1123-2
35. S. Amsterdamski: *Between Experience and Metaphysics.* Philosophical Problems of the Evolution of Science. Translated from Polish. [Synthese Library 77] 1975 ISBN 90-277-0568-2; Pb 90-277-0580-1
36. M. Marković and G. Petrović (eds.): *Praxis.* Yugoslav Essays in the Philosophy and Methodology of the Social Sciences. [Synthese Library 134] 1979 ISBN 90-277-0727-8; Pb 90-277-0968-8
37. H. von Helmholtz: *Epistemological Writings.* The Paul Hertz / Moritz Schlick Centenary Edition of 1921. Translated from German by M.F. Lowe. Edited with an Introduction and Bibliography by R.S. Cohen and Y. Elkana. [Synthese Library 79] 1977 ISBN 90-277-0290-X; Pb 90-277-0582-8
38. R.M. Martin: *Pragmatics, Truth and Language.* 1979 ISBN 90-277-0992-0; Pb 90-277-0993-9
39. R.S. Cohen, P.K. Feyerabend and M.W. Wartofsky (eds.): *Essays in Memory of Imre Lakatos.* [Synthese Library 99] 1976 ISBN 90-277-0654-9; Pb 90-277-0655-7
40. Not published.
41. Not published.
42. H.R. Maturana and F.J. Varela: *Autopoiesis and Cognition.* The Realization of the Living. With a Preface to 'Autopoiesis' by S. Beer. 1980 ISBN 90-277-1015-5; Pb 90-277-1016-3
43. A. Kasher (ed.): *Language in Focus: Foundations, Methods and Systems.* Essays in Memory of Yehoshua Bar-Hillel. [Synthese Library 89] 1976 ISBN 90-277-0644-1; Pb 90-277-0645-X
44. T.D. Thao: *Investigations into the Origin of Language and Consciousness.* 1984 ISBN 90-277-0827-4
45. Not published.
46. P.L. Kapitza: *Experiment, Theory, Practice.* Articles and Addresses. Edited by R.S. Cohen. 1980 ISBN 90-277-1061-9; Pb 90-277-1062-7

# Boston Studies in the Philosophy of Science

47. M.L. Dalla Chiara (ed.): *Italian Studies in the Philosophy of Science.* 1981
ISBN 90-277-0735-9; Pb 90-277-1073-2
48. M.W. Wartofsky: *Models.* Representation and the Scientific Understanding. [Synthese Library 129] 1979 ISBN 90-277-0736-7; Pb 90-277-0947-5
49. T.D. Thao: *Phenomenology and Dialectical Materialism.* Edited by R.S. Cohen. 1986 ISBN 90-277-0737-5
50. Y. Fried and J. Agassi: *Paranoia.* A Study in Diagnosis. [Synthese Library 102] 1976 ISBN 90-277-0704-9; Pb 90-277-0705-7
51. K.H. Wolff: *Surrender and Cath.* Experience and Inquiry Today. [Synthese Library 105] 1976 ISBN 90-277-0758-8; Pb 90-277-0765-0
52. K. Kosík: *Dialectics of the Concrete.* A Study on Problems of Man and World. 1976 ISBN 90-277-0761-8; Pb 90-277-0764-2
53. N. Goodman: *The Structure of Appearance.* [Synthese Library 107] 1977
ISBN 90-277-0773-1; Pb 90-277-0774-X
54. H.A. Simon: *Models of Discovery* and Other Topics in the Methods of Science. [Synthese Library 114] 1977 ISBN 90-277-0812-6; Pb 90-277-0858-4
55. M. Lazerowitz: *The Language of Philosophy.* Freud and Wittgenstein. [Synthese Library 117] 1977 ISBN 90-277-0826-6; Pb 90-277-0862-2
56. T. Nickles (ed.): *Scientific Discovery, Logic, and Rationality.* 1980
ISBN 90-277-1069-4; Pb 90-277-1070-8
57. J. Margolis: *Persons and Mind.* The Prospects of Nonreductive Materialism. [Synthese Library 121] 1978 ISBN 90-277-0854-1; Pb 90-277-0863-0
58. G. Radnitzky and G. Andersson (eds.): *Progress and Rationality in Science.* [Synthese Library 125] 1978 ISBN 90-277-0921-1; Pb 90-277-0922-X
59. G. Radnitzky and G. Andersson (eds.): *The Structure and Development of Science.* [Synthese Library 136] 1979 ISBN 90-277-0994-7; Pb 90-277-0995-5
60. T. Nickles (ed.): *Scientific Discovery.* Case Studies. 1980
ISBN 90-277-1092-9; Pb 90-277-1093-7
61. M.A. Finocchiaro: *Galileo and the Art of Reasoning.* Rhetorical Foundation of Logic and Scientific Method. 1980 ISBN 90-277-1094-5; Pb 90-277-1095-3
62. W.A. Wallace: *Prelude to Galileo.* Essays on Medieval and 16th-Century Sources of Galileo's Thought. 1981 ISBN 90-277-1215-8; Pb 90-277-1216-6
63. F. Rapp: *Analytical Philosophy of Technology.* Translated from German. 1981
ISBN 90-277-1221-2; Pb 90-277-1222-0
64. R.S. Cohen and M.W. Wartofsky (eds.): *Hegel and the Sciences.* 1984
ISBN 90-277-0726-X
65. J. Agassi: *Science and Society.* Studies in the Sociology of Science. 1981
ISBN 90-277-1244-1; Pb 90-277-1245-X
66. L. Tondl: *Problems of Semantics.* A Contribution to the Analysis of the Language of Science. Translated from Czech. 1981
ISBN 90-277-0148-2; Pb 90-277-0316-7
67. J. Agassi and R.S. Cohen (eds.): *Scientific Philosophy Today.* Essays in Honor of Mario Bunge. 1982 ISBN 90-277-1262-X; Pb 90-277-1263-8

# Boston Studies in the Philosophy of Science

68. W. Krajewski (ed.): *Polish Essays in the Philosophy of the Natural Sciences.* Translated from Polish and edited by R.S. Cohen and C.R. Fawcett. 1982
ISBN 90-277-1286-7; Pb 90-277-1287-5
69. J.H. Fetzer: *Scientific Knowledge.* Causation, Explanation and Corroboration. 1981
ISBN 90-277-1335-9; Pb 90-277-1336-7
70. S. Grossberg: *Studies of Mind and Brain.* Neural Principles of Learning, Perception, Development, Cognition, and Motor Control. 1982
ISBN 90-277-1359-6; Pb 90-277-1360-X
71. R.S. Cohen and M.W. Wartofsky (eds.): *Epistemology, Methodology, and the Social Sciences.* 1983. ISBN 90-277-1454-1
72. K. Berka: *Measurement.* Its Concepts, Theories and Problems. Translated from Czech. 1983 ISBN 90-277-1416-9
73. G.L. Pandit: *The Structure and Growth of Scientific Knowledge.* A Study in the Methodology of Epistemic Appraisal. 1983 ISBN 90-277-1434-7
74. A.A. Zinov'ev: *Logical Physics.* Translated from Russian. Edited by R.S. Cohen. 1983 ISBN 90-277-0734-0
*See also* Volume 9.
75. G-G. Granger: *Formal Thought and the Sciences of Man.* Translated from French. With and Introduction by A. Rosenberg. 1983 ISBN 90-277-1524-6
76. R.S. Cohen and L. Laudan (eds.): *Physics, Philosophy and Psychoanalysis.* Essays in Honor of Adolf Grünbaum. 1983 ISBN 90-277-1533-5
77. G. Böhme, W. van den Daele, R. Hohlfeld, W. Krohn and W. Schäfer: *Finalization in Science.* The Social Orientation of Scientific Progress. Translated from German. Edited by W. Schäfer. 1983 ISBN 90-277-1549-1
78. D. Shapere: *Reason and the Search for Knowledge.* Investigations in the Philosophy of Science. 1984 ISBN 90-277-1551-3; Pb 90-277-1641-2
79. G. Andersson (ed.): *Rationality in Science and Politics.* Translated from German. 1984 ISBN 90-277-1575-0; Pb 90-277-1953-5
80. P.T. Durbin and F. Rapp (eds.): *Philosophy and Technology.* [*Also* Philosophy and Technology Series, Vol. 1] 1983 ISBN 90-277-1576-9
81. M. Marković: *Dialectical Theory of Meaning.* Translated from Serbo-Croat. 1984 ISBN 90-277-1596-3
82. R.S. Cohen and M.W. Wartofsky (eds.): *Physical Sciences and History of Physics.* 1984. ISBN 90-277-1615-3
83. É. Meyerson: *The Relativistic Deduction.* Epistemological Implications of the Theory of Relativity. Translated from French. With a Review by Albert Einstein and an Introduction by Milič Čapek. 1985 ISBN 90-277-1699-4
84. R.S. Cohen and M.W. Wartofsky (eds.): *Methodology, Metaphysics and the History of Science.* In Memory of Benjamin Nelson. 1984 ISBN 90-277-1711-7
85. G. Tamás: *The Logic of Categories.* Translated from Hungarian. Edited by R.S. Cohen. 1986 ISBN 90-277-1742-7
86. S.L. de C. Fernandes: *Foundations of Objective Knowledge.* The Relations of Popper's Theory of Knowledge to That of Kant. 1985 ISBN 90-277-1809-1

# Boston Studies in the Philosophy of Science

87. R.S. Cohen and T. Schnelle (eds.): *Cognition and Fact.* Materials on Ludwik Fleck. 1986 ISBN 90-277-1902-0
88. G. Freudenthal: *Atom and Individual in the Age of Newton.* On the Genesis of the Mechanistic World View. Translated from German. 1986
 ISBN 90-277-1905-5
89. A. Donagan, A.N. Perovich Jr and M.V. Wedin (eds.): *Human Nature and Natural Knowledge.* Essays presented to Marjorie Grene on the Occasion of Her 75th Birthday. 1986 ISBN 90-277-1974-8
90. C. Mitcham and A. Hunning (eds.): *Philosophy and Technology II.* Information Technology and Computers in Theory and Practice. [*Also* Philosophy and Technology Series, Vol. 2] 1986 ISBN 90-277-1975-6
91. M. Grene and D. Nails (eds.): *Spinoza and the Sciences.* 1986
 ISBN 90-277-1976-4
92. S.P. Turner: *The Search for a Methodology of Social Science.* Durkheim, Weber, and the 19th-Century Problem of Cause, Probability, and Action. 1986.
 ISBN 90-277-2067-3
93. I.C. Jarvie: *Thinking about Society.* Theory and Practice. 1986
 ISBN 90-277-2068-1
94. E. Ullmann-Margalit (ed.): *The Kaleidoscope of Science.* The Israel Colloquium: Studies in History, Philosophy, and Sociology of Science, Vol. 1. 1986
 ISBN 90-277-2158-0; Pb 90-277-2159-9
95. E. Ullmann-Margalit (ed.): *The Prism of Science.* The Israel Colloquium: Studies in History, Philosophy, and Sociology of Science, Vol. 2. 1986
 ISBN 90-277-2160-2; Pb 90-277-2161-0
96. G. Márkus: *Language and Production.* A Critique of the Paradigms. Translated from French. 1986 ISBN 90-277-2169-6
97. F. Amrine, F.J. Zucker and H. Wheeler (eds.): *Goethe and the Sciences: A Reappraisal.* 1987 ISBN 90-277-2265-X; Pb 90-277-2400-8
98. J.C. Pitt and M. Pera (eds.): *Rational Changes in Science.* Essays on Scientific Reasoning. Translated from Italian. 1987 ISBN 90-277-2417-2
99. O. Costa de Beauregard: *Time, the Physical Magnitude.* 1987
 ISBN 90-277-2444-X
100. A. Shimony and D. Nails (eds.): *Naturalistic Epistemology.* A Symposium of Two Decades. 1987 ISBN 90-277-2337-0
101. N. Rotenstreich: *Time and Meaning in History.* 1987 ISBN 90-277-2467-9
102. D.B. Zilberman: *The Birth of Meaning in Hindu Thought.* Edited by R.S. Cohen. 1988 ISBN 90-277-2497-0
103. T.F. Glick (ed.): *The Comparative Reception of Relativity.* 1987
 ISBN 90-277-2498-9
104. Z. Harris, M. Gottfried, T. Ryckman, P. Mattick Jr, A. Daladier, T.N. Harris and S. Harris: *The Form of Information in Science.* Analysis of an Immunology Sublanguage. With a Preface by Hilary Putnam. 1989 ISBN 90-277-2516-0

# Boston Studies in the Philosophy of Science

105. F. Burwick (ed.): *Approaches to Organic Form.* Permutations in Science and Culture. 1987  ISBN 90-277-2541-1
106. M. Almási: *The Philosophy of Appearances.* Translated from Hungarian. 1989  ISBN 90-277-2150-5
107. S. Hook, W.L. O'Neill and R. O'Toole (eds.): *Philosophy, History and Social Action.* Essays in Honor of Lewis Feuer. With an Autobiographical Essay by L. Feuer. 1988  ISBN 90-277-2644-2
108. I. Hronszky, M. Fehér and B. Dajka: *Scientific Knowledge Socialized.* Selected Proceedings of the 5th Joint International Conference on the History and Philosophy of Science organized by the IUHPS (Veszprém, Hungary, 1984). 1988  ISBN 90-277-2284-6
109. P. Tillers and E.D. Green (eds.): *Probability and Inference in the Law of Evidence.* The Uses and Limits of Bayesianism. 1988  ISBN 90-277-2689-2
110. E. Ullmann-Margalit (ed.): *Science in Reflection.* The Israel Colloquium: Studies in History, Philosophy, and Sociology of Science, Vol. 3. 1988  ISBN 90-277-2712-0; Pb 90-277-2713-9
111. K. Gavroglu, Y. Goudaroulis and P. Nicolacopoulos (eds.): *Imre Lakatos and Theories of Scientific Change.* 1989  ISBN 90-277-2766-X
112. B. Glassner and J.D. Moreno (eds.): *The Qualitative-Quantitative Distinction in the Social Sciences.* 1989  ISBN 90-277-2829-1
113. K. Arens: *Structures of Knowing.* Psychologies of the 19th Century. 1989  ISBN 0-7923-0009-2
114. A. Janik: *Style, Politics and the Future of Philosophy.* 1989  ISBN 0-7923-0056-4
115. F. Amrine (ed.): *Literature and Science as Modes of Expression.* With an Introduction by S. Weininger. 1989  ISBN 0-7923-0133-1
116. J.R. Brown and J. Mittelstrass (eds.): *An Intimate Relation.* Studies in the History and Philosophy of Science. Presented to Robert E. Butts on His 60th Birthday. 1989  ISBN 0-7923-0169-2
117. F. D'Agostino and I.C. Jarvie (eds.): *Freedom and Rationality.* Essays in Honor of John Watkins. 1989  ISBN 0-7923-0264-8
118. D. Zolo: *Reflexive Epistemology.* The Philosophical Legacy of Otto Neurath. 1989  ISBN 0-7923-0320-2
119. M. Kearn, B.S. Philips and R.S. Cohen (eds.): *Georg Simmel and Contemporary Sociology.* 1989  ISBN 0-7923-0407-1
120. T.H. Levere and W.R. Shea (eds.): *Nature, Experiment and the Science.* Essays on Galileo and the Nature of Science. In Honour of Stillman Drake. 1989  ISBN 0-7923-0420-9
121. P. Nicolacopoulos (ed.): *Greek Studies in the Philosophy and History of Science.* 1990  ISBN 0-7923-0717-8
122. R. Cooke and D. Costantini (eds.): *Statistics in Science.* The Foundations of Statistical Methods in Biology, Physics and Economics. 1990  ISBN 0-7923-0797-6

# Boston Studies in the Philosophy of Science

123. P. Duhem: *The Origins of Statics*. Translated from French by G.F. Leneaux, V.N. Vagliente and G.H. Wagner. With an Introduction by S.L. Jaki. 1991
ISBN 0-7923-0898-0
124. H. Kamerlingh Onnes: *Through Measurement to Knowledge*. The Selected Papers, 1853-1926. Edited and with an Introduction by K. Gavroglu and Y. Goudaroulis. 1991 ISBN 0-7923-0825-5
125. M. Čapek: *The New Aspects of Time: Its Continuity and Novelties*. Selected Papers in the Philosophy of Science. 1991 ISBN 0-7923-0911-1
126. S. Unguru (ed.): *Physics, Cosmology and Astronomy, 1300-1700*. Tension and Accommodation. 1991 ISBN 0-7923-1022-5
127. Z. Bechler: *Newton's Physics on the Conceptual Structure of the Scientific Revolution*. 1991 ISBN 0-7923-1054-3
128. É. Meyerson: *Explanation in the Sciences*. Translated from French by M-A. Siple and D.A. Siple. 1991 ISBN 0-7923-1129-9
129. A.I. Tauber (ed.): *Organism and the Origins of Self*. 1991
ISBN 0-7923-1185-X
130. F.J. Varela and J-P. Dupuy (eds.): *Understanding Origins*. Contemporary Views on the Origin of Life, Mind and Society. 1992 ISBN 0-7923-1251-1
131. G.L. Pandit: *Methodological Variance*. Essays in Epistemological Ontology and the Methodology of Science. 1991 ISBN 0-7923-1263-5
132. G. Munévar (ed.): *Beyond Reason*. Essays on the Philosophy of Paul Feyerabend. 1991 ISBN 0-7923-1272-4
133. T.E. Uebel (ed.): *Rediscovering the Forgotten Vienna Circle*. Austrian Studies on Otto Neurath and the Vienna Circle. Partly translated from German. 1991
ISBN 0-7923-1276-7
134. W.R. Woodward and R.S. Cohen (eds.): *World Views and Scientific Discipline Formation*. Science Studies in the [former] German Democratic Republic. Partly translated from German by W.R. Woodward. 1991
ISBN 0-7923-1286-4
135. P. Zambelli: *The Speculum Astronomiae and Its Enigma*. Astrology, Theology and Science in Albertus Magnus and His Contemporaries. 1992
ISBN 0-7923-1380-1
136. P. Petitjean, C. Jami and A.M. Moulin (eds.): *Science and Empires*. Historical Studies about Scientific Development and European Expansion.
ISBN 0-7923-1518-9
137. W.A. Wallace: *Galileo's Logic of Discovery and Proof*. The Background, Content, and Use of His Appropriated Treatises on Aristotle's *Posterior Analytics*. 1992 ISBN 0-7923-1577-4
138. W.A. Wallace: *Galileo's Logical Treatises*. A Translation, with Notes and Commentary, of His Appropriated Latin Questions on Aristotle's *Posterior Analytics*. 1992 ISBN 0-7923-1578-2
Set (137 + 138) ISBN 0-7923-1579-0

# Boston Studies in the Philosophy of Science

139. M.J. Nye, J.L. Richards and R.H. Stuewer (eds.): *The Invention of Physical Science.* Intersections of Mathematics, Theology and Natural Philosophy since the Seventeenth Century. Essays in Honor of Erwin N. Hiebert. 1992
ISBN 0-7923-1753-X
140. G. Corsi, M.L. dalla Chiara and G.C. Ghirardi (eds.): *Bridging the Gap: Philosophy, Mathematics and Physics.* Lectures on the Foundations of Science. 1992
ISBN 0-7923-1761-0
141. C.-H. Lin and D. Fu (eds.): *Philosophy and Conceptual History of Science in Taiwan.* 1992
ISBN 0-7923-1766-1
142. S. Sarkar (ed.): *The Founders of Evolutionary Genetics.* A Centenary Reappraisal. 1992
ISBN 0-7923-1777-7
143. J. Blackmore (ed.): *Ernst Mach – A Deeper Look.* Documents and New Perspectives. 1992
ISBN 0-7923-1853-6
144. P. Kroes and M. Bakker (eds.): *Technological Development and Science in the Industrial Age.* New Perspectives on the Science–Technology Relationship. 1992
ISBN 0-7923-1898-6
145. S. Amsterdamski: *Between History and Method.* Disputes about the Rationality of Science. 1992
ISBN 0-7923-1941-9
146. E. Ullmann-Margalit (ed.): *The Scientific Enterprise.* The Bar-Hillel Colloquium: Studies in History, Philosophy, and Sociology of Science, Volume 4. 1992
ISBN 0-7923-1992-3
147. L. Embree (ed.): *Metaarchaeology.* Reflections by Archaeologists and Philosophers. 1992
ISBN 0-7923-2023-9
148. S. French and H. Kamminga (eds.): *Correspondence, Invariance and Heuristics.* Essays in Honour of Heinz Post. 1993
ISBN 0-7923-2085-9
149. M. Bunzl: *The Context of Explanation.* 1993
ISBN 0-7923-2153-7
150. I.B. Cohen (ed.): *The Natural Sciences and the Social Sciences.* Some Critical and Historical Perspectives. 1994
ISBN 0-7923-2223-1
151. K. Gavroglu, Y. Christianidis and E. Nicolaidis (eds.): *Trends in the Historiography of Science.* 1994
ISBN 0-7923-2255-X
152. S. Poggi and M. Bossi (eds.): *Romanticism in Science.* Science in Europe, 1790–1840. 1994
ISBN 0-7923-2336-X
153. J. Faye and H.J. Folse (eds.): *Niels Bohr and Contemporary Philosophy.* 1994
ISBN 0-7923-2378-5
154. C.C. Gould and R.S. Cohen (eds.): *Artifacts, Representations, and Social Practice.* Essays for Marx W. Wartofsky. 1994
ISBN 0-7923-2481-1
155. R.E. Butts: *Historical Pragmatics.* Philosophical Essays. 1993
ISBN 0-7923-2498-6
156. R. Rashed: *The Development of Arabic Mathematics: Between Arithmetic and Algebra.* Translated from French by A.F.W. Armstrong. 1994
ISBN 0-7923-2565-6

# Boston Studies in the Philosophy of Science

157. I. Szumilewicz-Lachman (ed.): *Zygmunt Zawirski: His Life and Work.* With Selected Writings on Time, Logic and the Methodology of Science. Translations by Feliks Lachman. Ed. by R.S. Cohen, with the assistance of B. Bergo. 1994 ISBN 0-7923-2566-4
158. S.N. Haq: *Names, Natures and Things.* The Alchemist Jābir ibn Ḥayyān and His *Kitāb al-Aḥjār* (Book of Stones). 1994 ISBN 0-7923-2587-7
159. P. Plaass: *Kant's Theory of Natural Science.* Translation, Analytic Introduction and Commentary by Alfred E. and Maria G. Miller. 1994
 ISBN 0-7923-2750-0
160. J. Misiek (ed.): *The Problem of Rationality in Science and its Philosophy.* On Popper vs. Polanyi. The Polish Conferences 1988–89. 1995
 ISBN 0-7923-2925-2
161. I.C. Jarvie and N. Laor (eds.): *Critical Rationalism, Metaphysics and Science.* Essays for Joseph Agassi, Volume I. 1995 ISBN 0-7923-2960-0
162. I.C. Jarvie and N. Laor (eds.): *Critical Rationalism, the Social Sciences and the Humanities.* Essays for Joseph Agassi, Volume II. 1995 ISBN 0-7923-2961-9
 Set (161–162) ISBN 0-7923-2962-7
163. K. Gavroglu, J. Stachel and M.W. Wartofsky (eds.): *Physics, Philosophy, and the Scientific Community.* Essays in the Philosophy and History of the Natural Sciences and Mathematics. In Honor of Robert S. Cohen. 1995
 ISBN 0-7923-2988-0
164. K. Gavroglu, J. Stachel and M.W. Wartofsky (eds.): *Science, Politics and Social Practice.* Essays on Marxism and Science, Philosophy of Culture and the Social Sciences. In Honor of Robert S. Cohen. 1995 ISBN 0-7923-2989-9
165. K. Gavroglu, J. Stachel and M.W. Wartofsky (eds.): *Science, Mind and Art.* Essays on Science and the Humanistic Understanding in Art, Epistemology, Religion and Ethics. Essays in Honor of Robert S. Cohen. 1995
 ISBN 0-7923-2990-2
 Set (163–165) ISBN 0-7923-2991-0
166. K.H. Wolff: *Transformation in the Writing.* A Case of Surrender-and-Catch. 1995 ISBN 0-7923-3178-8
167. A.J. Kox and D.M. Siegel (eds.): *No Truth Except in the Details.* Essays in honor of Martin J. Klein. 1995 ISBN 0-7923-3195-8
168. J. Blackmore (ed.): *Ludwig Boltzmann* His Later Life and Philosophy, 1900–1906. 1995 ISBN 0-7923-3231-8
169. R.S. Cohen, R. Hilpinen and Qiu Renzong (eds.): *Realism and Anti-Realism in the Philosophy of Science.* Beijing International Conference, 1992. 1995 (forthcoming) ISBN 0-7923-3233-4
170. I. Kuçuradi and R.S. Cohen (eds.): *The Concept of Knowledge.* The Ankara Seminar. 1995 (forthcoming) ISBN 0-7923-3241-5

# Boston Studies in the Philosophy of Science

171. M.A. Grodin (ed.): *Meta Medical Ethics*. The Philosophical Foundations of Bioethics. 1995 ISBN 0-7923-3344-6

*Also of interest:*
R.S. Cohen and M.W. Wartofsky (eds.): *A Portrait of Twenty-Five Years Boston Colloquia for the Philosophy of Science, 1960-1985.* 1985   ISBN Pb 90-277-1971-3

*Previous volumes are still available.*

KLUWER ACADEMIC PUBLISHERS – DORDRECHT / BOSTON / LONDON